Guide to the Management of Hazardous Waste

Guide to the Management of Hazardous Waste

A Handbook for the Businessman and the Concerned Citizen

J. William Haun, Ph.D.

Fulcrum Publishing
Golden, Colorado

Library of Congress Cataloging-in-Publication Data

Haun, J. William.
Guide to the management of hazardous waste : a handbook for the businessman and the concerned citizen / J. William Haun.
p. cm.
Includes bibliographical references (p.) and index.
ISBN 1-55591-065-3 (trade pbk.)
1. Hazardous wastes—Management. 2. Factory and trade waste. 3. Hazardous wastes—Law and legislation—United States. I. Title.
TD1030.H38 1991
363.72′87—dc20 90–85268
CIP

Printed in the United States of America
0 9 8 7 6 5 4 3 2 1

Fulcrum Publishing
350 Indiana Street
Golden, Colorado 80401

CONTENTS

How to Use This Book

The subject of hazardous waste, indeed all wastes, in modern society is complicated. Readers will not be surprised that a book on hazardous waste may also be complicated. This section shows how a particular reader can select those subjects and levels of detail that are useful to that reader. The author has conscientiously attempted to minimize the use of technical terms and jargon that overload all communication in this field. Unfortunately, if citizens are to have any idea what their government is talking about, they have to know at least some of the terms in use. Therefore, you will find some of them here, with explanations of their meanings. Glossaries in the Appendix will help.

The book is in three parts, each with selected material to meet specific needs for information and action by members of several communities of readers: business owners and managers, citizens whose lives are actually or potentially affected by management and mismanagement of hazardous wastes, and public officials whom constituents expect to know and understand the issues and the government response to the issues.

PART I

The Introduction is for all readers. Reasons for the recent increase in concern about the management of hazardous waste are discussed, with some insights into the nature of our society and how that bears on the problems considered in the book. This chapter supplies the common background of understanding that all citizens must grasp as they begin discussion of this major social problem.

In Chapter 2, "A Businessman's Guide to Life under the Hazardous Waste Laws and Regulations," the main message is that you'd better believe that the laws and regulations are real and are being enforced. Failure to comply with all the environmental requirements is a serious matter. Mismanagement of hazardous waste can bankrupt your company and deprive its managers of their liberty and their fortunes.

While much of the material is written for the independent business person who lacks staff and resources to comply and thus must personally deal with these matters, managers of large businesses will also profit by reading this material. Recently the trend has been to eliminate corporate staff specialists in this area and to "delegate" the responsibility for environmental compliance to divisional and plant-level management. This means in many cases that no one will have time to learn enough to ensure compliance. More than one company has found that this delegation entrusts the "garbage man" with decisions involving costs far beyond a division manager's authority.

Chapter 3, "A Citizen's Guide to Action on the Problems of Hazardous Waste," provides information for the individual in his or her role in the society. It aims first at the person or family whose personal well-being is somehow jeopardized by a release or a threatened release of some material. Useful information includes suggestions on how to interpret the language of public officials. Better yet, you'll find information on how to discover who's in charge and how to express your concerns. If necessary, the citizen can follow the suggested ways to demand action from either public or private entities and organizations.

The remaining three chapters of Part I discuss in some detail the identification of hazardous waste, which may surprise in what it *does not* include. Some of the basic terms and methods used in toxicity measurement are considered in some detail to help unravel the confusion between "toxic" wastes and other hazardous wastes. Finally, the reader can profit from a discussion in simplified form of the developing methods of risk assessment and risk management, which form the basis of the whole regulatory system for toxic and hazardous materials.

PART II

This part describes the major methods of hazardous waste management. Because business managers have an obligation to specify the methods used to manage and dispose of their wastes, they need to know at least the rudiments of these methods even if they assign the task of implementing decisions to technical specialists.

Concerned citizens can turn to these chapters for help in following public and private decisions on waste management and for a general appreciation of the complexity of the technology and the difficulty of obtaining official permission to locate, build, and operate any waste management or disposal facility.

PART III

This part begins with a review of some basic chemistry that is accessible to anyone who remembers a few fundamentals from a high school chemistry course. It is included to further a prime objective of this book: to reduce the sense of mystery and the consequent fear that permeates public concern about hazardous wastes. The use (and frequent misuse) of chemical names in the press and electronic media creates a false aura of mystery. "Halogenated solvents," for example, are just old familiar substances that have been in common use for most of this century. Technical ways for expressing the amount of acid in water simply make it easier to be fully understood. This chapter is by no means comparable even to an elementary textbook in chemistry, but it can assist in understanding wastes better.

Two chapters discuss the fundamental Federal laws applying to past and present waste management and disposal. Each chapter is considerably shorter than the law that it discusses and is couched so far as possible in nontechnical language.

Businessmen should take the time to read these chapters through. The focus is on the complexity and severity of these laws and the regulations developed to implement them. Business managers must understand that many of the commonly accepted principles of civil liberty do not apply to companies or even to individuals accused of violation of some highly technical requirements. The higher courts have spoken on many of these subjects, and it is too late and too publicly unpopular to expect relief.

Interested and concerned citizens can find much that is fascinating in these detailed discussions.

SOME OBSERVATIONS

Some years ago, the author was involved in a church-setting discussion about clean air in one of our major cities. After considerable talk in which the data for that particular city showed that emissions from automobiles were primarily responsible for the problems, a participant expressed a common reaction. A fine lady who was always in the forefront of the fight for human dignity and individual rights defined the solution: "The government simply has to make most of those people stop driving those cars!" Other participants pointed out that, in a democratic society, developing rules that are fair and reasonable within the Constitutional framework is a difficult and expensive effort. Her reaction was: "Well, if you have to get into all that abstract stuff about freedom and individual rights, then just forget about it. You'll never solve the problem!"

So it goes in the field of toxic and hazardous substance management. Basic fairness to the population as a whole demands that we develop and follow sound, safe disposal practices for the wastes that must be produced after applying the utmost wisdom to the problems of reducing wastes to the barest possible minimum. Unswerving application of coercive legislation and severe criminal penalties may be needed to deal with serious and willful actions. However, the solution to our environmental problems lies not in the courthouse but in the hearts and minds of an informed and vigilant citizenry. Top corporate and government management persons, alongside "ordinary" persons, must play their responsible parts as citizens of the nation.

A DISCLAIMER

To write intelligibly about the management of hazardous wastes is not possible without extensive reference to the regulations and laws that have been devised at all levels of government to control them.

This book, however, does not offer legal advice, nor should it be relied on by anyone who must know the application of law and regulations to a particular set of facts. The conclusive statements are the opinions of the author based upon many years of involvement as an engineer/scientist in the development and implementation of this complex body of law, regulation, and practice.

Guide to the Management of Hazardous Waste

Part I

Information You Can Use Now

Chapter 1

INTRODUCTION

BACKGROUND OF THE WASTE PROBLEM

As we near the end of the twentieth century of the Common Era, as anthropologists and philosophers prefer to say it, we seem to have plenty to worry about. It's not hard to get involved in a conversation about people's fears of nuclear war, about the drug crisis, the crime rate, or clean air in the cities.

Amid all these crises, and there is an endless number in the catalog, we keep encountering press reports and concerned conversation about the problems of "Toxic Waste" or "Chemical Waste" or "Industrial Waste" and how the world may end when the last of us expires under a pile of some reeking goo. Several surveys have found that schoolchildren are more concerned about a future dominated by these wastes than they are about dying in a nuclear war.

The business owner or operator today has the largest stake of all in the working out of a management system for hazardous wastes that is both safe and believable to the public. That vested interest in solutions comes from two facts: (1) business generates nearly all the hazardous waste, and (2) business is the only entity in the American society that can really do anything constructive about it.

No company or plant wants to make waste, much less hazardous waste. Wastes are usually considered to be evidence of inefficiency and poor management of processes and operations. That is often true in a simplistic sense, but it is also true that some wastes are unavoidable consequences of the production of otherwise desirable things.

One of the effective scenes in J. R. R. Tolkien's great fantasy *The Lord of the Rings* describes the results of the conquest of the beautiful Shire, home of the Hobbits, by the evil empire of Mordor. The streams are lined with "great, groaning" mills that spew endless amounts of filth into the once-pristine streams and air. Obviously, Tolkien had no real difficulty equating pollution with evil. It is interesting that no suggestion is ever made about what the "mills" may have produced other than "filth."

This dire scenario with all the blame and ill will falling on business tends to make business people angry and defensive. While this is understandable, it is not productive. At least in American and other Western capitalistic and mixed economies, your customer is finally the consumer, who is also you, your employees and neighbors, and with devastating simplicity, the voter!

What sense does it make to try to resist action that is clearly demanded by the people most important to your success? It helps to be wise, but it is necessary to be pragmatic and smart enough to stay out of trouble with your source of business.

Since it can be shown that practically all hazardous waste is produced by business and industry, and even more results from consumer use and misuse of the products sold to them, then it is clear that the business community is ultimately going to have to do something about it.

People cry out that government must control hazardous wastes and see to their safe management.

Government can only react, in a free society, by deciding what it will accept as good management of wastes and by using its persuasive and its coercive power to make business do it. This simple equation often escapes business people. Government works by forcing others to perform—it is conceivable that some public facility might be set up to dispose of hazardous wastes, as is already done with the ordinary municipal kind. It is just about inconceivable that such a move would be politically acceptable. Politicians avoid blame whenever possible—why would we expect political leaders to be willing to be blamed when a public facility which they authorized is thought to be endangering their constituents? How much easier to invoke the high moral ground that the polluter must pay—for the disposal of wastes from identified industrial sources and for any faults which the system might have or might develop.

So, when all is said and to be done, industry and business are the only part of the economic society which has the ability to manage hazardous waste. We might as well set about it and do it well, because if we don't, we will just continue to pay and pay until the public and their representatives are satisfied.

At any rate, no knowledgeable citizen of the United States can fail to have a concern and some opinions about hazardous waste and its management. The recent history of public opposition to just about any proposal to locate a waste treatment or disposal facility bears highly vocal witness to the fears the public has about such places, even if they are told repeatedly that only "non-hazardous" materials will be handled there.

Reacting as always after the fact, Congress has passed strong, punitive, and confused legislation that demands action when there is no intelligent basis for action. Administrators try in their fashion to carry out the will of Congress, writing endless tomes of regulatory language that only a skilled bureaucrat can penetrate.

The outcome is that ordinary people's fears and concerns are apt to disappear in an impenetrable thicket of experts, lawyers, and bureaucrats who tend to maintain that those same "ordinary citizens" can't possibly understand what's going on and that they may as well accept the decisions of the experts who run the system.

The primary thesis of this book is that reasonably literate people can become an informed citizenry that can and should participate fully in the decisions of government, institutions, and businesses that affect their lives. Government exists to find the best answer for all concerned, so not everyone can achieve his personal goals. We will try to face the fact that all life is full of risks of one kind or another. We cannot eliminate all risks, but we can, by gaining understanding, participate in the *management* of the risks to which we and our families may be exposed in modern life.

In this context, every person is a citizen and consumer. It has already been said that some citizens have special responsibilities as owners, managers, and operators of businesses. Most people are simply interested in the long-term well-being of themselves and their families and communities. A few will be interested in becoming deeply involved in the policy-making process, either as it affects their local concerns or at the national level.

A major problem in the design, implementation, and enforcement of laws controlling the management of hazardous waste and toxic materials is that there is no really common body of knowledge among all kinds of citizens around which consensus decisions can be fashioned. The author hopes that this book can at least begin the process of development of that common understanding that is always necessary to the appropriate functioning of the democratic process.

Many observers conclude that Rachel Carson's book *Silent Spring*, regarding the impact of pesticides on birds and other animal life, was responsible for awakening the current environmental movement. Obviously, plenty of people were ready to be awakened, even mobilized, to defend the "natural" environment from damage, whether by evil intent or simple carelessness and insensitivity.

Under this stimulation, citizens of industrial countries, particularly those in the United States, have seen and heard press stories many times repeated about hazards to their health from sources usually unknown and often considered not really understandable. Political campaigns and even whole careers have been built on the need for environmental protection. It is not always clear whether the environment is to be protected or whether people are to be protected from hazards in their environment. As it turns out, these two forms of protection are really the same thing. In general, most students of these matters conclude that anything that threatens the human environment eventually threatens humans—and as we shall see, choices must be made between our comfort and our survival.

Unclean air and smog are not new; they were documented hazards in Europe even before the Industrial Revolution. Describing smog in bureaucratic language as "fair" or "poor" or even "unhealthy" air quality and measuring it with an air quality index seems to make it even more mysterious and therefore more threatening than the old pea-soup or even "killer" fogs that plagued most great cities until recent times. The last killer episode in the developed world may have occurred in London in 1952, when 4,000 more people died than would normally have been expected to in the time period of the fog. In the United States, the memory of the 1948 killer fog in Donora, Pennsylvania, still haunts the search for better air, perhaps even more than it should, given the changes in our sources of energy and therefore of air pollutants since that time fifty years ago.

Clean water was a prize of great public value centuries before modern concerns arose. First-rate engineers of the ancient world, the Romans built elaborate and extensive aqueduct systems to bring good water to their cities. The ruins of these enormous structures, like those of the *quanats* of the ancient Arabians, remind us that people are willing to invest great amounts of time and money to achieve the goal of a clean and dependable supply of an essential of life.

Waste water was also of major concern to people from early times. Rome's *Cloaca Maxima*, still in use, was literally the Greatest Sewer. It collected and removed the waste of the city to the nearest river where the dirty water was abandoned to nature and to the concern of anyone downstream, just as we did in the modern world until very recently.

During the past century, the scientific basis for understanding water-borne disease was discovered. The conquest of those diseases through sewage collection and treatment and the purification of drinking water is a great public health achievement. In the developed world, the great plagues of typhoid, hepatitis, and others have vanished, leaving a few folk tales about "Typhoid Mary" as their only traces in society.

The sanitization and purification of foods and drugs since 1913 made sense to the citizen because he had a rudimentary understanding of the germ origin of disease and folk knowledge about poisons and "bad" food. We felt that these threats to our well-being, so effectively presented in the press and political speeches, were at least understandable and that the mitigating action made sense. Basic sanitation was becoming a way of life anyway among most Americans during the first years of the twentieth century.

It seems clear that we have come a long way in the control of threats to the public health. Why then do we experience fear and helplessness whenever we get into conversation with our fellow citizens about the subject of wastes?

That the fear and concern exist is hardly arguable. The morning paper will tell you about something in your community that is causing concern. It may be lack of landfill space, the need for new places to put wastes of every kind, even the objections of someone to every proposed way of handling wastes.

When these matters reach the comic pages you know they are important. Recently, in "Calvin and Hobbes," Calvin was discussing his need for a really scary costume for Halloween. After considering and rejecting several, he finally decided to go as the scariest thing he knew—a barrel of "toxic waste"!

When the press is dominated by television and the news of the whole world is allowed thirty minutes including commercials, finding problems is a major industry and a productive one at that. Going on to the interpretation of the meaning and significance of a problem is a much more challenging job. Detailed research and seeking the best knowledge from health professionals and others who may know what they are talking about are only rarely of interest or within the allowed-time assignments of daily news reporters. Even magazine and book-length authors have constraints.

A result of this time pressure on the news is that many reports jump quickly to a conclusion that can support a bold headline or a "Tune in Tomorrow" based on a few words from an interviewee and a dramatic picture.

The typical news report results in capsulizing the very real and grim story of people exposed to chemical wastes in New York under the simple rubric "Love Canal" or the identification of major problem sites as the "Valley of the Drums" or the "Stringfellow Pits." The names become symbols far separated from their reality.

When a citizen reads or hears or sees a video portrayal of "another Love Canal," he will understandably feel some revulsion and caution, if not active fear, and take whatever action he can to remove any threat that may be presented by the image. The fact that very few people know much about Love Canal in no way prevents this instinctive reaction when the symbol is so associated with bad situations. This same instinctive reaction to cues serves us well in avoiding many dangers, from staying on the proper side of the centerline of a road to avoiding the contents of a bottle with the skull and crossbones or "Mr. Ugh" on its label.

In this book, as we said earlier, we shall attempt to provide a basic understanding of these complex problems, their beginnings, and some solutions. At the scientific, technical, and legal levels, the Byzantine complexity of the issues surrounding hazardous waste and its management confuses everyone. Finding ways of dealing with the effects of these wastes when they escape control sometimes seems impossible. The combining of all this is likely to intimidate all of us, including the experts.

We may have to oversimplify, and for the most part we will avoid the temptation to retreat into jargon but only, we hope, toward the objective of wider understanding, to arm the citizen with knowledge needed to participate effectively in decisions affecting him/her and our descendants in an ever-more-complicated world.

REDEFINING WASTE

Waste in Society

The term "waste" has much meaning for modern society—all bad. Who among us has not been admonished by parent, teacher, or employer for wasting time, or talent, or energy, or money? A feature of every home or office is the wastebasket, into which all kinds of useless (for the moment) things are cast. Innocent objects, these wastebaskets, delights to small children, a frequent source of fires in offices and commercial establishments.

Wastebaskets have an interesting feature which will have more meaning to us later—they get filled up and have to be emptied. When this happens, someone in society has to deal with the contents not only of the innocent plaything of the creeping child, but of thousands of others which may be more suspect in their contents.

We use the word waste in many other ways. Politicians frequently assault "waste and inefficiency" in government in general, or in the Defense Department, or among contractors. Tracts of land having no visible productive use are usually called "wasteland," although one person's wasteland may be another's "fragile wilderness," since productive use is defined by the needs or objectives of the user.

Nonetheless, when Newton Minow (one-time chairman of the Federal Communications Commission) made a widely reported speech in which he described commercial television as a "vast wasteland," he communicated to a large audience at a gut level that keeps his memorable phrase alive as a part of our language.

In manufacturing industry, waste is the subject of nearly every sentence written or spoken about a process or system. Manufacturing staff meetings are called to discuss "down-time" (wasted time while machinery or personnel are out of service), waste reduction, and efficiency of the production system. Efficiency is a peculiar word which attempts to accentuate the positive by measuring and reporting the amount of the inputs that were *not* wasted, therefore used in the production of salable goods. The whole profession of industrial or manufacturing engineering is founded on the idea that all human activities are inherently prone to waste and that special skills and knowledge are valuable in reducing these inefficiencies.

At another level, we all know that wastes are an essential feature of life. The most skillful maker of clothing knows that at the end of the project there will be scraps for which no earthly use can be found. Children learned to play with blocks because no carpenter can come out precisely even in his wood supply. Making a metal part for a home model produces trimmings and filings and the like, and so does the production of a part for the real machine that the model reproduces.

We don't like to think about it, and we rarely talked about it until television commercials were found to sell laxatives, deodorants, and the like, but waste production by our own bodies is essential to life. The way in which the waste systems function can bring us feelings of good health or of severe illness. It is simple good science that moves our physician to request urine and feces specimens for evaluation. Even the archaeologists practice fecal analysis but mask the unpleasant implications by calling feces "coprolites," presumably to connote sanitation by time. A visit to the dialysis unit of a modern health center is a sharp and permanent lesson in the misery which results when our kidneys fail to remove effectively the waste products of cellular life. The body drowns in waste and will eventually die.

All these bodily wastes are repulsive to us. Even in countries where they are valued as recycled matter in agriculture, handling of them is assigned to the lowest social class. Significantly, for many generations most human wastes have been disposed of in holes in the ground, possibly following the example of some of our animal co-ancestors, possibly to make it hard for an enemy to find us. Today's hole may connect to a sewer or septic tank, but we still have a sense of finality when something disappears "down the drain."

Interestingly, the waste products of life are not always repulsive and instantly rejected. Yeast cells, mankind's friend (and occasional enemy) for thousands of years, consume carbohydrates (starch and sugar in the main) and produce alcohol as a waste product of their metabolism, mostly ethyl alcohol but often others, together with a rich variety of chemicals: aldehydes, ketones, ethers.

Since it happens that the short-chain alcohol "ethanol" is the only one of its series the human body can metabolize in any significant amount without severely toxic effects, people early found pleasure in the chemical soup the yeasts produce. The chemicals produced in association with alcohol in fermentation are called "congeners" and are responsible for the poetic and lyrical flights of language used to describe wine, beer, and whiskey, among thousands of beverages around the world. Aldehydes and esters are not romantic, but a great wine has "nose," "character," and "body," all contributed by the congeners or original raw material. Ethyl alcohol has no taste of its own.

Bread is made in a similar way, with yeasts providing not only alcohol and congeners for flavor but also the carbon dioxide gas that forms the bubbles in raised dough. Just about anyone can tell the difference between bread raised by fermentation and that raised by chemical action of baking powders.

All in all, it is amazing how waste materials in the right places become not only inoffensive but even essential and admired. How many gardeners swear by sheep manure as the only way to successful floriculture? Waste wood chips and bark are welcome around home and nursery. Soaps have been made for millenia from inedible fats otherwise wasted. Even today, special detergents are made from chemical derivatives of tallow. The products are not soap but are often better for a particular purpose.

With this sense of the need to evaluate wastes for their useful aspects and finding their place in our world, we can turn from living things to inanimate ones. Let's consider industrial waste, its origins and its nature.

Industrial Waste

We have seen how life processes take in raw materials (nutrients) and use them in ways that sustain life, generating energy and tissue and reproducing itself in myriad ways. Wastes are inevitable products of life, and we can sometimes find value in those wastes. In many ways industrial processes imitate life systems. Raw materials like petroleum, ores, metals, agricultural commodities, water, and air are subjected to chemical and physical conditions that change their forms into new ones more useful for some purpose.

Starch in a beer fermenter, when treated with malt enzymes, becomes sugars that yeast can convert into alcohol. In the chemical plant, oil is cracked to ethylene with large energy input in the form of heat; then the ethylene reacts with water in sulfuric acid to produce ethyl alcohol. The alcohols are chemically identical—but each contains impurities characteristic of its method of synthesis. The alcohol made from ethylene cannot legally be sold for beverage uses. The reasons for this radically different treatment of the same chemical are political, not scientific or technical. In fact, both varieties of alcohol are synthesized, that is, made from some other material. In one case, the synthesis is carried out by enzymes in a yeast-cell culture, in the other by chemical reactions in a powerful acid solution.

The point of this discussion is that a chemical material may be produced, generated, or synthesized in many ways. The origin of the material may be important but usually only because of associated impurities or perhaps political or even esthetic concerns. If we are to understand the *effects* of chemicals, we need to concentrate on their *identity*, not their origin. In this way we can begin to make sense of the issues.

Wastefulness and Waste—A Necessary Connection?

We have a cultural bias against waste that may lead us to believe that wastes result only from wasteful practices or shoddy operations. A related belief holds that all wastes can be recycled. In some biological (therefore "natural") systems, wastes are inevitable companions of life. We shall see that to a large extent this results from laws of nature and that these laws apply in nonliving systems as well.

It is a familiar observation that in physical matters most changes are not reversible. By this we mean that if we saw a board in two pieces, no amount of ingenuity or glue will ever put it back together exactly as it was. Even if the wood is carefully split, small pieces become disordered and can never be restored to their positions.

On the other hand a teeter-totter that is unbalanced so one end sinks to the ground can be restored to its balanced position by a push of the feet—adding a little energy. Thus, given the need for some outside energy, the position of the teeter-totter is described as a reversible system. One can make either end high or low by just the right application of energy. If both ends are equally heavy, it will remain in whatever *equilibrium* position one chooses. The word equilibrium is derived from this picture: equal weight (librium) on each side.

In the early days of the science of chemistry, researchers observed that some common compounds were extremely stable and resistant to change in the range of reaction conditions that were attainable in crude apparatus. As chemical science progressed, it became possible to isolate the elements, some of which made up these extremely stable compounds. Water, for example, could be broken down into hydrogen and oxygen, both, as it turned out, very reactive elements. Experimenters found that these two could react violently to remake the water from which they came.

Other compounds were easier to break down, and it was found that most chemical reactions form equilibrium mixtures of raw materials and reaction products that can be made to proceed to complete reaction only by some applications of the laws of chemical equilibrium. We can increase the amount of one reactant beyond the theoretical amount needed, we can increase temperature or pressure, we can add a "catalyst" to the brew. In general, though, chemists know that no reaction proceeds to completion—that is, no reaction in laboratory or plant can be made to use up exactly its original raw materials and to produce nothing but the desired product.

In a practical sense, some exceptions exist: explosive reactions are an example where evidence of an equilibrium state is hard to find. Biological systems use the special properties of complex molecules called enzymes to bring about reactions that would be forbidden because of their energy needs in nonliving mechanisms.

Nonetheless, much of the art of industrial chemistry and chemical process engineering is devoted to finding ways to maximize the amount of a desired product made in a reactor while minimizing leftover reactants and unwanted co-products.

Another complicating matter in making desirable chemicals is the practical limits of purification of the input chemicals. Most industrial chemicals, for economic and technical reasons, contain many impurities—compounds other than that on the label.

We joke a lot about "Murphy's Law," an inspired way of expressing our frustration with the real world: Anything that can happen, will, and in a way to maximize the trouble we will encounter. Well, in the case of the soup in a chemical reactor, Murphy was right. Everything reacts with everything, impurities and all.

What this means is that all chemical reaction systems have to be followed by separation and purification steps that can remove the unused input materials and the myriad possible impurities so that the final product can be most useful for its intended purpose. In this simple idea lies the explanation for much "hazardous chemical waste" that is of concern to us today as participants in the "Chemical Century" so glowingly proclaimed in the third and fourth decades of the twentieth.

A review of the tables at the end of the chapter on "How to Recognize a Hazardous Waste" shows how this works. Under the list of "Specific Wastes That Are Hazardous," we find the terms "stillbottoms" or tars listed among Organic Chemicals many times. These are purification wastes that result from the kinds of problems we have just discussed. Those who want to find out more about these kinds of things can refer to the chapter in this book entitled "Chemistry of Waste Materials" or to a standard text in chemistry.

Many other wastes are made as results of purification operations. Simple degreasing of machine parts before painting, for example, often uses trichloroethylene (TCE) as the solvent to remove the grease and oil. TCE is expensive, so most factories and shops want to use it and reuse it. Simple distillation seems to be the answer: boil off the TCE and leave the oil, grease, and dirt behind. Unfortunately, when the goo in the still is boiled down, it eventually begins to decompose before all TCE is distilled from the residue, releasing, among other things, hydrochloric acid, very dangerous stuff indeed, and very restricted in the amounts that can be released into the air.

The result is that something remains to be disposed of: spent solvent or stillbottoms containing TCE, one of the most feared chemicals of the age, if one judges by press clippings. It is hard to remember that we began using TCE to improve safety of the workers by replacing the highly flammable gasoline and kerosene that were once common in these shops.

Thus we see that waste products have a degree of inevitability; they are necessary to life itself. Some wastes are useful if not good, some are merely in the way, and some may be dangerous. The example of recovery of degreasing solvent shows that even recycling can produce difficult problems of final disposal of residues. Thinking about waste is not a pleasant avocation, yet ignoring it is not possible in a world that has just over five billion inhabitants.

Since waste is inevitable and (we argue) is always with us, why then is so much fear and revulsion caused by the mere threat of "hazardous waste" in our community or neighborhood? What is a "hazardous" waste anyway? Like many simple questions, these have complex answers. The whole of our chapter "How to Recognize a Hazardous Waste" is devoted to these answers, but some brief descriptions are possible as preface to the following discussion.

At the simplest level, hazardous waste is anything so labelled by some government agency. The Federal Environmental Protection Agency, state environmental protection agencies under many names, and even some counties and municipalities designate some materials as hazardous waste—you can look it up!

One problem with these designations is that you have to know a lot about the subject to look anything up in the tables. The lists are based on a set of assumptions that may be, but usually are not, made clear. For example, since hazardous wastes are regulated under the laws applying to solid wastes, one needs to know that liquids, even gases, are classifiable as "solid waste" under these laws and regulations.

In addition to the lists of mostly esoteric chemical substances, the definition of hazardous waste is based on "characteristics" that, when present, make a waste material hazardous for legal and regulatory purposes. Whether possession of one or more of these properties actually makes a substance hazardous may be open to debate, but to the extent one believes the tests, there is no doubt that a substance must be managed as a hazardous waste unless the owner or generator wants to risk the penalties in the laws.

The four characteristics that define hazardous wastes not to be found on a specific list are:

1. Ignitability—the property of easily being set afire

2. Corrosivity—the property of eating through metal

3. Reactivity—the property of exploding or otherwise releasing dangerous amounts of energy or fumes, etc.

4. Toxicity—defined by a wide variety of tests, chief among which is the Toxicity Characteristic Leaching Procedure, itself the subject of much controversy

We will examine the definitions and develop an ability to recognize hazardous waste in another chapter. For the moment, it is important to understand that these definitions abound in absurdities when applied to the real world. For example, the popular soft drink you enjoy on a hot day has a crisp bite because of the acid taste contributed by the carbon dioxide or citric acid used to make it fizz. The same characteristic (acidity) might classify the drink a hazardous solid waste if you poured it on the ground. On another hand, garbage and other materials from households and residential complexes may contain hazardous materials but can generally be gotten rid of in municipal waste collections. The public is usually not considered ready for strict controls, so political considerations make it unlikely that such wastes will ever be successfully restricted at the point of use. Attempts are being made to regulate "consumer hazardous waste," but few government jurisdictions have made significant progress in that direction.

Making sense out of the regulations may well be beyond any possibility in any book like this one, but we will make a try at it later. At this point, we need to examine the ideas and intent behind the growing concern about the hazards and dangers of waste management.

Historically, the tendency of mankind has been to get rid of its embarrassments by burying them in the ground. Everything from human bodies to broken clay pots to "unclean" ritual objects has been thrown into a hole or down a cliff or into a river, lake, or ocean since time began.

Archaeologists rejoice in the discovery of a midden where household wastes of ancient people were dumped. Much can be learned about people by what they discard. Lately, continuing revelations about the life of early Colonial Williamsburg have come from the results of excavations of the old privies. After all, if you were a little boy or girl or a servant who broke the favorite china or spoiled something in the house, where could you hide the evidence and not be discovered? Down the hole! No one will dig it out—at least for several generations.

Wastes therefore mirror the society that produces them. It should come as no surprise that a modern industrial society produces a modern industrial waste, even if much of that waste actually comes to the point of its final disposal at the hands of a consumer.

As we have pointed out, too, along the way to the manufacture of these new consumer goods, useless and unrecoverable by-products are bound to be made.

The twentieth century has been well named the Chemical Century; thus much of the waste of this time is of chemical-industry origin. Indeed, since all life is based on chemical systems, we find that the study of our environment and of our impact on it ultimately finds expression in chemical and biological terms.

The very word hazardous implies some kind of danger, a word that is used to describe something menacing. An icy sidewalk or a sharp knife are clearly hazardous but only if one steps on the walk or brings the sharp edge of the knife into contact with something it can cut. Hazards are remote, like lightning on the horizon, or immediate, like the oncoming vehicle in your lane, bearing down at high speed. We evaluate hazards all the time and make judgments as to their meaning to us, often whether the hazard is great enough to require action to change our behavior or plans, to avoid or reduce the risk to an acceptable one.

Few persons will choose to become sky divers, knowing the hazard of death and injury involved. Millions fly on commercial airlines despite diligent reporting in the press of every accident or near-miss. Indeed, rational analysis of risks is rare and mercifully so, for a careful analysis might lead to paralysis of our ability to function at all. Who is willing to calculate his or her chances of surviving an auto trip to a movie?

We tend to accept the risks of life with equanimity if not neglect. We cross the street between crosswalks; nearly a third of all adults smoke cigarettes; we use power tools and engage in thousands of other risky pursuits, often as recreation. In doing so, we believe we can manage the risks by looking carefully about, by taking precautions, by making judgments about our actions, by managing our affairs.

The difference between the Grand Prix driver out on the track at two hundred miles per hour and the young hotshot changing lanes on the freeway to prove his skill lies in training, experience, and technical backup. More important, Mario Andretti on the track or road-course is participating in a society of *choice*—that is, his fellow drivers choose to share the road under these conditions; the other drivers exposed to our show-off on the freeway did not really choose to enjoy his company.

Acceptance of risk is related to choice. Those risks or hazards that we choose to accept are no longer frightening because of our conviction that we are in control. People who relax and are comfortable in the heaviest traffic or the most dangerous work may be terrified as passengers or observers. Psychologists find that most fear of flying in commercial aircraft results from the feeling that one's safety is in the hands of anonymous pilots whose arcane skills can only be imagined.

Choice is an essential means of controlling our lives. We can eliminate travel hazards by staying home.

We can decide not to smoke, maybe not to have sharp knives around, to have someone in to mow the lawn. The issue is, what kind of existence remains after we have stopped taking risks? A life without risk is a life without exposure, exposure to real things, to experience, to stimulation, to other people and their world. Few will choose such a life, preferring a reasonable acceptance of the "normal" mishaps.

The problem of hazardous waste is that we feel no element of choice. No one willingly accepts contact with such forbidding stuff. Hardly anyone, that is, since workers who produce, store, transport, and dispose of such materials presumably do so as an alternative to seeking other jobs.

No, the fear arises out of concern that if hazardous waste is anywhere near us, it must inevitably reach out and expose us to some adverse effect. Adjectives like insidious, hidden, and especially unknown, are applied. Even when remarks are made acknowledging the necessity of proper disposal sites, the same commenters almost universally add that, of course, their community or neighborhood is, for a variety of good reasons, not a good choice.

Among professionals working in waste management, this set of attitudes is called the NIMBY syndrome—*Not In My Back Yard*. Sure, you've got to dump the stuff somewhere, but NIMBY! It is exactly the same reaction that opposes the building of a new airport because of the noise, but feels abused when forced to drive many miles to a remote location to catch a flight.

The easy choice seems always to give the problem to someone else. This has reached such a point that State governments are actually writing and enforcing regulations that simply require that hazardous wastes be taken out of the State.

When no one wants hazardous waste, then maybe we ought to stop producing it! Can't these nasty people quit producing this nasty stuff? Like many simple questions, these have complex answers.

The complications arise from something we will call the Law of Unintended Consequences. This law just says that everything is connected to everything else; it is usually simplified to the more familiar saying—there's no such thing as a free lunch.

In our modern (or postmodern) industrial society and free market economy, each of us makes choices every day that have consequences beyond our vision and often our understanding. The sum of these individual choices is a style of life, of convenience, choice, and quality of products that most of us are not willing to give up. So it is with the creation of wastes, some of which are hazardous.

In *Environmental Quality and Residuals Management*, a 1979 book written for Resources for the Future (a research group concentrating on resource management and environmental quality), Allen V. Kneese and Blair T. Bower analyzed the "residuals" produced by a number of common industrial processes. "Residuals" are all those things other than the desired salable product that are produced along the way to making that product. A humble example from this excellent work will show the effects of our choices on the amounts of waste materials produced.

Toilet tissue is a modern essential, and the market demand is for soft, white tissue. Claims are made for squeezability, for strength, for gentleness, and other compelling virtues, but let's look at the comparison between white (bleached) and plain, light brown, or gray (unbleached) tissues.

Table 1 shows a comparison of the residuals resulting from producing each product by the classical Kraft process starting with softwood logs. Not all of these residuals will be released to the environment—most will be captured in some sort of control system—but these materials have to be managed somehow.

Table 1

Residuals Resulting from the Production of One Ton of Air-Dry Tissue, in Pounds

Residual Type	Bleached Tissue	Unbleached Tissue
Gaseous		
Chlorine	1.2	0
Chlorine Dioxide	0.6	0
Sulfur Dioxide	25.6	12.1
Hydrogen Sulfide and Organic Sulfides	25.5	23.2
Particulates	58.5	52.7
Liquid		
Dissolved Inorganic Solids	263	22
Dissolved Organic Solids	244	41
Suspended Organic Solids	113	107
Suspended Inorganic Solids	4.5	4.1
Five-Day Biochemical Oxygen Demand	147	31
Solid		
Inorganic Solids	82.0	73.7
Organic Solids	0	0

Source: Allen V. Kneese and Blair T. Bower, *Environmental Quality and Residuals Management* (Washington, D.C., Resources for the Future, 1979), 65–66.

There are many processes for producing pulp and tissue, and reference to Kneese and Bower's book will show a wide range of residuals produced by some of the alternatives. The selected information in Table 1 does point out that some waste quantities produced are very different for the white bleached tissue as compared to those produced with the less acceptable gray-brown variety.

The big difference is found in the Five-Day Biochemical Oxygen Demand. This particular analytical test measures the amount of sewage treatment required for the plant effluent—nearly *five times* as much for the white tissue as for the same tonnage of unbleached tissue. Worse yet, in biological treatment of the waste, about one-half pound of organic sludge is produced for every pound of biological oxygen demand eliminated. Thus the bleached (and softer) tissue will result in a five-fold increase in sewage sludge, which must be disposed of also.

Whether these residuals are "hazardous waste" in the legal and regulatory sense is really not germane to the point of this discussion. What is important is that *the market decision to prefer white toilet tissue* results in the need to manage two to five times as much waste as might be *essential* to the production of a supply of functional toilet tissue. Some of that waste may be hazardous or at least have to be treated as hazardous under the regulatory scheme.

Many millions of times each day, consumers vote with their dollars in the market to bring about results comparable to those we followed in the toilet tissue case.

As another example, many people prefer soft, washable acrylic fibers for clothing and other fabrics, replacing wool, which has to be dry-cleaned and has other perceived problems. Wool is grown on the bodies of sheep, whose animal wastes we urban residents never see and that are rarely treated in any way. The wool is cleaned and prepared by ancient technologies that most of us might think of as traditional or friendly, maybe messy but not threatening. Residents downstream from a wool-scouring plant might not be so easy to persuade, but that's another matter for now.

Acrylic fibers are made in the main from acrylonitrile, a known-to-be toxic and possibly cancer-causing chemical derived from propylene and hydrogen cyanide; and methacrylic acid, derived from similar stuff. Manufacturing the raw materials results in waste products found in the tables of hazardous waste: stillbottoms, etc., that may or may not be recognizable to the uninitiated as connected with such common articles as your favorite, soft acrylic sweater, but can be clearly tied in by a little research.

Similar arguments apply to the manufacture of polyester fibers and fabrics and to the various forms of nylon available. Similarly, the production of the raw forms of most of our "modern materials" has predictable consequences. The chemical industry, consequently, is among the leaders in the management of hazardous residuals—they have to be!

Do the advantages of lower prices, personal comfort, and ease of cleaning justify the need to manage the wastes produced in manufacturing the raw materials for the consumer products? Who can say? The market decision has been for the synthetics by many thousands of tons per year, although the wool producers are making a comeback, often by treating their "natural" product with chemicals of similar character to those used in the synthetics.

These few examples are intended to show that each decision we make in our daily lives has consequences, most of which are only dimly understood, if at all, by most of us.

Failing a mass decision to return to some imagined time in which "synthetic" materials were not known and only "safe, natural" products were needed or used, we must accept the reality of the waste problem. Table 2 lists some common consumer items and some of the chemical substances that are used in their manufacture. Wastes from the original production of the chemical raw materials and from manufacturing the articles themselves can enter the waste stream in many ways. Some of the wastes generated in response to our demand in the market will be hazardous if we come into contact with them, and that is the story of this book in its most compact form. Walt Kelly was attempting to express one aspect of this truth when he had Pogo the Possum say, "We have met the Enemy, and he is us!" Pogo's simple truth has become a rallying cry for the environmentalist movement, but few really grasp its deeper meaning.

In the United States a fundamental assumption of the society is that an informed citizenry will make sound decisions for the governance of the country. Public education and the free press are intended to be major contributors to the achievement of this state of good government. In the debates of two hundred years ago about the form of government under which we now live, the primary source of disagreement was, at root, over whether this principle of trusting the citizens to rule would really work.

How in the world can a citizen become "informed" in the way envisioned by Jefferson when faced with a field like hazardous waste? Even the language of any dialogue seems alien and often incomprehensible: dumps have become Waste Management Facilities or Superfund Sites. The opening paragraph of a proposed regulation in the Federal Register recently contained twenty-one capital-letter acronyms by actual count! Only a skilled bureaucrat with a prodigious memory could read it and have any real clue as to its meaning or even its subject.

Table 2

Some Common Modern Materials and Their Major Chemical Inputs

Article	Input Chemical	Toxic	Cancer causing
Acrylic Fiber			
	Acrylonitrile	X	X
	Methyl Methacrylate	X	
Polyester Fibers			
	Xylene	X	X
	Terephthalic Acid and Esters	X	
Urethane Foams			
	Benzene	X	X
	Nitrobenzene	X	X
	Aniline	X	X
	Toluene	X	
	Dinitrotoluene	X	
	Phosgene	X	
	Diisocyanates	X	X
Plastic Pipe			
	Vinyl Chloride		X
	Acrylonitrile	X	X
	Styrene	X	X
	Benzene	X	X
	Butadiene		X
Laundry Detergents			
	Benzene	X	X
	Toluene	X	
	Organic Phosphates	X	
	Inorganic Phosphates		
Nylon Stockings and Lingerie			
	Benzene	X	X
	Ammonia	X	
	Cyclohexane		
	Adipic Acid		
	Hexamethylene Diamine	X	
	Caprolactam	X	

Repeating our earlier words: it is a premise of this book that a reasonably literate citizen *can* become informed on the main aspects of hazardous waste problems to an extent that will have real value in participation in public decision making as it affects him and his family and his community.

We will discuss factors involved in determining risk, the place of relative risk assessment in making judgments, and the sources of information available to anyone willing to make the effort.

Members of the general public can choose to participate in the development of solutions or not, and we will discuss how one can get involved. Citizens who happen to be responsible for the management of industrial and commercial wastes have no choice—they must become involved and knowledgeable or be prepared to face heavy penalties in both legal sanctions and public opprobrium. We will discuss the ways in which the businessperson can become informed and play a proper role in the management of this great public concern.

Experts *can* be encouraged to speak clear English—even lawyers and engineers. Rational discussion is possible even today, if understanding begins to replace fear and hostility and if questions are clearly based on knowledge and thought.

Finally, it must be accepted that hazardous wastes, like dirty air and dirty water, are the result of the actions or inactions of many people, often, although not always, driven by a perceived imperative to supply a demand for some material or some thing in our market economy. The American consumer may feel betrayed that no one ever told him or her that synthetic fabrics contribute to the generation of hazardous wastes, but they contribute no more than hundreds of other kinds of desirable products do. If the reaction to gaining this understanding is a search for scapegoats and an attitude that someone must pay, then the real need to develop a national consensus on ways of managing wastes will not be met, and the angry confrontations will continue while dangerous wastes continue to be *mis*managed to the detriment of all.

Chapter 2

A Businessman's Guide to Life under the Hazardous Waste Laws and Regulations

Throughout this book, all readers are considered to be citizens with an interest in understanding the problems of hazardous waste management and in playing a responsible part in the national effort.

Business owners, managers, and those who form part of the business system are citizens, too, with a difference. That difference arises from the natural output of larger quantities of wastes of all kinds from manufacturing and service businesses. There is another, important difference: the laws and regulations are applied more strictly to business operations than they are to homeowners and commercial and residential property owners and managers.

The thesis of this chapter is that business leaders cannot afford to stick their heads in the sand but must make at least a reasonable effort to define their responsibilities and take action. Persons familiar with the extent of the obligations of most companies in this area watch with fascination the current level of corporate takeovers. In the time and with the kinds of resources devoted to one of these battles, the acquiring stockholders cannot possibly understand even the simplest concepts about these open-ended liabilities arising from the requirements of Federal environmental regulations.

Since all the liabilities go with the property and the company, the acquirers may acquire hidden obligations that can wipe out much of the worth of the acquisition. Some notable examples can be cited, such as Occidental Petroleum's acquisition of Hooker Chemical Company just in time to feel the brunt of public opprobrium and costs of the Love Canal site.

This chapter is intended primarily as a general guide to the regulations and laws on hazardous waste management as they apply to small and medium-sized manufacturing and service businesses and to their professional advisers—particularly legal counsel, consulting engineers and chemists, and real estate consultants or brokers.

Many executives of large companies could profitably study this chapter as a brief summary of their own responsibilities in waste management. Absorbing this information should make them more appreciative of the work of those specialists whom large companies employ to handle these important business functions.

YOUR ATTENTION, PLEASE

You Must Believe

The major problem most business people have with the hazardous waste laws is that they may know in an abstract way that the laws exist, but they do not really believe what they have heard about them.

Whether managers of big business or owner-operator entrepreneurs, most business people frequently react to attempts to describe their personal responsibilities and their exposure to liability under these laws and rules with skepticism bordering on and rapidly becoming angry denial that such rules can actually be in existence, much less enforced.

Recently in a consultation with professional-level personnel who are responsible for developing and promoting a compliance program for a large corporation (profits near $500 million per year), top management indicated that it wanted to put more emphasis on this important area of "social responsibility" but that no increase in "head count" would be authorized to accomplish the increased emphasis, which was only vaguely defined. This common jargon simply indicates that if someone can be found within the company to do some work instead of their current assignment, then the work can be done. This attitude in a company with many thousands of employees whose top management level does not even know what hazardous wastes it produces or what happens to them is, regrettably, not unique.

Companies that produce and market hazardous substances have always paid attention to this area of regulation. Their practice may not always be in accord with their public stance, but at least they have responded.

Many companies that use hazardous chemicals in maintenance, manufacturing, or sanitation, and in quality-control laboratories are run today by executives whose background in marketing, sales, finance, and/or law has left them completely uninformed about today's complex manufacturing plants and processes.

Most small businesses are owned and operated by people whose natural talent is entrepreneurial—following nontraditional and unusual ideas has provided their business opportunity. The last thing they want to do or even have anything to do with is a slavish adherence to a detailed bureaucratic procedure for dealing with waste—they see it as a waste of valuable time in a typical day-and-night work schedule.

This discussion was originally intended to focus on the needs of the small business, and that will remain the primary focus. Whether large companies choose to do so or not, they can devote manpower and other resources to their compliance responsibilities. In the small entrepreneurial company, the ultimate responsibility for everything falls on the owner-operator. If violations occur, his personal assets and often his personal liberty are at stake in an enforcement action. The chief executives and other operating officers of large corporations *believe* that they are not exposed to personal liability for violations that occur in their organizations, even if they have failed to provide resources to the persons who have been delegated the responsibility to do the job. Recently, some prosecutors and judges have set out to impose penalties on officers of public companies to correct this misunderstanding.

Whatever the size of the company or business, the risks associated with either ignorant or willful failure to comply with the hazardous waste laws have the potential to destroy or seriously affect the viability of a business. As we shall see, the liabilities can be open-ended and not bounded by any limits that provide effective protection, even for the largest companies. This is the fundamental reason that one cannot obtain insurance protection for the cost of environmental impairment that results from normal business operations as opposed to a "sudden unexpected" incident like a truck accident or an equipment failure. The insurance underwriters simply have no way of estimating the magnitude of the financial penalties and cannot calculate a rate for the insurance.

The logic seems inescapable: these rules generate liabilities that are incalculable in literal terms; the public has demonstrated its demand for proper behavior of businesses; compliance is possible by the application of effective management and resources. A weak and ineffective program, no matter how much rhetorical support it may get, cannot protect the company from excessive liability. Financiers, stockholders, directors, and employees need to assess their company's response to these programs and insist that the response be enough.

A final word in this regard. The estimated liability of all American corporations resulting from the requirements of the Comprehensive Environmental Response, Compensation and Liability Act (CERCLA, see Chapter 13 on Superfund) is $600,000,000,000. That's nearly half of the annual Federal government budget and is several times the total annual profits of all publicly held corporations.

Despite the huge size of this potential liability, one seeks in vain any provision in the books and accounts of the corporations, unless they are actually in the process of a really major cleanup. The accounting profession simply ignores the problem under a judgment that few of these liabilities have been defined by technical work or in contracts or court orders. In the absence of hard data, the billions in contingent costs are simply lost in comments on the financial statements that there are no "known liabilities that are believed likely to have material effects on the financial condition of the Company" or some similar language. In a world where $600 billion can be ruled "not material," what businessman or woman can justify significant expenditures for determining compliance?

For the smaller business, a simple real estate transaction—say the purchase of a site for a new plant—can expose the company to unlimited liability for cleaning up contamination left behind perhaps long ago by previous owners who are now impossible to find.

As we shall see, in this field government agents can decide that you are responsible and dictate what you must do to remedy the situation, or can clean it up themselves and send you the bill. If that does not get your attention, then there is little hope for the future.

A Law with Very Long Arms

As just discussed, there is a problem in getting the attention of obvious candidates for regulation under the hazardous waste laws and regulations. There is even more difficulty in getting the attention of people who provide services to business. Corporate lawyers who specialize in counseling small businesses must understand the need for compliance and the high price of failure. The impact on the activities of industrial and commercial real estate brokers, on financiers and banks and other similar services is not readily apparent, but in fact, no one related to the business system is unaffected by the requirements of the principal environmental laws.

We will discuss real estate transactions more fully a little later in this chapter, but a few questions may get our attention to some of the issues:

1. Is a broker liable if he fails to disclose a past industrial use of property that may limit its usefulness or subject the buyer to large costs for cleanup?

2. When an attorney reviews an abstract of title, is he obligated to provide an opinion on any historic uses that could have created contamination?

3. When creditors foreclose or create a bankruptcy, who is responsible for any obligations under the hazardous waste rules?

Similarly, failure of a supplier to comply properly with his obligations to provide information on the hazardous nature of a product sold to a business can create some very expensive liabilities. One can go on and on with this tedious recital of all the problems associated with failure to comply rigorously, but the point has been made. The business owner, his suppliers, financial institutions, and professional advisers all have major stakes in the success of the compliance program.

The really frightening part is that they all frequently depend on the best efforts of the lowest-paid members of the manufacturing organization—the service and janitorial staff. Effective training and supervision of these invisible persons is the first essential after one decides to "get with the program." Business analysts are fond of the entrepreneurial manager who is willing to "bet the company" on the outcome of a business decision. Today, you may "bet the company" on the performance of someone you don't even know exists.

A COMPLIANCE PROGRAM

Identify Your Waste

The first step in compliance has to be finding out what kinds of waste are produced in your operation, whatever that operation might be. Remember, many businesses that are not engaged in manufacturing create substantial quantities of wastes that *may* be hazardous and that in any case need intentional management.

The key to this is the dedication to change from a casual policy toward wastes as something to be "thrown away" to a policy that develops and communicates your intent as to how you and your people are to manage wastes whenever and wherever they are generated. Even the nonhazardous wastes need to be managed purposefully, because you may someday have to justify your conclusion that they are not hazardous.

Identification of wastes is in principle a simple task, in practice a difficult, often boring, usually time-consuming and sometimes expensive process. You must have on hand a copy of the Hazardous Waste Rules for your State, county, or city. You must study and understand them before this process begins. If you cannot understand these complex rules, seek advice from the agency that promulgated them; they will often provide at least a service of answering questions about specific compounds. For a large operation or one in which the key personnel have no time for conducting this evaluation, a qualified professional consultant in waste management must be retained. *Even honest mistakes can protect only against criminal liability for violations. Civil penalties are often imposed even when a good-faith effort has been made but without the needed trained personnel.*

There is simply no substitute for the observation of the activity by a person who understands the production process, who notes when waste is created, how much is made, why it is made, and what happens to it immediately after it is made as well as later when it must be moved or otherwise gotten out of the way. Do solvents drip on the floor when a washed part is taken out? Who wipes it up and with what? Where do the rags or other absorbents go?

Does your machine leak or overflow occasionally? When a contractor's personnel finish a job, what do they do with the leftover paints, solvents, cleaning compounds, caulking compounds, adhesives, and a thousand other things?

You need to know the sources of waste in your operation and to have some idea about the quantities. Be sure to insist that the various wastes be kept separate so you can know whether mixing them is sound practice or simply creates more hazardous waste than is necessary.

Waste Reduction Source, the newsletter of the Minnesota Technical Assistance Program, put it this way: "If you mix chicken salad with chicken manure, the only possible product is chicken manure." This excellent program, supported by the Minnesota Pollution Control Agency and the U.S. Environmental Protection Agency, is the source of many good ideas expressed in this discussion.

So, the first step results in an *inventory* of wastes, their kinds (solid, liquid, gases in bottles or containers), what part of the operation generates them, and their quantities. Don't forget to check the dumpster out back and the warehouse corners for the stuff the foreman doesn't know what to do with. Find out what's going down the sewers, especially the floor drains, and measure again the amounts of material being recycled or shipped out for recovery. What happens to paint filters and paint removers used to clean filters and booths?

A key part of the *waste inventory* is an inventory of all materials that enter the operation at the plant, the job site, the shop. Both OSHA (Occupational Safety and Health Administration) and many States require that you have on hand in readily accessible places the manufacturer's Material Safety Data Sheets (MSDS) listing any hazardous materials that might contribute to the production of hazardous wastes.

The senior managers at the generator site must read the MSDS's that are on file. If you cannot understand what these papers are telling you, imagine how difficult it may be for the ordinary worker to comprehend the hazards that can exist, much less to plan his/her work to control the hazards to which he or she may be exposed in the course of routine work.

The person in charge must read and understand the Material Safety Data Sheets because most businesses buy supplies almost exclusively on the basis of the performance, in some particular application, of the material as supplied. Few buyers have any idea what chemicals are to be found in common supplies like cleaning compounds, machine coolants, or special solvents. Often these were marked only with trade names for identification, which triggered the imposition of regulations under the general rubric of "Employee Right to Know," which now require much information to be placed on containers and to be available in the workplace.

The supervisor has the final responsibility in these matters to translate the "right to know" into the "right to understand"—a much more difficult objective.

Classify Your Waste

With the inventory complete and organized well enough to describe the wastes, their quantities, and their probable constituents (based on a review of the hazardous characteristics, if any, of the materials used in the process that produced the waste), you need to divide the wastes into classes in order to decide how to manage them.

There are two classes of hazardous waste. The first to check for on your inventory is the "Listed Wastes." Generally, these are generic classes of waste materials like chlorinated solvents (trichloroethylene, methylene chloride, etc.) that have been used and are "spent," that is, cannot be used further without treatment; or wastes from specific kinds of manufacturing processes, like residues from recovery stills for certain solvents. There is also a long list of wastes from chemical manufacturing that may take a Ph.D. chemist to read. Each of the listed wastes has a number that is essential to its later disposal. An auto body repair shop will usually have wastes bearing numbers like F002—Waste Paint Remover, F005—Waste Paint Thinners, or D008—Paint Arrestors Containing Lead.

Now, with your own knowledge of your operation, and preferably with the assistance of a professional expert, review the listed wastes as they apply to your location. Remember, while the Federal rules are relatively easy to find in the Code of Federal Regulations Parts 261 and 265 at your local library, they are not complete for most locations in the country. It is likely that a State list and not unlikely that a county or city list will be more extensive and *more stringent*. Those additional lists are generally the ones that you *must* follow. To locate the applicable ones, call or write your State Hazardous Waste Office (see the listing in the Appendix) to obtain copies.

Begin with identification of all the wastes on your inventory that are exempt from regulation as hazardous waste. Some regulatory agencies publish lists of exempt waste, but the mere absence of a waste from a list cannot be taken as a sure indication that it does not require management as a hazardous one. Such materials need to be carefully reviewed in the light of any special requirements, of the possible need to be classified as "characteristic" waste, for example. Such bad actors as polychlorinated biphenyl (PCB), asbestos, and a few others are regulated under special Federal rules that are not the same as those for most hazardous wastes. However, some of the states require these kinds of materials to be reported and otherwise treated as hazardous wastes under their rules. Such unique features demand that each business get help on the rules in its own location—a general work of this kind can even be misleading if we try to get into details. Typical of some exempt wastes are:

- Nonhousehold rubbish like unusable paper and board, wood, and some plastic scrap

- Scrap metal that is recycled
- Used oil that is recycled in some cases, after evaluation
- Arsenical-treated wood wastes generated by end-users of such wood
- Some chromium-bearing wastes in which the chromium exists primarily in the trivalent form with no hexavalent chromium

A quick look at this list, which is not by any means complete, reinforces the need for expert advice and a careful review of the possible components of any waste stream.

Set aside on the inventory any wastes that you may have that are exempt. In general these can be disposed of in sanitary landfills if the landfill operator will accept them. If there is a problem with acceptance, then your expert review may be of some value in persuading the landfill.

You will find it worthwhile to review the rules on empty containers that were used for hazardous materials. At least in small operations, these containers are often the only residues of the hazardous raw materials, the main part of which can be shipped as a part of the final product.

The general rule for containers that have been used for any hazardous material (not acutely hazardous, which we will discuss below) is that the container is empty if all the removable material has been taken out of the container by the use of common practice *and* if at least one of the following conditions has been accomplished:

- No more than one inch of residue remains on the bottom
- No more than 3.0% of residue remains in a container having a total capacity of 110 gallons or less
- No more than 0.3% of residue remains inside a container having a total capacity of more than 110 gallons

Please be sure these rules are applicable in your jurisdiction—some may be different. The Department of Transportation rules can cause confusion with your transporter. Discuss this in advance. The application of these rules is discussed in Chapter 12 under the heading "Transportation of Hazardous Waste."

If you have containers that have contained *acutely hazardous waste*, then the containers have to be triple-rinsed with an appropriate solvent, with each rinse using an amount of solvent equal to at least 10% of the capacity of the container being rinsed. This method only breeds more hazardous waste unless you can use the rinsings in the process. Generally, the rules allow you to find and use an "equally effective" method of cleaning or to use an inner liner in the container that can be handled as solid waste. Any such equally effective method should be approved by your regulatory official.

Having set aside any exempt wastes and with the fate of empty containers decided, then comes the difficult question: is any of your waste hazardous or extremely hazardous, and if so, how much of it do you generate in a month?

Preferably with the help of your expert adviser (who may be a State Technical Assistance Counselor), find which, if any, of your wastes are Listed Wastes in your State or locality. Often these will not be the same as those listed in the Code of Federal Regulations, so you need your local list. Most small businesses need to look for used chlorinated cleaners and solvents, like carbon tetrachloride, freon, methylene chloride, perchlorethylene, and all the related ones, of which trichloroethylene is by far the most common.

Many other solvents and adhesive bases, like alcohol, esters (ethyl acetate), ketones (MEK), methyl alcohol, toluol or toluene, and xylene are Listed Wastes or, because they may comprise most of a waste, make that waste subject to regulation as the listed solvent.

Electroplaters have their own range of chemicals and residues to be evaluated, while operators of the ordinary auto body shop and machine shop will often produce more hazardous waste than they realize.

Once more, review the lists in the light of what you do and get help to interpret them as they may apply to you. Make up a complete listing of all Listed Wastes and the quantities produced each month. This time, add to the list the locations in the plant where these materials are stored, what kind of containers they are stored in, and any labels or marks that have been applied to the containers or the area of storage. You'll need this information in a little while.

A potentially more troublesome problem arises from the second class of hazardous wastes—those that are not found on lists but are classified hazardous because they possess one of four "characteristics" that the law and the rules define. These characteristics have already been defined in the Introduction, but we should look at them again in the practical context. They are:

Ignitability. For liquid wastes, this characteristic is possessed by any liquid having a "flash point" in excess of 140° F (60° C). The flash point is determined by a prescribed laboratory method that requires special equipment and training. For most substances, the flash point can be found in the manufacturer's Material Safety Data Sheet. Solid waste is also ignitable if it can *spontaneously* catch fire and burn so persistently that it represents a hazard. To show how tricky this can be, fine steel wool that has been used to rub down wood or metal with drying oils like linseed and tung oil will often self-ignite in contact with the air and must, unless treated, be considered an ignitable hazardous waste. Even the traditional "oily rags" we have all been taught to keep in sealed containers for safety can be a problem unless they are handled in accord with a safe practice.

Reactivity. This characteristic, for which the EPA has not been able to develop a specific laboratory test, is possessed by materials which are unstable in themselves or which may react with other wastes or air or water to create an explosion, a stubborn fire, or to generate large amounts of noxious or toxic gases and fumes. In some regulations the class is called "oxidizers" or some similar term, but this is really too limiting.

Corrosivity. The characteristic is based on the strength of acidity or alkalinity which the product possesses. Generally, any water-based waste having a pH of less than 2 or more than 12.5 is hazardous under this characteristic. Acids of this concentration are able to corrode steel at rates in excess of 1/4 inch per year, and any solution that can do that is considered hazardous for corrosivity. Concentrated alkalies are destructive of organic materials including human skin.

Toxicity. This characteristic is based, for Federal purposes, on a procedure called the Toxicity Characteristic Leaching Procedure. This is defined by a complex and strictly defined laboratory procedure which extracts a sample of the material with water or a dilute acid in a specially designed extraction vessel, filtering through special filters, and analyzing the liquid extract. This test and its interpretation are both changing and are somewhat controversial until further testing and validation have been conducted. In the present method, any concentrations of certain metals and a few organic materials that exceed 100 times the National Primary Drinking Water Standards are considered to define a hazardous waste. Because of dissatisfaction with this test, a number of states have adopted their own definitions of the characteristic of toxicity.

Other Characteristics. The individual States, having authority to set their own requirements if they are at least as strict as the Federal ones, have adopted a bewildering variety of characteristics which can define wastes and the management systems that must be followed in handling and disposing of them. Once again, the only way to know whether your wastes are hazardous or not is by reference to your state and local regulators. In some states, notably California, it is unlawful to dispose of many substances as a matter of policy. Often the State need not prove that harm to people or the environment might result from unregulated disposals. Use your expert!

Using the characteristics to decide whether your wastes are hazardous is a job calling for intimate knowledge of the operation that generates the wastes and for expert judgment as to whether any of the non-Listed Wastes and nonexempt wastes on the inventory are likely to exhibit the properties that make them Characteristic Wastes.

Working with accurate knowledge of plant operations, experts can usually perform screening to help eliminate some wastes from further consideration on the basis of the chemical constituents and of the history of the material. Those wastes that cannot be eliminated in this screening effort will need to be tested. In general, these tests will have to be performed by a qualified independent laboratory, often one that is certified by State or Federal authorities for the particular tests that are to be run.

With the results of the laboratory tests in hand, your inventory of wastes is complete, and the ones that have hazardous characteristics should be added to it. Remember to list the quantity of each kind of waste produced (generated) each month. This is also a good time to recheck and ensure that all containers are labeled in accordance with regulations. As a minimum, each container should have an approved label that clearly indicates that the contents are hazardous waste and that has a proper description of the waste—a waste number or Department of Transportation classification—and the date on which the waste was placed in storage. Labeling is a subject in itself which we will discuss somewhat more fully later.

Determine Your Generator Status

If any of your wastes are classified hazardous by any of the lists or characteristics we have just discussed, your *facility* is a generator of hazardous waste. The amount of waste generated is of no significance in this determination. *Any* hazardous waste production makes the facility a generator and subjects the facility to the regulations.

The quantities of various classes of waste produced and/or stored determine which regulations apply, and this is important, as we shall see. However, if any hazardous waste is stored, treated, or disposed of on the property on which the facility is located, you are required to obtain an EPA Generator Identification Number.

The Identification Number is obtained by filing EPA Form 8700-12 with the Regional Office which is responsible for your location. For a copy of the form, see the chapter on the Resource Conservation and Recovery

Act (RCRA). You will also find a listing of the Regional Offices of the U.S. Environmental Protection Agency in the Appendix. The law required all generators to file these notices by August 18, 1985, and any generation, treatment, storage, or disposal after that date without having filed the notification and obtaining a number is unlawful.

In the interests of maximizing compliance with the regulations, the EPA has provided some easing of enforcement action for facilities that file later. But it is essential that the notice be filed *as soon as you (the operator) become aware that the requirement affects your facility*. It is undoubtedly better to file as soon as you know you must, rather than try to avoid the rules. The latter course has too much risk because too many people need the Identification Number to do business with you, and sooner or later the moment of truth will arrive. Besides, it's against the law and could make you subject to prosecution, with jail time and huge fines of up to $25,000 per day of violation real possibilities.

In order to file the Notification, you must state how much hazardous waste of what kinds you generate or expect to generate. The crucial question is: more or less than 1,000 kilograms (2,200 pounds) each month? The regulations are different and increasingly more demanding for generators of 100 kilos (220 pounds), less than 1,000 kilos, and more than 1,000 kilos per month. All must file their Notification and all must comply with some level of controls, but there is some simplification for the smallest-quantity generators. The Form 8700 itself does not provide for a classification of a generator of less than 100 kilos per month, because for most purposes, these smallest of the Small Quantity Generators are labelled Conditionally Exempt SQG's. The condition on the exemption is that their waste not be classifiable as Acutely Hazardous Waste in accordance with the RCRA rules. If it is so classified, the generator is subject to all the rules of management of hazardous wastes that apply to all generators. A number of states, including California, Louisiana, Minnesota, Rhode Island, West Virginia, and others provide no exemptions from the full requirements on the management of hazardous wastes for Small Quantity Generators. One hundred kilograms per month amounts to about one-half of a 55-gallon drum of all-liquid waste per month.

Packaging, Storing, and Transporting Wastes to Disposal

Once it is determined that your facility is a generator of hazardous waste and you have applied for your EPA Identification Number, you must find a way to dispose of your waste that is safe, effective, and within the rules of your State or locality.

Generally, waste materials which can be recyled within the same process or facility are not subject to any regulation. This is obviously the best solution to the problem. Unfortunately, few plants are lucky enough for this situation to prevail, at least for all of their waste. Solvent recovery stills are a good solution for a large user of most solvents and degreasers, but they do end up with a residue in the still that cannot be eliminated easily and that usually is among the listed hazardous wastes. A good example is found among dry cleaners who installed solvent recovery systems under pressure from air pollution rules and who now produce stillbottoms which are Listed Wastes. The old practice of exchanging used solvents for new from a supplier has just about died out because the solvent distributors are often unwilling to assume the burden of disposing of the residues.

In some states, even used materials intended for internal recycling must be reported on the semiannual or annual reports which must be submitted. More about these reports later.

Whatever the situation, packaging, storage in-plant, and transportation to a (usually) different site for disposal are all essential to compliance with the laws. Few small businesses are equipped with personnel or machinery to treat and dispose of their own hazardous wastes. If this is your choice, read carefully the chapter on the Resource Conservation and Recovery Act, obtain good professional help, and apply for a Treatment, Storage, and Disposal Permit. This process alone will likely lead to a careful reexamination of the wisdom of choosing this option. Some businesses will find it worthwhile, particularly electroplaters and some other metal finishers who can use established technology that is recognized as able to achieve "detoxification" or "delisting" of specific wastes containing heavy metals. Most such businesses will find there are cost-effective and easily operated package treatment units available to them that can save much hassle *and* make a small return on the value of the recovered metal.

At this point, you must make a major decision. You need to select a service company to transport your waste to a disposal facility. More important, most small businesses need to establish a relationship with a service company that does more than haul.

Remember, the generator of a hazardous waste is responsible for that waste and for its effects on the environment *forever*. This is not a metaphor for some long time; it is the literal language of the law. The transporter and even the disposal facility have their own responsibilities and accountabilities, but if they fail to achieve protection of the public for any reason, the original generator is accountable for the damage and for its abatement whenever and wherever it happens. Therefore, your choice of both transporter and disposer needs to be based on knowledge and deliberation, not on sales presentations, much less casual choice.

Help is available.

The State or local agency which regulates hazardous waste disposal in your area may have a list of licensed haulers and often will know what final disposal facility is available for your area and your kind of waste. Not every disposal facility can handle all kinds of waste since its license or permit limits it to specific kinds and often restricts the amounts allowed to be handled.

Many cities and counties license all refuse haulers who operate in their jurisdiction. A check with the official responsible for licensing may result in a list of possible ones. Your trade association or even neighbor companies in an industrial park or district may know of haulers with good reputations who already serve your area. Ask, find out as much as you can.

Most small businesses, as we have said, need a service that is more than simple transportation. A good transporter will know what disposal facilities are logical candidates for your waste, how to contact them, how the disposer and the transporter prefer the wastes to be packaged, and can help you understand how to classify your wastes for shipment under the Department of Transportation rules and fill out the Uniform Hazardous Waste Manifest for each shipment.

For a small-quantity generator, the length of time wastes may be stored on-site without a Treatment, Storage, and Disposal Facility Permit is affected by the distance to the final disposal site. The Federal rules allow for accumulation of wastes up to 180 days without need for a storage permit, provided the total amount does not exceed 6,000 kilograms (13,200 pounds) and that an employee is on-site or on-call at all times to handle a hazardous waste emergency. If shipment must be for more than 200 miles (not unusual today), then wastes can be accumulated up to 270 days, but subject to the same quantity limit of 6,000 kilograms. As of March 1, 1986, all hazardous waste must be disposed of in a properly licensed facility, and none may be placed in sanitary landfills or similar garbage-disposal units.

For the smallest generators (Conditionally Exempt and generating 100 kilograms or less per month), quantity in storage is limited to 1,000 kilograms and length of time for storage without a storage permit is 180 days if 200 miles or less and 270 days if more than 200 miles from the disposal site. Disposal must be accomplished at a permitted hazardous waste disposal site, which will ordinarily require preparation of a Uniform Hazardous Waste Manifest to accomplish the transportation to the final site. Under the Federal program, the record-keeping may be less demanding than for the larger-quantity generators, but most state programs give no relief from these demands. Special rules apply if the waste is acutely hazardous, and you should refer to your consultants and professional advisers on the requirements.

Regular generators of more than 1,000 kilograms/month of RCRA waste can accumulate only up to 90 days without a storage permit.

The amount of storage you will need is determined by the amount of waste you produce, but even more by the hauler's schedule. If you can arrange a pickup every week at reasonable costs, little storage is needed—a significant advantage for the generator. For a long haul to a disposal site 200 or more miles away, the 180-day or even the 270-day period may be economical.

With your inventory of hazardous wastes in hand, including the amounts, types, and package sizes and types in which you keep them, you must contact one or more of the transporters or service companies which have turned up through the contacts suggested above.

The first essential is: does the transporter have his own EPA Identification Number? If he lacks this, there's no need to go any further with him. Try someone else.

The second essential is: does the transporter provide service for the kinds of wastes you produce? To which disposal facilities? What is the schedule of pickup and shipment to the disposal facility?

What services or advice is the transporter prepared to give in addition to basic transportation? Can you get counseling on the size and type of packaging needed to be acceptable for transport and to meet the final disposer's needs for its handling and processing system? In short, is the transporter prepared to work with you to achieve your goals as well as his own? If not, then maybe you need to look at someone else.

If the chosen company still appears to be one you wish to work with, then remember that *the generator is responsible for the performance of the transporter*. You must diligently inquire into all aspects of the transporter's background, reputation, and performance to reduce your own liability. Some important questions are:

- What permits/licenses/registrations does the transporter have in effect? For example, he may need an EPA Identification Number and State licenses for all states through which your waste will pass.

- Can the transporter show Certificates of Insurance complying with at least the minimum requirement of all jurisdictions entered?

- Are the company's drivers qualified in accord with the rules issued under the Motor Carrier Safety Act and applicable State rules?

- Is there evidence that the drivers are trained for safe handling of hazardous waste?
- Do they know emergency response procedures and have a written plan to follow in the event of some accident or emergency?
- Do they know the proper Department of Transportation (DOT) shipping names, United Nations (UN) numbers, hazard classes, and shipping labels for the range of materials that they haul?
- Have they the necessary papers and training to follow proper loading procedures?
- Have they been trained in proper placarding procedures?
- Are they trained in the use of hazardous waste manifests? Can they assist their generator customers if needed? Will they refuse a shipment if it is not properly manifested and might be in danger of refusal at the disposal site?

It should be clear that these and other questions which should occur to you are far more detailed and even intrusive into the business affairs of the transporter than would ordinarily be reasonable or even allowed by a company which you hire to provide a product or service. That does not alter the situation—the facility in which the waste is generated is responsible for the actions of the transporter to a degree that is simply not found in ordinary business relations, and this responsibility cannot be transferred by the generator to the transporter. The generator must rely on the transporter to perform without direct supervision, and his only protection is an attempt to assure himself that the transporter will do the job right.

Selecting a Hazardous Waste Management Facility—Final Disposal

Selecting a hazardous waste disposal facility is the responsibility of the generator. This comes about even though you may seek help in identifying possible facilities and your transporter may prefer to work with one or more and strongly recommends them. Under the basic principle of the hazardous waste laws, the generator is forever responsible for his waste. Thus it is incumbent on the generator to do everything in his power to select a waste disposal facility that can be relied on to perform its function properly so that the generator will not have to pay again in the future for disposal of wastes that he thought were gone forever. It is the management of the TSDF (Treatment, Storage, and Disposal Facility) which performs the functions for which the generator is held accountable. This accountability cannot be contracted away. Even though the TSDF is held to be responsible for any mismanagement, the original generator will have to pay its share of any remedial costs which are not covered by the disposer's funds in the future—and this can be a very large liability.

When one or more disposal facilities are identified as candidate sites, it is best that the generator visit the site and find out for himself what management techniques will be applied to his waste if it is accepted by the facility. Since in many cases the disposal facility is likely to be some distance away, this is sometimes expensive and difficult to arrange. It may be that the generator's representative will not understand what the disposal site's proposals entail. If this happens, it is essential to bring in the expertise of your professional adviser on waste disposal—at additional expense. When you finally ship your wastes, you will be required to declare on the Manifest that you have chosen the treatment, storage, or disposal methods which the disposal facility will apply to your wastes. Clearly, to make such a choice intelligently requires at least a review of the options available with the TSDF operator.

If you are unable to visit the site or to understand what's going to happen to your wastes, then there are some sources of information. Your hauler must have some information about the site and its history of violations and other problems. A call to the hazardous waste regulatory office in the state where the facility is located should turn up any reports of problems with the site. Ask to speak with the person in the state agency who inspects the site you are considering. Tell them your interest and ask about the site's current status and past record. Is it currently subject to any enforcement actions? If not, is it in compliance or operating under some agreement to comply in the future? Is the facility listed on the National Priority List? (See Chapter 13 on Superfund.)

If you receive satisfactory answers to these questions, then others having to do with your direct relationship with the site will come to mind.

- How do you get your waste to the site? Does the service include pickup and hauling?
- If access is by means of contract haulers, does the site have one or more preferred ones in your area?
- Does the facility serve other customers in your area? Who are they? Contact them to get any experience that they may have with the service.

- Will the facility provide you with copies of their permits covering the types of waste you intend to send?

- If you become a customer, what sampling of wastes before shipment will be required? Will the waste be sampled and analyzed on receipt, and in either case, what fees are charged for analyses? Can you provide analytical data generated by an independent laboratory and avoid reanalysis charges?

- What are the charges for the service? Is there a minimum charge for any shipment, or is the charge based on the amount and type of waste in each load?

- Can the disposal facility provide you with special services at need, like the packing and handling of "lab packs" for disposal of many small containers of hazardous substance that can be generated in cleaning out a chemical stockroom or laboratory?

- How much lead time is necessary for arrangements for a shipment? Few small businesses have a regular schedule for generation of controlled wastes, so pickups may not always be arranged on permanent schedules.

Finally, when the arrangement is set, obtain certificates of insurance showing how the facility's coverage protects you in the event of insolvency, bankruptcy, death of an owner, or any other perils that can befall a business. Do not expect to obtain, directly or indirectly, insurance protection for you in the event of mismanagement of your wastes—you cannot avoid this liability by insurance in most cases. To nail down all these agreements, it is essential to obtain a contract for the service which spells it all out in formal language so there are no misunderstandings.

Shipping Manifests—Not a Simple Assignment

In order to accomplish the "forever" connection between a generator facility and its hazardous waste, each shipment of waste, whether to a disposer, a treater, or a recycler, if it leaves the original generator's property, must be accompanied by a properly executed, multipart form called the Uniform Hazardous Waste Manifest. It has always been the hope that the Federal form 8700-22 (in whatever version is currently approved) would be adopted for all shipments. At least some of the States, however, insist on more information than the Federal form requires; so there are a number of versions around. Be sure to check with your State regulators for the current form or get it through your transporter, who has a major stake in getting this document right. In some cases, when shipping certain wastes to or even through another State, you may need to use that State's Manifest or to provide special information not called for on your State's version.

Preparing a Manifest for a shipment of hazardous waste calls for special training, information, and skills that are not customarily found in the shipping department. Special training, renewed annually, is needed.

Before starting, the generator's EPA Identification Number and a Document Number for this particular Manifest must be entered. Both of these numbers are essential for tracking the Manifest and therefore the shipment it accompanies. The ability to track is required by law, and any failure is punishable by heavy legal penalties. Any generator will someday be grateful for this bureaucratic requirement if it is faithfully done. Some of the State agencies provide forms that are serially numbered when you get them. Use of such numbers helps comply with record-keeping requirements that are burdensome, as we shall see. A copy of a Federal Manifest is included in Chapter 12 on RCRA.

Preparation begins with the generator's name, mailing address, and telephone number. This telephone number is likely to be the point of contact with the generator in the event of any emergency during transportation of the waste, and someone should *always* be ready to respond if it is called. *Always* means 24 hours a day and 7 days each week.

Next come the name(s) and USEPA Identification Numbers of each of the transporters who will handle the waste on its way to the Treatment, Storage, and Disposal Facility. *All* transporters must be listed, so you must find out from your transporter, if you do not already know, what other companies will handle your waste and which one will make final delivery to the TSDF.

The next section names the "Designated Facility." This is the one you have contracted with for treatment or disposal of your waste, with *its* EPA Identification Number. You cannot legitimately send a shipment of hazardous waste forth from your facility unless you have made an arrangement for a place for it to go for treatment or disposal (in rare cases, for long-term storage).

The meat of the Manifest is the list of materials being shipped. This requirement calls for the special skills already mentioned, for many reasons.

First, the listing calls for the "Proper Shipping Name" which is based on Department of Transportation rules which are listed in Volume 49 of the Code of Federal Regulations under Parts 171 through 177; the Hazard Class, a carefully defined term; and an Identification Number derived from a United Nations–agreed system for worldwide use. Finding the information to do this can be confusing and difficult. Training of waste management personnel in this area is expensive and may not be productive if your wastes are consistently the same over long periods of time. Your hauler or waste disposal facility can help; indeed, the hauler cannot legally transport your goods unless they are properly labeled and manifested. Information and assistance may be obtained from your State waste management agency, your State transportation agency, the Materials Transport Bureau of the U.S. Department of Transportation, or your trade association. It is impossible to reproduce here the vast amount of information needed to approach this part of the task from the regulations alone. The small business simply must seek help.

For each waste listed on the Manifest, the number and type of containers being shipped are listed. A few examples are: metal drums, barrels, kegs (DM); fiberboard or plastic drums, barrels, kegs (DF), tank trucks (TT) and so on. (The letters in parentheses designate the material from which the kegs, etc., are made.) The total quantity of waste described under each designation must be listed, including the weight of the container if it is to be discarded. The unit of quantity used must be stated. On State-compliance manifests a State Waste Number or other special designation or description may be required.

Be sure to check your State's Manifest form for special requirements such as a State Generator's Identification, State Identification Numbers for transporters, and special information requirements for the disposal facility.

The listing of wastes on the Manifests is to some extent an art that is adapted to the peculiar circumstances of each location, hauler, and disposal facility. Additional information about the components which make a particular waste hazardous may be needed if the Proper Shipping Name is generic and not informative. Special needs in the event of emergencies may have to be shown on the shipping papers. One recent article recommended that the appropriate page of the Department of Transportation Emergency Response Guide for the hazardous material number (UN/NA Number) of the waste be copied and attached to the shipping papers. A well-trained driver for a transportation company will have a copy of the current Emergency Response Guide in his vehicle at all times.

The regulations require that the Manifest be "signed by hand," demanding that a person sign his name to the form *and to the declarations printed on it*. There are two of these:

- The first declares that the shipment represented by the Manifest is in every way described, packaged, marked, and labeled in accordance with all the regulations that apply to it.
- The other declares that, unless exempted by the regulations, the generator has a "program in place" to reduce the volume and toxicity of waste generated to an economically practicable minimum level and that the generator has "selected the method of treatment, storage, or disposal currently available to *me* which minimizes the present and future threat to human health and the environment."

The regulations make it clear that these Declarations are signed under penalties of perjury under Federal and State law. This is discussed more fully elsewhere in this book, but these are serious undertakings which make it wise for only a responsible manager to sign. These are not routine shipping documents, as must be repeated over and over again.

The Manifest as a Tracking and Recording Device

The Uniform Hazardous Waste Manifest is usually printed as an eight-part form; that is, it has eight copies of the blank form bound together in register and with duplicating medium bound in so that an ordinary typewriter or computer printer can be used to print all the required copies at once. The distribution may or may not be indicated on each page but is generally as follows:

- The first five copies are given to the transporter (usually the truck driver) when the shipment is delivered to him. He signs in space 17 on the form as Transporter 1, acknowledging the receipt of the material.
- Copy 6 is usually sent to the State regulatory agency having jurisdiction.
- Copy 7 may be a spare which can be sent to a second State regulatory agency if required by their rules.
- Copy 8 is retained by the generator for at least three years, some say forever, and is later matched with Copy 1 when that copy is returned by the TSDF as discussed below.
- If the transporter passes the waste to another transporter for final delivery to the TSDF, Transporter 2 acknowledges the receipt in Space 18 on the form and keeps a copy for his own records.
- When the shipment finally arrives at the Treatment, Storage, and Disposal Facility *and is accepted*, the operator signs the form as a receipt in

Space 20, with any exceptions to the Manifest shown in Space 19. The last transporter keeps a copy for his records.

- The TSDF distributes copies of the form for accepted wastes as follows:
  - Copy 1 to the generator. If the generator does not receive this copy within 35 days, he must ask the TSDF for a status report and file an Exception Report of the circumstances with the State regulatory agency and the EPA Regional Office within 45 days of the failure to receive the receipted copy.
  - Copies 2 and / or 3 to the affected State regulatory agency.
  - Copy 4 is retained by the TSDF in accordance with its own record-keeping requirements.

The idea behind all this is that the generator and the affected agencies will match up the receipted copies of the Manifest with the originals to ensure that all shipments are fully accounted for and that no illegal disposal has occurred. The horrendous paper load which all these shipments bring about makes it seem unlikely that the paper trail will actually turn up many violations. After the fact tracking of shipments may be helpful in law enforcement. Generators of any significant number of shipments are advised to apply good data management techniques which will turn up the exceptions and make it possible to track all shipments within a few hours of finding a need. State offices will likely at least attempt to obtain a resolution of every Exception Report, and none should be ignored but checked out and resolved by the generator as soon as possible.

Congress is considering a national plan for tracking every shipment of hazardous materials, including hazardous wastes, in the country by use of centralized data input and electronic data management systems. If this system is ever implemented, it boggles the mind with its complexity, although it may eventually include the reports on waste handling as a subset of the main data bank. We shall see how far this dream gets!

Data Management and Reporting

As the responsible agent for his hazardous waste for all time, the generator has some heavy record-keeping obligations. In general, copies of each Manifest, all test and analytical results on wastes shipped off-site, and copies of all routine reports described below must be retained for three years under the Federal rules, while some States require longer retention times. *These must be considered essential business records and kept in safe storage away from the ordinary perils of the business property.* Whoever is responsible for the integrity of the records should seek legal advice on the adequacy of storage versus the need to refer occasionally to the records.

Several classes of information must be retained for the "active life of the facility." Among them are exception reports resulting from nonreceipt of Manifest copies, required summary reports to the State and the EPA, records of employee training, the Contingency Plan for the facility, and the EPA Identification Number.

Finally, each generator is required to file reports on his total hazardous waste management activity with regulatory authorities at intervals that vary widely depending on the agency. Federal reports are required every two years from generators of more than 100 kilograms per month. This requirement can be triggered by one month when the waste generated exceeds the exempt level. Many States call for reports annually, usually on a schedule that fits the state's fee structure. Oh, yes, many States have significant fees to be paid by each generator, transporter, or other manager of hazardous waste—usually to cover the State's administrative costs for the program.

Contingency Plan—A Must for All Generators

Federal law and regulations require each generator to prepare and maintain a Contingency Plan for dealing with emergencies arising out of the management of hazardous waste in a facility. Normally, these Plans are kept in the plant or facility and not submitted to the regulatory agency. It is not unusual for an inspector to ask to see the Plan at a routine inspection, and if an emergency occurs and there is no Plan in evidence, its absence is considered a serious violation of the rules.

Generally, every generator, even the smallest, must plan to comply with a broad requirement to manage its wastes in a manner that prevents problems like fires, leaks, spills, etc. The facility must have adequate fire extinguishing and spill control equipment on hand, with some form of internal communications so that the discovery of an emergency situation can be readily reported. In some States and communities, it is required that local "first responder" fire and emergency units be informed as to the kinds and amounts of hazardous wastes generated at the facility.

This notification requirement has been superseded by the new rules issued to implement the new Title III of Superfund, usually called the Community Right to Know Act, which demands reporting to the community in complex and demanding ways all manner of hazardous materials on hand and in use in facilities. However, it is often true that old requirements stay on the State and local books, sometimes for years, even after new ones are overlaid on them; so it pays to check it out.

Larger generators must meet more specific requirements, and the means of compliance must be spelled out in the Contingency Plan, particularly describing in detail who is to do what and when and providing the information needed so that the actions can be taken without having to leave the scene to look up anything.

Typically, a generator must be prepared to respond to an emergency by:

- Designating a person as Emergency Coordinator who must be on-site or on call *at all times*. This requires that vacation and sick-time relief must be specifically planned for and that the personnel must be available when needed. The small business owner who takes a well-earned vacation must appoint someone to fill in.

- Providing a telephone which is readily accessible near the location of the potential emergency but where a user does not have to be exposed to the emergency to make a call.

- Posting by the telephone in clear and readable form at least the following:

    - The name and telephone number of the emergency coordinator (note that this may have to be changed from time to time)

    - Location of fire and spill control equipment and fire alarms

    - Telephone number of the fire department serving the area

- Providing training of employees in waste handling and emergency procedures as required by the regulations—more about this later.

When an incident does occur, the Emergency Coordinator must respond by doing at least the following:

- Extinguishing fires or calling on a fire department to do so

- Containing any spilled materials and/or cleaning up contamination by approved methods

- Reporting any fire, explosion, or spill to the National Response Center (800) 424-8802 and to your State Spill Reporting Center or Hot

Note: The states of Michigan and Minnesota suggest a form such as that shown in Figure 1 be used for this posting.

Figure 1

HAZARDOUS WASTE EMERGENCY

EMERGENCY COORDINATOR	Name	____________
	Phone	____________
ALTERNATE	Name	____________
	Phone	____________
FIRE DEPT.	Phone	____________
HOSPITAL	Phone	____________
POLICE	Phone	____________

Fire alarm is located: ____________

Spill control equipment is located: ____________

Fire extinguishers are located: ____________

National Response Center: 1-800-424-8802
Minnesota Pollution Emergency: 1-800-296-8100

MICHIGAN DEPARTMENT OF NATURAL RESOURCES DNR

Line with at least the following information to the extent available:

- Name, address, and EPA Identification Number of the facility

- Time and type of incident (spill, fire, explosion, etc.)

- Quantity and type of waste involved (estimates, if necessary)

- Extent of any injuries

Note: It is more important to make a first report promptly than it is to wait and get firm data on the amounts of spilled materials or wastes generated in the cleanup. Failure to report promptly an incident involving a reportable quantity of a hazardous substance may be a felony under Federal law. Prompt reporting means as soon as possible under the circumstances. Delays of a few hours have been held to be excessive. The natural tendency to get more data should be resisted—an update report can be made after the emergency is past.

Personnel Training

One of the more difficult requirements imposed on a small business that is a generator of hazardous waste is the specific training which is demanded for every employee who has any responsibility for the management of hazardous waste. The employee's association with wastes may be as simple as moving drums from the point of generation to storage, and if marking or labeling is also part of this job, the training needs are greatly increased.

One of the tricky parts of the regulations is that some of the Occupational Safety and Health Administration's general safety training is incorporated by reference into the EPA and State regulations.

For the Federal requirements, the general requirement to train employees of a generator is buried in the Code of Federal Regulations Volume 40, Part 262.34. The specific content and character of a training program is described in the same Volume in Part 265.16, and goes on in some detail.

In general, it seems impractical for a small business to get involved in the development of in-house training capability. The requirements are complicated, and people who devote full time to this area of education have a tremendous advantage in their ability to keep current and adjust their course content as needed.

The starting point for all training programs is one general requirement that each company has to do on its own or with the help of some outside professional. That is the development of the list of activities of each employee who deals with hazardous wastes so that the training need for each person can be defined.

The Federal program assumes that there are or can be developed formal job descriptions for all employees. In many small businesses this may not be true. Somehow, you must develop enough of the basic data on job content so that, with the help of your trade association, State industrial assistance office, or purveyors of training courses, you can decide who is to be trained, in what subject matter, and how.

Many organizations offer training courses in thousands of subjects of potential interest to business. Training in hazardous waste management is relatively new, and it may be hard to find reliable courses that are offered in places and ways that are useful to the wage employees you need to reach.

Probably the best two sources of information and help in this task are your State regulatory office and your trade association, preferably in its local chapter organization, where businesses can set up cooperative programs to share the burdens of these training needs.

One measure of the importance the government places on this requirement is the regulation that requires retaining the documentation of employee training for three years after the employee has left the company—a long retention time indeed. The assumptions behind these rules seem to be that all business operates in basically the same way—with the elaborate job analysis and training departments characteristic of large corporations somehow available to the small business far from the cities where these services are the stuff of daily life.

In any event, there is no real alternative to providing training for your employees doing the waste management task for your company. Typical courses may last from one day to a full week and can be expected to cost around $100 per day, plus any travel expense and often wages paid for the time devoted to training. This investment must be carefully planned and executed so that it produces real benefit to the company in reduced errors and avoidance of enforcement actions. Remember that serious violations of these regulations can result in prison sentences and large financial penalties—upwards of $25,000 per day of violation!

Finally, don't forget to train yourself if you manage your own small business. The buck stops with the boss, and you'd better take the time to learn all that is taught your employees, even if you think you have better things to do. If the miserable problems of waste management get out of control in today's world, you're out of business anyway.

Waste Minimization and Alternatives to Solid Waste Generation

If your facility is regulated as a generator and ships hazardous waste off-site for disposal, you have to declare (on the Uniform Hazardous Waste Manifest) that you have at least thought about the possibility of treating, detoxifying, or reducing the amount of waste you generate. This obligation assumes that the analysis goes at least far enough to develop an economic reason for any decisions made—including the one to do nothing.

Chapter 10, in this book, entitled "Waste Management: Minimization and Recycling," gives some leads on developing a waste minimization program and can help in the process. Several strategies, particularly the avoidance of careless mixing of wastes, can be helpful and save trouble and money.

One of those strategies which should be evaluated by the small business is the possibility that, by proper treatment, significant amounts of hazardous wastes may be managed properly through the use of public sewers.

We must hasten to add a caution: sewering of wastes other than normal sanitary discharges is always subject to controls, the enforcement of which is in the hands of the owners/operators of the sewage treatment works where your discharge is eventually handled. The rules are strict and can be enforced by several levels of government, so care is needed even before beginning.

A list of candidate wastes for discharge to the sewers should be prepared with complete analytical and physical property data for each. A review of the waste streams with the staff of the sewage collection and treatment agency can then be arranged, at which your proposals can be considered. The agency must consider many factors, including their own regulations, Federal and State rules under the Clean Water Act and other laws, and the effects of your proposed discharge on the safety and life of the sewer system and treatment plant. In addition, many chemicals can pass through a treatment works and cause water quality problems in a receiving stream.

The process of clearing your idea with the sewage treatment plant may seem complicated. But for an operation that produces large volumes of liquid wastes that can be readily treated by neutralization or other simple means, the savings can be great, particularly since disposal of liquid in landfills has been substantially prohibited. In some recent cases, Federal prosecutors have filed felony charges against business operators who have discharged hazardous waste to sewers in violation of the sewer system's permit and operating rules. In such cases, conviction can lead to fines of hundreds of thousands of dollars and imprisonment of the responsible owners or executives. You must clear these matters with your sewer authority!

In a similar way, perhaps you can obtain approval to burn some of your waste oils in your heating boiler by following the necessary procedures which can be worked out with your State regulators. Used lubricating oil from your own plant, being reclaimed or recycled, is not a hazardous waste in most jurisdictions, but is subject to its own rules, which are intended to encourage its use for productive purposes. In certain cases, even wiping cloths from furniture finishing can be burned in a boiler furnace which was fired with waste wood from the factory. Don't try this without checking with your regulatory agency, but much is possible along these lines.

The important thing is that the waste you never generate does not contribute to your classification as a large, intermediate, or small quantity generator and may keep your facility from being subject to the most restrictive parts of the regulations. Besides, you may save a lot of money!

Polychlorinated Biphenyls (PCB's)

PCB's are a class of chemical compounds that have not been produced in the United States for many years. Despite this, their durability and resistance to natural and man-caused degradation has kept them in the environment in nearly the whole world. The argument over the actual toxic effects and safe levels of exposure of humans and environmental subjects is irrelevant to this discussion. Let us agree that many people, most notably the governmental authorities, are convinced that these materials (more than 100 compounds lumped together) are dangerous and must be controlled at a very low level of concentration and exposure.

The Federal regulations on PCB are based on the Toxic Substances Control Act and are different in structure and procedures from those under the hazardous waste rules. However, some State programs treat PCB as a hazardous waste and require combined reports with the RCRA-type wastes.

PCB's were introduced in commerce in fluids that had to be resistant to degradation at high temperatures, had certain desirable electrical properties, and were fire resistant. They were chosen as cooling fluids for electrical transformers, high-temperature reactors, and for some hydraulic power systems. Today they are not used in any new equipment and have not been for a number of years.

The most likely place that a business may find PCB in its facility is in older electrical equipment, particularly in transformers and capacitors. Many plants have capacitor banks installed to correct a power factor problem. If the capacitors are mounted on the power supply system owned by your utility, the chances are that the regulations have been met by the utility's own crews.

The best course of action for the small and medium-sized business seems to start with a consultation with a specialist from your electrical utility. In many cases, the utility representative can advise you whether any of your equipment is likely to contain PCB and can help sample the material for analysis in a qualified laboratory. If the equipment is in use (energized), then only qualified electrical maintenance personnel should be involved in this.

When and if any PCB fluid or PCB-contaminated equipment is found, plans should be made for its removal to one of the few licensed disposal facilities in the country.

The large national hazardous waste management firms will provide this service for substantial fees. If the equipment has to be replaced, several of the larger electrical equipment suppliers provide a complete service for draining and rinsing equipment and disposing of the waste fluids. Pole-mounted equipment that is relatively old may well be shipped to the incinerator and fed as a unit to the destruction system, which is usually based on a rotary kiln like those used for lime and cement manufacture at extremely high temperatures.

When PCB is found in your facility, report it at once to the regulatory authority having jurisdiction and follow their procedures faithfully. Under no circumstances should a small or medium-sized business which lacks specialized environmental staff attempt to manage PCB or dispose of it in any way without assistance from the specialists we have mentioned. The exposure to public censure and concern is enough hazard; the financial penalties can be incredible.

Underground Storage Tanks

A full discussion of the Leaking Underground Storage Tank Program is in Chapter 13 on Superfund. The focus in that program is on petroleum storage tanks, since they are believed to be by far the most numerous. Several million have been reported under the rules which demand reporting of those tanks in service after January 1974.

It goes without saying that any business which has gasoline, fuel oil, or other petroleum-product storage tanks covered by the regulations must follow them as generally indicated in the Superfund chapter. We shall discuss the effect of these tanks on certain property transactions later.

Underground storage tanks that are used for storing wastes or hazardous substances are covered by the hazardous waste management rules under RCRA. They are treated as storage facilities and may, depending on the length of time wastes or other materials are stored, require a permit as a storage facility. The literal requirements for hazardous waste storage systems regarding placement away from flood danger, groundwater monitoring, and all the others apply and must be followed. If you have any such tanks, contact your regulatory agency to find out what you must do to comply with their rules.

Inspections by Government Agents

Small businesses tend to be run by strong-minded and independent people who don't like strangers wandering through and around their businesses and property. Because of this feeling, it is vital that each operator realize that the environmental laws provide for government personnel to have access to your property at "reasonable times" for the purpose of determining your compliance with the laws and regulations. In most cases the inspectors today are likely to be State employees, but in a few they may be Federal personnel or even contractor's people hired by the government to do the inspections.

Business operators might as well get used to this idea. In a few cases, the courts have held that the inspector has to get a search warrant before entering your plant against your will, and the use of contractor's personnel as inspectors has been challenged successfully a few times. In general it's not worth the hassle for the ordinary business. Those who want to fight to the end on the principle of preventing government invasion of their rights are to be admired. But they'll likely lose in today's courts, where ideas of the balance between individual rights of businesses and the right of the public to be protected against whatever is today's peril are often reflected in decisions to open your doors to the world.

The best way to deal with a government inspector is by being cooperative and expecting the same of the inspector. You or your people should greet the person claiming to be an inspector and politely ask for his/her identification and evidence of authority to inspect your plant or property. This might be a badge or other formal identification with an agency, but its acceptance on your part should be dependent on a statement of the inspector's jurisdiction and the intent of the inspection. You are entitled to this information so that you can provide access to those plant areas and personnel that the inspector will require to accomplish his objective. A courteous request that the inspector provide you with the name and phone number of his supervisor with whom you can check his credentials is perfectly in order and should be done if there is any doubt about the bona fides of the person.

Cases have been experienced when inspectors were found wandering around in open areas of the plant without properly identifying themselves in the office before entering the property. You need not tolerate this behavior but should ask such persons to come into the office and be properly identified before continuing. If they are not properly identified and authorized, they are mere trespassers.

During the inspection, be sure the inspector is accompanied by management personnel who can take notes and answer questions. The inspector is obligated to follow all safety and sanitation rules which are in effect and should be required to do so. Taking of photographs in your facility is generally subject to your approval in advance and should be allowed only if your personnel are equipped to take duplicate shots. If the plant handles flammable and/or explosive materials, you can and should prohibit the use of flash equipment for photography.

Similarly, while the inspector is generally entitled to take appropriate samples, plant personnel should always obtain a split of the same sample, which should be retained for future reference.

Finally, ask for a report at the end of the inspection, first an oral one on-site if possible, otherwise a copy of any written report to the inspector's superiors as soon as possible after it is submitted.

Inspections are clearly authorized by the laws in these matters, but they are to be conducted within reasonable limits on the time they are conducted and with safeguards for civil liberties. The USEPA has a limited force of criminal investigators who have the same authority as sworn peace officers as do U. S. marshals. If you deal with these officers, the matter is potentially a criminal one and is subject to the safeguards of Constitutional procedure which apply to all police activity.

BUSINESS TRANSACTIONS THAT ARE AFFECTED BY THE HAZARDOUS WASTE LAWS AND REGULATIONS

In the introductory paragraphs of this chapter, we suggested that the hazardous waste laws and regulations had to be understood by small business and by professional advisers and providers of specialized services to those businesses. In particular, it is necessary to consider the dramatic effects of the existence of the hazardous waste laws on several kinds of business transactions.

Specifically, we shall consider the new relationships between parties to business transactions like buying and selling of industrial real estate, mergers and acquisitions, purchase of assets, and formerly routine financing activities including secured loans, mortgages, and bankruptcy.

First, we must once again warn that this is not a book of legal advice. The discussion that follows is intended to show the business owner-operator some of the issues that should be taken up with legal counsel and other professional advisers when contemplating the kinds of transactions and decisions which are discussed. A major point is that professional advisers who are fully cognizant of the current state of the rules and court decisions in these areas are needed, and maybe this brief discussion will help in selecting and employing the engineers, attorneys, and other specialized helpers who may be wanted.

The basis of the issues raised is that the hazardous waste laws are founded on the concept that the owner of property is responsible for any contamination of the soil, water, or other damage to natural resources that exists on the property or results from migration of hazardous substances from the property. More important, the Environmental Protection Agency has great flexibility in deciding which, in a chain of ownerships, it will take enforcement action against to remedy the problem, if any. This can be called the perfect example of the "deep pocket" theory of liability—the EPA will pursue anyone who has or had an interest in the property whom it believes has the financial resources to pay for the achievement of the public interest in cleaning up any contamination.

For years, the EPA has pursued this course consistently in carefully chosen cases until it has a record of court approvals of these policies. Further, at the last amendment of Superfund (the Superfund Amendments and Reauthorization Act of 1986), much of this structure was built into the Federal statute. States are required also to have at least the same authority to impose legal sanctions on violators of these acts.

Thus, the ownership of property that has been contaminated at some time imposes obligations not only on the current owner, but on all those who may have some ownership interest or control now or in the past. The responsibility may be assumed, but cannot be avoided by contract, if the agency is to be believed. The principle is defined in Section 107 of the Superfund (CERCLA) Act.

In recent years, companies have often purchased land for ordinary use only to find that there is contamination that results in an order from a government agency to clean it up. Remedial actions to treat contaminated soil, to restore groundwater, or to replace leaky tanks and piping can cost more than the value of the property. While it may be possible to recover some of these costs from the seller, that is chancy at best and often takes years of litigation, while the government demands action now and will sue to recover its costs in a forum where the advantages are all on its side.

These newsworthy cases aside, in recent times just about anyone who can be held to have an ownership interest in property can get caught up in the government's search for a responsible party who can afford to pay. Thus, lenders who foreclose on a security interest in property, no matter how temporary their ownership, have been held responsible. Bankruptcy has been found to insulate a current owner from Superfund liability in a few cases, but creditors who take over real property in bankruptcy settlements can also get caught. Even banks who have made ordinary business loans secured by the company's assets can find themselves, in foreclosure, the responsible owners of a Superfund site—not quite what they had in mind as security. Sometimes an unsecured loan is a better risk.

The recital of those at risk can go on and on: the title insurance company, the real estate broker who may have failed in his duty to disclose significant facts, even an attorney who renders a title opinion without research into environmental liabilities can be in trouble.

How does the businessperson protect against these potentially disastrous events? There is no real option short of an assessment of the property before completing any transaction. The Superfund provides for a so-called "innocent landowner" defense against liability, but this is hedged with a requirement for diligent inquiry on the part of the one claiming the defense *before the fact*, so it is likely to provide little help for a business which can be presumed to be operated by knowledgeable managers.

Some of the States have adopted laws that require, in effect, that State approval be given in advance before title to industrial real estate can be transferred, with the State agencies authorized to withhold approval until the site is cleaned up to their standards. More States are considering this form of regulation modeled on New Jersey's Environmental Cleanup Responsibility Act. Many have requirements that any enforcement action against a site be recorded in the deed for the delectation of future owners.

In response to all this, a number of individuals and professional organizations now provide an environmental assessment service designed to provide all parties with the information needed to appraise their liabilities, if any, in transactions involving land or other real assets. One of the major areas of this practice is in the appraisal of environmental risks in consummating mergers and acquisitions, where the acquiring company can inadvertently assume major undefined financial responsibilities not always shown on a balance sheet and operating statement. Remember earlier when we pointed out that the accounting profession had found no way of listing some $600 billion of future Superfund liability as material to the financial statements of their audit clients? This problem is the flip side of that issue.

Every environmental assessment is unique. The questions asked and the kinds of answers needed must be developed for the particular situation. There are, however, some general kinds of information that must be gathered and evaluated.

First, a picture of the site itself must be developed. Its general setting: rural, urban, isolated, or in a congested area with many neighbors. Who are the neighbors, and what kinds of activities do they conduct? What physical factors affect the site? Is it in a flood plain, situated over "vulnerable hydrogeology," or subject to unusual natural events like high winds and heavy rains resulting in large runoff from the land that may be contaminated because of long intervals between storms?

Second, the history of the site should be developed in detail, beginning as early in the history of the region as possible. It is hard for a modern company to cope with its responsibility for soil and water contamination which was left over from some pole-treating yard that was purposely located far from town seventy-five or more years ago, but you'd better find out. Local lore can turn up stories of famous fires and explosions on and near the site which can give clues to the present, and don't forget that environmental enforcement actions, monitoring data, and reports are available at your State or local agency. By law, these are public records, but it may take a lot of persistence to find them and review them.

Third, evaluate the information in hand and make a judgment whether hazardous materials might have been manufactured, used, or stored on the site in the past. If the likely past existence of some such materials can be inferred, then you will know what to look for in further work if it's needed.

Fourth, make a thorough evaluation of the entire property, buildings, storage facilities, particularly tanks and piping whether above or underground. Look for signs of poor maintenance, for spills or unusual piles of rubbish or unusual soil or vegetation areas. Open pits are always suspect, and their purpose and history should be followed up even if they appear to be empty today. Floors showing an accumulation of old spills over structural cracks can indicate subsoil contamination that will be hard to measure.

Inside, check for nonobvious hazards like exposed and friable asbestos and old electrical equipment which may contain PCB's, along with nooks and crannies in the storage and warehouse area where small amounts of hazardous materials may have accumulated through neglect over the years. Abandoned or closed plants often have unknown liquids and solids left in tanks and trenches; these should be approached carefully until the material is identified.

Fifth, try to find records of water wells in the neighborhood. Usually the County Engineer or the Health Department will have a listing of all well permits ever issued for a developed area. Many wells are not used for potable water supply but for cooling and other purposes, so they continue to operate even if there is a public water supply. It is possible to obtain samples or even health department test records of analyses which can give an indication of any groundwater contamination in the area, without the expense of drilling monitoring wells, at least until more is known.

Sixth, while checking the outdoor areas of the property, trace the surface drainage and the condition of any creeks, ditches, or other drainage that serves the site. A knowledgeable eye can often, without fanfare, get a pretty good idea of the water quality in such drainage channels and determine whether testing is needed.

Finally, be prepared to spend sufficient funds to test suspicious soil properly by extraction or purge-

and-trap methods for at least the list of priority pollutants. In the same way, well-water samples need to be analyzed and perhaps soil and groundwater sampling carried out before the transaction is closed.

Priority-pollutant scans by a reputable laboratory using proper sampling and Quality Control/Assurance methods are expensive, with each sample costing upwards of $1,000 per analysis. Similarly, the services of the environmental professionals needed to plan and interpret the work properly will cost a good deal, from a few thousand dollars for a small and simple site to tens, perhaps hundreds of thousands, when real difficulty is encountered on an important site. In one case, a company had planned to acquire some industrial-development land from a municipal agency to expand its employee parking space. A quick walk over the site by an independent professional whose office was within a short drive of the large-city site showed the presence of some odd-looking waste piles which were tested and found to be blast furnace dust containing prohibited concentrations of heavy metals, particularly lead. A careful review produced an opportunity to buy only the uncontaminated part of the site for a parking lot, leaving the remainder as a problem for the municipal agency which was peddling contaminated land without a care.

The moral of this story is that even those who you might think are most concerned about compliance with environmental laws sometimes find it convenient to avoid the subject unless a buyer speaks up and insists on hard data before committing. In particular, there are numerous tales, many documented, about companies buying or being offered redevelopment land that is heavily contaminated in its soil or groundwater. A forewarned business will look these over very sceptically before accepting the opportunity to build or expand.

In closing this chapter, it is incumbent on us to urge that any material business transaction today is potentially affected by the current and past state of compliance of the property and its owners with a host of rules and regulations, most of which did not exist within the life experience of senior managers (which time will surely cure). When conducting a site appraisal for hazardous materials contamination, the study should be extended to an Environmental Compliance Audit so that the fabled "surprise" that is anathema to managers can be avoided.

No news is not good news in these days—it is essential to know both the bad and the good.

See the Appendix for the list, "State Agency Offices Responsible for Hazardous Waste Management," with addresses and telephone numbers.

Chapter 3

A Citizen's Guide to Action on the Problems of Hazardous Waste

RESPONSIBLE CITIZENSHIP IN THE MANAGEMENT OF WASTE

The simplest responsibility of citizens wishing to help solve the problems of waste management lies in regulating personal behavior to ensure that in our daily lives we contribute to sound management practices and set an example for those who may be less informed but want to do the right thing. This chapter is mostly addressed to the "ordinary" citizen living his/her life at home and in the immediate community. Business people are citizens, too, and will find use for some of the ideas discussed here. Still, since businesses often have a larger impact on the waste streams to be managed, they have their own Guide to Action in a separate chapter.

We can start with an understanding of household wastes and the need for their proper management. The real hazard to society that might be posed by discarded household and consumer chemical products ordinarily found in the regular trash collection and handled in landfills is a controversial subject. Some reports place the total amount of such materials in commercial and domestic wastes at 0.35% to 0.40%, which does not seem like much, while others guess that the levels are one-tenth of these concentrations. Whatever the facts turn out to be, and we probably will never know precisely what they are, even small quantities each day or each week from each of millions of households and commercial facilities have the potential for placing a lot of hazardous materials into sanitary landfills. Those wastes will, over time, either degrade or escape into the environment, usually through the groundwater, somewhere, sometime.

So, the first place the citizen can have an impact on the problems of hazardous waste management is in his own and his family's behavior. Two aspects of this behavior are susceptible to direct action, as opposed to indirect intervention in the political and regulatory system. We will discuss the latter a little later, but first, let's talk about measures we can take today and into the future that will have a positive and beneficial effect on the nation's hazardous waste system, if enough of us follow them.

Just about every home in the United States has in its cupboards, shelves, garden sheds, and garages, significant amounts of a staggering variety of chemical substances. A good first step is the location and identification of as much of this as you can find.

Where to look? Try this list:

- Basement
- Kitchen
- Garage
- Utility room
- Storage shed
- Laundry room
- Bathroom cabinets

What to look for? Long-abandoned or forgotten packages, particularly if the labels are missing or unreadable and the products are caked, hardened, or otherwise appear useless. In particular, if you don't know what it's for, then assume it's no good and a candidate for disposal. Obviously, good, clean packages of materials that you or your family use regularly or intend to use are not to be classified as waste—not even the regulations require that.

At this stage of the search, the materials should be left where they are and listed on an inventory or slip of paper that can be carefully preserved. Collecting any mass of these items in a small area is best done *after* you have a plan for getting rid of those you don't want.

Some classes of products for which you should look are:

Cleaning and Sanitation Products:
- Drain and toilet cleaners
- Heavy-duty floor cleaners and wax removers
- Oven cleaners
- Disinfectants
- Cleaning solvents and spot removers
- Bleach and ammonia

Hobby Supplies:
- Glues and glue thinners
- Photographic chemicals
- Fixatives, waterproofers
- Paints, inks, and solvents for hobby work
- Lacquer and lacquer thinner
- Shellac and thinner

Health and Beauty Products:
- Obsolete and left-over prescription drugs
- Unused over-the-counter drugs and food supplements
- Aerosol products—hair spray, deodorants, etc.
- Fingernail polishes and removers

Gardening and Pest Control Products:
- Obsolete and banned pesticides (DDT, Clordane, etc.)
- Sprays and dusts for plants or soil
- Ant and rodent killers
- Pet supplies—flea powder, wormicides
- Weed killers and other herbicides—some are now banned

Automotive and Power Equipment Products:
- Antifreeze
- Starting fluids, solvents
- Cleaning products, tar removers, waxes, etc.
- Brake and transmission fluids
- Gas and oil additives
- Used motor oil
- Old and unused gasoline or grease removers and solvents
- Used batteries of any type, battery acid

Paint and Building Products:
- Latex and oil-based paints
- Paint thinners, strippers, and solvents
- Lacquers, stains, varnishes, and shellac
- Spray cans
- Wood preservatives and mildewcides for paints
- Acid etchants
- Latex for concrete additives
- Caulking compounds, asphalt, roofing compounds

Disposal of Household Hazardous Waste

Now that we have located and listed the hazardous wastes found around your home (cabin or vacation home, too!), the next problem is: what in the world am I going to do with all this stuff? The answer depends on where you live and how conscientious you want to be. Remember the cardinal rule: there's no such thing as a free lunch or an easy solution to a waste disposal problem. More to the point, few areas have a decently organized system for the homeowner to use for disposal of small quantities of hazardous waste, even if it is highly dangerous stuff. Many if not most private waste haulers, however, are fully informed and greatly restrict the freedom of their drivers to collect anything that fits their own description of a prohibited material for their particular disposal-facility license. Be prepared for your hauler to refuse to accept a collection of old bottles and cans or a bucket that looks suspicious. Don't yell at the driver—he's not responsible for the problem and can get into serious trouble if he picks up something he shouldn't.

Some waste management authorities offer a service to the general public for disposal of small amounts of hazardous waste. Generally, this takes the form of a place where you can deliver your material and leave it for disposal, perhaps after payment of a fee. The trouble is that even where this system is in place, the receiving facilities may be open only a few days a year. In one case, the disposal facility was open on one given day in September that year, not much comfort when you're spring-cleaning in May!

The only way to proceed is by reviewing your list of waste materials. Some have designated disposal points required by law. In every community, for example, some service stations are designated as receiving points for waste lubricating oil. A posted notice to the public is required, and in general, these facilities must accept any used oil brought to their station.

In most communities, there is a similar system for disposing of old automotive-type storage batteries. The combination of sulfuric acid electrolyte with the lead inner structure of the battery makes these common pieces of junk both dangerous and valuable for recycling—an ideal combination. Look in your telephone book for recycling stations or battery dealers, many of whom will take your battery off your hands. Some places will charge a small fee, although some will pay *you* a few bucks for an old battery.

As for the rest, pragmatism must take over. A call to your local waste disposal authority or waste disposal firm may turn up a location and time when you may dispose of your accumulated wastes. Maybe some of the materials that are not useful to you can be used by your neighbor, your church, your relatives, or friends. Try to be sure that whatever you give away is not just moving to a new place of rest. Don't give it away unless someone will actually use it correctly.

Small containers, like fingernail polish bottles, empty aerosol cans, and even near-empty pesticide containers of reasonable consumer-quantity size can probably be handled in the regular garbage-and-refuse collection and handling system as well as any other. Because of small amounts and high potential for mischief, prescription drugs should probably be washed down a drain where they'll do the least harm. At any rate, try to dispose of the materials through some legitimate and official system.

There are some important don'ts.

Don't bury the stuff in your backyard.
Don't put liquid chemicals in the trash.
Don't put used oil or batteries in the trash.
Don't burn leftover chemicals.
Don't mix chemical wastes together.
Don't dump chemicals in a storm sewer—they will show up again, likely in a lake or stream in your neighborhood.

To sum up, one of the many responsibilities of today's citizen is the considered and informed disposal of his own wastes. To assist us in that endeavor, we can develop some helpful habits. Among these are attempts to find a nonhazardous alternative in the first place, limiting purchases of such supplies to the smallest amount practicable, and sometimes using human effort instead of chemicals. "Elbow grease" may not always be pleasant to apply, but at least it's nontoxic.

Life-style Choices as a Way of Helping with the Problems

Earlier in the book we discussed the impact of consumer choice on the creation and magnitude of certain large-scale hazardous wastes. We pointed out, for example, that white toilet tissue is a far greater generator of wastes during its manufacture than is a less-white tissue that might function well enough.

Similarly, the choice of synthetic fibers for textiles has a price in that substantial amounts of hazardous wastes must be managed as a result of the manufacture and use of the raw materials, fibers, dyes, and finished fabrics. Nearly every consumer product, dazzling as its performance and usefulness may be, presents some elements of choice that affect the total waste stream.

Clearly, the ever-growing popularity of disposable products designed for one or only a few uses offers choice. We do not argue that the choices are easy or even that the same choices are appropriate for people in different circumstances. Disposable diapers, for example, are obviously consumers of large amounts of resources for a purpose that can and still is met effectively by the use of cloth diapers. For a hardworking parent with a small child who is not always getting enough parental time, maybe the use of disposables to save the time and hassle of laundering is the right personal choice—many parents evidently think so.

Choices are not always obvious. The use of expanded polystyrene food containers and drinking cups has crept up on the world because of their fantastic convenience. From the beginning, the production of the polystyrene essential to these products has generated relatively large amounts of hazardous stillbottoms and other by-products that must be managed as high-tonnage hazardous wastes. Only recently, the long life of the discarded containers in the environment and the slow release of fluorocarbon gases used in the manufacture of the expanded plastic have been publicized and their impacts on wildlife and the ozone layer of the stratosphere been brought out as reasons for reconsidering their use. The alternative is the use of vast quantities of paper made from wood and finished with sophisticated chemical coatings to achieve satisfactory performance. This form of consumption has its environmental price also, and it is very difficult for an individual citizen to sort all this out.

About all that can be suggested here is that readers seek out knowledge of the effects of these choices wherever they can. Responsible popular scientific publications occasionally publish information in the field, and participation in some of the better environmental activist groups will help. Do not be discouraged—we have spent two centuries developing the technological ability to permit these choices and cannot realistically expect to understand how to deal with them as a society of nonspecialists without some effort and some more extensive education on the subject.

YOU HAVE A RIGHT TO BE INVOLVED

One of the most ancient rights of people in society is the right to address the ruling power about grievances. In the early days, the grievance might arise in personal life or out of the actions of governments or at least one's superiors as society was then organized. Kings and lesser officers held levees or public meetings just to make themselves available to render judgments between persons or between persons and society. In this regard, perhaps the most famous story of justice in the Old Testament arose: King Solomon ruled that a baby should be destroyed to settle a dispute, knowing (or hoping, some may say) that the real mother would place her infant's life ahead of her own interest.

As populations increased and the burden of judgment became too heavy or too inconvenient for the sovereign, judges were selected to listen to the people and render just decision in accord with the customs (common law) of the people and the rules laid down by the king (statutory law).

When the government of the United States was designed, the founders were concerned that the historic rights of the English people to have access to and impact upon their government should be preserved. This concern was expressed in the Declaration of Independence, and later in the First Amendment, the "right of the people peaceably to assemble, and to petition the Government for a redress of grievances" was guaranteed.

The need felt by the Founding Fathers to state explicitly this particular right of the people was based on a sentiment still heard wherever citizens gather to exercise their fundamental right to criticize their government (also part of the First Amendment): Nobody ever listens to me! My opinion doesn't count!

The fact is, in this country, if your voice isn't heard, you're probably not making enough noise or are making it in the wrong places.

Your Right to Be Heard

So, the first principle of citizen involvement is the realization that you must be heard as a matter of right, not because some government employee allows it. No one in a position of authority can rule you out of order because your opinions are not important. As long as what you have to say is relevant, couched in nonabusive language and limited in length to the same time allowed others to speak, you need not beg to have your say—you must be afforded the opportunity in a public forum.

The courts, as a society of lawyers, have complicated the process for the citizen by introducing the need for a licensed spokesperson who lives within the subsystem we call the legal system. Nonetheless, the Constitutional requirement of "due process under law" is intended to accomplish the same objective. Sometimes, due process can become an end in itself; then the citizen turns to the political system to ensure that the courts are fair.

All of the regulatory procedures established under the recent environmental laws include requirements for the affected public to be informed about the activities of the official agencies, whether in granting licenses, considering enforcement action, or any other procedures.

The most common requirements are for public notice in "newspapers of general circulation" and for public hearings or information meetings to be held in the area affected by the proposed action. Many, if not all, State environment and health-related agencies offer the possibility of placing one's name or that of an organization on a mailing list. Lists may be maintained on a subject basis: all notices dealing with water pollution, for example, or those whose subject is solid waste. Once on these lists, one will automatically receive the required notices as well as others in which the agency believes the listed persons will be interested.

Personal Involvement Is Worthwhile

The second principle of citizen involvement is that it is worthwhile if the subject is of real concern to you as an individual. Taking on the role of a spokesperson or representative of a group may complicate your ability to speak and be heard. Some jurisdictions may regulate such activity and classify it as lobbying, but none can interfere with the principle that a citizen can represent him- or herself on a matter that is of concern to that citizen.

Personal involvement in these matters is worthwhile for two kinds of reasons, one broad and general, the other pragmatic politics. Whether elected or part of the bureaucracy, public officials cannot last long in office if the public turns against them. We have all had experience with arrogant or simply uncaring public employees who seem to be immune to complaints or even attempts by their political supervisors to correct their attitudes. In public hearings and public information meetings, however, this attitude is rarely expressed by environmental and health regulators. Of all public officials, they are aware that strong public support of their efforts is essential. In most such public discussions, courtesy and a certain forbearance are almost always present.

Pragmatically, the old cliché about squeaking wheels works, and works better than usual in environmental actions. Most agencies have two internal procedural tracks for making decisions. Noncontroversial matters go smoothly through a (frequently) short process to decision. For example, once a permit is classified as noncontroversial, it is very difficult to invoke the procedural rules to guarantee a more extensive publication of the facts and a fuller airing of the decision-maker's information base.

Citizen involvement at two levels of controversy (described below) will trigger a more comprehensive response by a regulatory agency. The first may look easy, but it is often difficult in practice. Typically, an agency will publish in a paper and/or to its mailing list a proposed action, like an intent to issue a permit for a landfill or for a discharge of waste into public waters, giving a date for the proposed action and some descriptive material on the nature of the action. The agency then says that it intends to act without a public hearing unless some number (10, 20, maybe 1,000) of persons files petitions within a set period of time, requesting a full and public hearing. If the required number of letters in the proper form is not received before the deadline, the action is confirmed, and that's that. Finding the required number of "objectors" within the time limit may not be easy.

A time-proven way of ensuring full care in action by the regulatory agencies is involvement of the local press, if possible the national press, better yet, television. No regulator is willingly going to stand accused of ignoring public sentiment when it is demonstrated on the evening news by a photo opportunity or a few sound bites that catch the eye of the video camera. Following this sort of action is not always one's own cup of tea, and it is not easy to gain the kind of exposure that Greenpeace, for example, gets for its often-outrageous ways of expressing its objections to something. Nonetheless, such tactics often ensure that the regulatory decision, while not necessarily better, will at least be arrived at in public. In the long run, citizen involvement can and does work to ensure that the final decisions of the regulatory agencies are in accord with the laws and the rules generated under those laws.

Most important, in our system those who take no part in the process have no basis of complaint when the outcome is not what they want. More and more, justice (in the United States at least) is defined as the action of authorized officials who have followed a rigorous procedure designed to ensure that they have all the facts before decision is rendered. If an official is lawfully entrusted with making a decision (like issuing a license for a landfill), usually no appeal is possible if "due process" has been followed. This is the meaning behind the limitation of the rights of citizens to bring suit against the regulators to those matters in which the administrative decision-maker is granted no discretion under the law. Where the law and regulations place the choice of action or no action in the hands of an informed decision-maker, that decision-maker is immune to a citizen's suit except for a general obligation to avoid "arbitrary and capricious" actions—the meaning of which is very limited in law.

One important limitation on personal appearances and involvement in public meetings needs to be expressed. Several of the laws provide for and require a proceeding called an "evidentiary hearing" before an Administrative Law Judge in certain cases. These matters often involve the presentation of large amounts of technical and scientific information, some of which may be of controversial quality or validity, and are usually held in the course of major decision making, like, for example, the decision to ban DDT from use in the United States. This is a forum for lawyers and experts that is as demanding procedurally as any complex lawsuit. Few, if any, private citizens will have a role to play in such hearings.

IF YOU WANT TO BECOME INVOLVED

Few citizens become involved in the process of environmental decision making until they or their families, neighbors, or community are affected, possibly threatened, by a situation. Of course, some people get involved out of a sense of responsibility, because of their own concern about a larger issue, or a hundred other reasons. These people are the exception, and this book is not really intended to be a handbook for public-interest spokespeople. It is intended to help a citizen who feels threatened or concerned or just mad as hell about something in the press or in his neighborhood's over-the-fence griping and who wants to find out some facts and to express his concern where it is most likely to do some good.

The first question one needs to ask is: why do I want to be heard? Do I want to know the facts about something? Is some action being taken that I want to know about? Try to formulate your question: "I read in this Sunday's paper that—" or "I was driving down Smith Road and saw some people in white suits digging around in a field near—" or "For the past few days a strong odor has been coming from the direction of—."

Remember that your State and local control agencies may well have many situations they are watching or trying to get corrected. To respond to any question from the public, they will need specific information on the location and the nature of any problem about which information is being sought.

The City Hall Crowd

The first place to contact in trying to find answers to questions of this kind is your City Hall. Because of their ability to get observations from twenty-four-hour police and fire patrols, city officials are usually pretty well informed. If not, the municipal people are as anxious as most citizens to know what's going on, and they may well have the clout to get answers that are not easily available to the individual.

Most municipalities of reasonably large population have someone on the staff who is assigned to the control of pollution and/or health problems. In small towns this may be the fire or police chief; in larger ones a specialist is often on the payroll. Nowadays, cities have many mandatory responsibilities under State environmental laws and just about have to have a trained person available to deal with these matters. At any rate, call or go to the local government first.

Don't be discouraged if the first or the fifth person you ask has no idea what you're asking about. If you can't find anyone who can give you answers or at least suggest whom to ask, you have learned something about your city and can begin to ask: why not?

If you make a contact at the municipal level, give them a little time to respond. Few such staff people can or will drop everything to get information for individual citizens unless there is a clear emergency. Real emergencies have their own dynamics, and in the case of a serious spill or discharge, the emergency response organization should have a public information official whose name and location you can find out from the State agency if the municipal staff does not know.

So, always start by "asking City Hall." If you cannot get a satisfactory response, then try other sources. These sources are more likely to respond to your requests if you make clear to them that you have already tried the local routes.

County and Regional Sources

Many States have chosen to assign major roles in the management of solid and hazardous waste to counties or to regional special agencies. The nature of the collection and disposal process has fostered regional planning where core cities are surrounded by suburbs with their own philosophies and capabilities that make cooperation difficult. There is simply no way of knowing the organization of these agencies without some research.

The simplest way of finding out the nature and location of the next-level government agency beyond your city is by asking the city staff. Some of them will see this as a way to get rid of you, others as a challenge to keep you away from the regional or State people. Chances are, however, that the city staff will be helpful.

Always call the office of the regional agency before sending any correspondence or visiting the published office location. Tell the operator who answers what kind of information you want. If the operator says he/she will refer you to an office for that information, see if you can get a name from the operator before the other phone begins to ring—if you lose the connection, you'll save time on another call. The telephone systems in many public offices are often unreliable, or if not, they are blamed for a lot of poor operator training.

Once again, if the person to whom you are referred refers you to still another source of information, try to get a name of a person as well as the name or number of an office. Many times, for reasons best known to the managers of these public agencies, different responses will be obtained from different individuals in the offices. Knowing whom your conversation or appointment is with ensures that you will get some consistency in the responses, and more importantly, recognizes the fact that government employees, like the rest of us, appreciate being known as individuals and often will be more responsive to repeated contacts by the same person over a period of time simply for this reason.

In most agencies, solid waste, hazardous waste, and emergency response actions are found in the same office within the agency. This simply mirrors the Federal organization, for the good reason that each of the offices receives considerable grant support from the Federal EPA, and support is easier to get for parallel activities.

Many, if not most, regional waste management agencies have operational as well as policy and emergency functions. That is, they often actually run the sewage treatment plants, the incinerators, and the landfills that service their area. Any information about an operating waste disposal facility should ordinarily be addressed to the office that is responsible for such facility operations in the area. If the complaint is about a facility that is not theirs, they'll happily inform you where to get more information. If it's one of theirs, keep trying. You may have to contact the top administrator's office or even the politically selected members of the governing body.

Political appointees and elected officials can usually be counted on to be impatient with a staff whose behavior generates calls from the public to them—and they will usually lean on the staff to get you off their backs. This does not always feel good to the one who is forcing people to respond—you the caller—but public officials know they are finally accountable in this way and will eventually respond. After all, this is the part of public service that Harry Truman was talking about when he advised those who couldn't stand the heat to get out of the kitchen!

The Role of the State Agency

In the years since the Federal effort in environmental management was institutionalized in the Environmental Protection Agency, there has been continual pressure for the states to take a larger role in the development of policy and in the practical process of regulation and administration of Federal programs.

This effort to get the states to play a larger role is usually based on the assumption that State officials and employees will be better informed on the local and regional conditions that bear on the development of all kinds of waste disposal and environmental issues. No one doubts that the concept is correct, but there have been well-founded doubts that every State government can or will make the long-term commitment to a professional organization that has the technical competence and political power to carry out the needed work in the intense political atmosphere of a typical State government.

As expected, the result has been somewhat uneven, but today most of the States do have basically competent and reasonably effective administrative and even technical organizations to develop and enforce environmental policy and to administer the laws and regulations from the Federal level that have been delegated to them by the Congress. Substantial Federal grants to support the State programs, usually in the form of matching funds, helped build this competence over the years.

The result is that, by act of Congress, the most complex programs mandated by Federal environmental law are actually administered by the State governments. In particular, permitting and monitoring of permittee performance under the Clean Air Act, Clean Water Act, Resource Conservation and Recovery Act, and the Hazardous and Solid Waste Act are delegated to States almost completely.

In order to carry out these heavy responsibilities, most of the States have adopted the Federal model by creating comprehensive environmental management agencies and giving them substantial powers to promulgate regulations and to act as enforcement officers to ensure that the regulations are obeyed. All the Federal acts that they must administer require that the State laws provide for standards of environmental quality and for penalties for violations of regulations that are *at least as strict* as those prescribed in the Federal statutes. In all cases, the States are permitted, but not required, to have their own environmental standards as long as those standards are *at least as stringent* as the Federal ones.

A few States still prefer separate organizations for the development and enforcement of regulations affecting different environmental media. Thus an Air Quality Board or a Waste Management Board may well be part of a State's administrative structure. Otherwise, such titles as Environmental Protection Agency, Pollution Control Agency, or Department of Environment, even Department of Environmental Conservation or Quality are encountered. It sometimes seems too bad that there is no standard nomenclature, but then, there's nothing "standard" about the States of the Union.

Whatever it is called, the State Environmental Agency is a very powerful, though often underfunded and understaffed, organization. Any citizen who is having difficulty finding out what's going on in his area or who cannot get information from the municipal officials on matters of concern can and should contact the State Agency and is entitled to expect that Agency, if not actually involved in any environmental issue, to know what official organization is involved and what the current status is.

By the way, State offices generally have a better reputation for prompt return of telephone inquiries than do the Federal agencies.

The most effective way of getting information from State agencies begins with the blue pages of your local telephone directory (or other public agency directory where blue pages are not the norm). Look for the key words: Environmental Quality, Pollution Control, etc. In many larger-area directories, the state listing may be broken down by major offices: Solid and Hazardous Waste, Water Quality, or whatever. If so, call the one that sounds likely and tell the person who answers what you want to know or talk about.

Chances are, that first person will not know what you want to know. There's just too much going on. Ask for a referral to someone or some office (preferably some individual) who might have the information or knowledge you seek. Often, the call cannot be transferred, and if you leave a message, try not to accept the statement that "someone will get back to you." Try for the name of the someone or at least the name and number of the office that will be asked to respond, so you can follow up.

Always allow a couple of days before following up, unless yours is a genuine emergency. If it is a genuine emergency, call one of the emergency or "hot line" numbers listed in the telephone directory, not an administrative office.

By the way, lest anyone think that the mail is a better way of contacting public agencies, a comment is in order. If a letter can be directed to a specific person within the agency or is a formal communication for the purpose of establishing a position in the case file or docket, then fine, use the mail. Experience says, however, that a letter blindly sent to an agency requesting information will, as it would in most large private organizations, be shunted to relatively low-level staff for decision as to how the agency will respond and may well get lost for weeks before any action is likely to occur. Call first and get a name or at least a specific office or title, as we have said repeatedly.

What to Do When You Find Someone to Talk To

The heading may seem silly until you try it. Many people who have little or no technical background feel bound to couch conversation with specialists in some version of the jargon of the field. Remember that the state employee or official on the line is another person with the same general language as your own and give him/her credit for the ability to understand plain English.

Typical openings would include: "I'm looking for information on the status of a new landfill I've heard about in River City." "I live in Maple Grove and have been worried about a bad odor that seems to be coming from—."

When the person responds, ask if there is any official document on your subject. If there is, ask that it be sent to you. Rarely is such a request refused or met with a charge for copying, mailing, etc. If that is the response, then ask what library or public reading room has a copy you can read. State and Federal laws require full public access to all information in the environmental field, and there are practically no secrets. A few narrowly defined "trade secrets" are given some confidentiality in the laws, but the amounts and types of wastes permitted to be managed are specifically excluded from the definitions of trade secrets.

If the situation you want to follow is an ongoing one, ask how to get on the mailing list to receive all information about it. In contested cases, this will often produce copies of all the papers filed in the case, whether by the government agency involved or by private proposers and/or opponents. Like all official proceedings, a huge pile of paper is almost always generated by all involved. Somewhere in there are the nuggets of information you want. Probably most important for the nonexpert participator are the notices of public meetings and hearings on the matters covered by your request. These are usually your best source of solid information unless you are familiar enough with the language and meaning of the written papers that may flood your mail box if a significant controversy is involved.

About the only other way to find out about public meetings and hearings is a careful daily reading of the "official" newspaper for your area. The Official Notices are not the stuff of exciting daily reading but often are the only kind of notice that is legally (not necessarily politically) required of a local or State agency. Typically, Federal laws and regulations require "publication of notice in a newspaper of general circulation" or some such language. Most EPA offices, however, will endeavor to get more prominent notices to the public by issuing news releases, etc., with the hope that the notices will end up in the editorial sections of the papers and not only in the smallest type in the unreadable columns on the same pages as the classified ads.

If a matter of real interest and concern is being considered, read the paper every day or week. No other procedure will always discover the notices you and your neighbors want and need.

PUBLIC MEETINGS—OPPORTUNITIES AND PROBLEMS

As we have discussed earlier, in principle all proceedings and decision making in environmental law and regulations are public. Since the purpose of the whole structure is the protection of the public, any other way of procedure would be ludicrous. The few exceptions to this involve matters of agency administration or of litigation in which a confidential talk with attorneys about strategy may sometimes be done in private. Note a distinction, however. At many of these meetings, the public is entitled to sit in and listen to the proceedings. The most common sign of this is that the press is on hand.

In the discussion that follows, the term "public meeting" is applied only to those kinds of meetings in which input from sources other than the regulatory agency or the proposer (or opponent) of an action can be introduced. At the risk of some oversimplification, public meetings held by regulatory agencies in environmental cases fall into three categories of decreasing degrees of formality and of difficulty for the average citizen to be heard: evidentiary hearings, public hearings, and public information meetings.

We mentioned earlier the evidentiary hearing, the most formal proceeding short of an adversarial trial in a court of law. It is almost inconceivable that any ordinary citizen would be involved in one of these except when represented by an attorney qualified in the practice of administrative law before the agency involved. If you ever get involved with one of these, rely on an attorney.

Next in degree of formality is the public hearing, intended to provide an opportunity for the ordinary citizen to speak or enter information into an official record intended to affect decisions. Strictly speaking, a public hearing is held after a public official or body has decided on a course of action or a decision. For example, a city council proposes to act on the recommendation of its staff to rezone a parcel of land. Before the vote is taken, it is customary (often required) for a public hearing to be held at which the details of the proposed rezoning are presented to the council and the public, and opinions are sought from the proposer of the rezoning and the public themselves as to the wisdom, legality, etc., of the action by the council. You may be required to arrange in advance to be on the agenda of the hearing. Time limits may be imposed on speakers. Sometimes priority is given in the order of appearance to those whose property is most affected by the change. The city's land-use expert and the city attorney usually advise the council in their area of professional skill, and others may be called on as needed.

In the case of a city council, the public hearing will usually be closed after the allowed time has been used or when all have spoken. The council will often proceed to a discussion in public of the pros and cons of the proposed action and may vote. If the matter is highly controversial, it may likely be taken under advisement and action postponed to another time—perhaps one when the room is less likely to be filled with concerned, sometimes irate, citizens.

Public hearings, dear to the hearts of legislators and writers of municipal ordinances, are a real pain to many public officials, who would prefer not to hear quite so much from their constituents. Hearings also have a well-earned reputation for being boring, time consuming, and filled with seemingly endless ranting by people who don't know what they're talking about. Consequently, hearings may be held at inconvenient times and places and often in intimidating rooms where only professionals tend to be comfortable.

This situation can be improved by citizens who want to participate. Your right to be heard is seriously compromised when the hearing is held only during working hours when most residents of a neighborhood are unable to attend. You can and should request an additional session at a more reasonable hour, citing facts from your own experience to support the need. If the place is difficult or dangerous or excessively expensive to reach in a reasonable time from your home or the site of the proposed action, request that it be moved or another session held. Federal guidelines usually require the hearing to be held in the area affected by the decision.

To whom do you address your request for an appearance or for a change in time, location, or format? All proper notices of hearings will include the name, address, and telephone number of an individual person or (more rarely) an office that is in charge of the hearing. Make your requests as soon as the meeting is announced, if possible—at a minimum, as soon as the problem is identified.

If no reasonable response is made to your request for changes in the hearing, you must appeal over the head of the person named in the notice. A letter to the head of the decision-making body is always in order. Send a copy of the letter to the editor of the local paper with a request that it be published. Other ways will occur to you, like a call to the consumer-action reporter at a local TV or radio station. Even a complaint to your State legislator (whom you should know about!) will often get some results. If not, be sure that an objection to the conditions is somehow entered into the record of the hearing, usually by including the objections in any statement made or by writing a letter to the presiding officer with a formal request that the letter be made a part of the record. This may be scant comfort as you waste a vacation day at a dull hearing, but your objection may serve as a means of attacking the action taken, if it is not one you desire and you wish to contest it in court or another forum.

The third, least formal and most varied in structure, among public meetings is commonly called a public information meeting. Public information meetings are usually called early in the planning process of a regulatory agency. It is common for the issue to be presented: "Our community is running out of landfill space—." "This agency is proposing to undertake a planning process to develop alternatives for the management of certain kinds of hazardous wastes—." The public is invited to advise the agency as to its ideas about the problem and any options that the agency ought to pursue. The discussion may proceed in an open-ended fashion, or there may be an attempt to limit it to a few subjects. There is really no prescribed format for all the possible useful forms of this kind of public meeting.

More unusual, but still frequently held, are public information meetings to announce new programs or policies that are thought to be important enough to merit more than a sparsely attended news conference or a press handout that may not be sufficiently newsworthy to get the "proper" exposure.

Public information meetings are often worth the time and effort if their subject relates to a matter in which you have a direct interest. At such meetings, where the atmosphere is less threatening than in a hearing, you can learn something about the officials who have responsibility for your area's regulatory structures. Listening to the presentations and the questions from the floor can be valuable preparation for your own appearances.

Participation in Public Hearings and Meetings

Many of us are reluctant to get up and say anything in a public meeting. Whether this reluctance stems from simple stage fright or from more complex causes, it needs to be overcome for the public hearing process to work at all. Obviously, a few paragraphs in this or any other book are not going to overcome some folks' natural difficulty in addressing an audience, whether in a hearing room or a church social, for that matter.

We can say that you need not be intimidated by any concern that what you have to say is not important—you are a citizen and have a right to your input if *you* think it's important. As we have said earlier, a nonspecialist member of the community need not attempt to use the precise technical language that may be heard from those specialists. You are perfectly free to call the subject of the hearing "this stuff" if you want, and you can demand that the information presented be couched in plain language. However, you may find that the experts are not capable of clear expression! Then and there, insist that someone be found who can speak English (or other applicable language).

One of the unusual effects of participation in a public meeting about something that deeply affects you and your neighborhood is that even experienced speakers are sometimes emotionally involved to a degree that makes it surprisingly hard for them to express themselves. It may seem unlikely, but the experience can shake your confidence.

The only means that is of real help to both inexperienced and usually facile speakers is preparation. Before you ask for the floor for a statement or comment, make some notes, and don't be afraid to refer to them. The more intimate the effects of the proposed action on your own life, the more important this preparation becomes. We all have freely expressed opinions on the deficiencies of the "government," but expressing these sentiments to the government as personified by our neighbors and friends—not unusual in an environmental public hearing—can be difficult.

Follow the motto of the Boy Scouts: "Be Prepared."

ENVIRONMENTAL ORGANIZATIONS

Environmental organizations come in every conceivable form and size. It is hard to imagine any citizen who has an interest in matters of social policy and public health who has not heard at least of the Sierra Club, National Audubon Society, National Wildlife Society, Friends of the Earth, Greenpeace, and many others. The very large number of such organizations found in the United States and indeed in the world cannot be counted here. The World Environment Center in New York publishes directories of many kinds, one of which provides background information and contact addresses for hundreds of environmental organizations around the world. In Washington, the National Wildlife Federation and the Global Tomorrow Coalition, among others, also provide useful directories. Many of the organizations have local chapters in larger cities and in college towns that can be found in the telephone directory.

Volumes can be written on just the major organizations in this field, but the real difficulty about characterizing them in this handbook is that thousands, formed strictly for the purpose of affecting a single issue or small geographic area, vanish as quickly as they appear when the issue that holds them together for some reason disappears or is resolved.

For the would-be citizen activist, there are two basic motivations for joining, or at least joining forces with, the organized form of the environmental movement: a general interest in the subject and the hope of enlisting help from knowledgeable people in dealing with an issue or a problem that affects you and your own interests.

Joining the "movement" is a relatively inexpensive way to tap into what the Harvard Business School calls the "ongoing stream of knowledge" about developments in the field. Most of the comprehensive organizations like those named earlier have publications, magazines, and newsletters that spread the word on environmental issues, at least on a national and worldwide scale, and tell you who's who in the field. You may have to evaluate which ones most closely reflect your own philosophy—it can be pretty hard to read some activist's views over and over when you would prefer more information and less argumentation. Nonetheless, enough groups cover a wide enough spectrum of opinion and argument that just about anyone can find a home if he really wants to. Besides, some of the best and least expensive magazines in the country are published by the great environmental organizations—well worth reading even if you do not always agree with the editorial policy!

For the individual citizen or a small group of citizens who are embroiled in a fight over an environmental issue, the environmental groups, particularly the larger ones, offer a resource that is often not appreciated. Most environmental disputes today, as in the past, develop when relatively small numbers of citizens decide to oppose some action of a government or government agency that they believe will adversely affect them, their environment, or their health. This is true even when the initiative for the proposed action comes from the private sector, since under today's regulations, few major private actions can take place without government approval. Approvals are often based on a recommendation that comes only after years of negotiation and modification that substantially commits the government agency to the project even before it is officially proposed.

In any case like this, whether the proposal is a public works project or a private development, citizens are really hard pressed to have their input considered. Participation in public hearings is fine for local zoning disputes and perhaps even for some large-scale development decisions and should not be overlooked. Very often, in local matters at least, the voices of concerned citizens can be effective.

It soon becomes obvious that the citizens' groups are outgunned and subject to a wearing-down process that is largely based on a belief that people have short memories and short attention spans in these matters. A municipal water supply issue, for example, may occupy the legal staff's time for many years, in some notable cases twenty-five or more, while the citizens are gradually led to believe in the inevitability of the action; they grow older and less passionate and finally do nothing in the ultimate decision.

If the large-membership environmental organizations can be interested in the issue, they can offset this advantage of government agencies. Many of them have permanent legal staffs who can be assigned to follow a case over years of administrative and litigative proceedings. Their interest lies in being repeatedly associated with public resistance to certain kinds of government and government/private actions, since this provides the public notice and the reputation that they need to continue to attract members. Of equal importance, winning a few gives these organizations clout where they really want it—in Washington. In Washington, the key question—"What have you done *for* me lately?"— is often made stronger by the related one—"What have you done *to* me lately!"

Give the Sierra Club and Friends of the Earth and the sister organizations credit. They rarely intervene in a strictly local dispute unless the issues involved are of significance to the organization and its philosophical base. These are institutions that were founded by and still tend to be led by people of deep convictions about what's right and what's not in man's relationship to his physical world. While their attention to an issue may be strongly influenced by the idiosyncrasies of the current leadership, it is almost always based on at least a vision of a better world.

Nonetheless, when seeking the help of environmentalists in working an issue, the active citizen's group must remember that the former are human organizations and have needs, too, as participants in citizen-initiated action. Their primary need is to be perceived as working the issue with a substantial body of public support. Lobbying in Washington on major issues of national and international policy can be effective because the Congress knows the large membership base and the long history of successful advocacy behind it.

Coming into a community to try to affect decisions at a community level is a different kettle of fish, however. Not even the Isaac Walton League or the National Wildlife Federation can long be perceived as "carpetbaggers for the environment" and maintain their effectiveness. In a pragmatic way, fighting unpopular causes needs popular support. It all depends on whom the cause is unpopular with.

So, when planning or organizing a public resistance (or, for that matter, a public campaign in support of right action), contact the local leaders of your environmental organizations, and perhaps see if you can attend a chapter meeting or arrange to meet a national office representative when he comes to your town. Subscribe to some of the publications. They will lead you to other resources that are available at meetings, seminars, bird walks, and in publications. If all else fails, write to the national head of the organization whose address is always to be found in the masthead of the publication.

The great resource that can be available in this way is hard to overstate. These people understand how to use the provisions of the law and regulations to help pay for the enormous expense of litigation that is often the only way finally to get the attention of public officials who are themselves highly practiced at using the courts to achieve objectives that they believe worthwhile, often based on a very parochial view of the world. In one recent case, a city council of a modest-sized but growth-minded city in the West appropriated seven million dollars for pursuit of a water-rights case with little debate; in another part of the country from which the water would be drawn, the opposing organizations are finding it difficult to gain enough citizen support to pay for the efforts of even one attorney to draw up papers in the required official form for the Water Court.

In this last case, hopes are high that the National Wildlife Federation will help the poorly organized opposition and at least turn the battle into something like a fair fight. To be effective, and most important, to keep the priority attention of the national leadership that the Federation staff must have to continue the fight, the local people must be prepared to show up and be visible or even vocal where that is permitted. Even a circuit judge will sometimes consider a controversy more seriously when his courtroom is packed with serious-looking citizens who obviously have more than a casual interest in the proceedings. It is not unusual for a judge to stop proceedings while he explains the matter to an orderly and deeply interested crowd in his court.

SUMMARY

Perhaps citizen action has been the strongest influence in the development of the new environmental ethic and the new environmental laws resulting since 1969, when the enactment of the new Clean Air Act and the National Environmental Policy Act announced a new era and a new ethic. Citizen action has expressed itself in almost every conceivable way, from teach-ins that even turned violent on college campuses to major political debates that can polarize public opinion and even determine the outcome of elections local and national.

The thesis of this chapter is that the need for continued public input has not weakened. Indeed, citizen input remains a strong element for effective resolution of local, national, international, and global issues.

Citizens can take action in their personal conduct to minimize their personal contribution to the management problems of hazardous waste, can teach their children, and can influence their neighbors to right actions.

Acting alone and through local organizations, informed citizens can assist public officials to make the often-unpopular choices that they must make to deal with the real hazards to human health and the environment. Informed citizens can help their local governments to avoid irrational and panicky actions driven by fear and bad information.

Above all, on the basis of knowledge, thought, and reflection on matters important to them, citizens can accomplish the real aim of government in these and all other times—to provide for the general welfare. In the case of providing a management system for hazardous waste, the general welfare is an elusive concept, but certainly it must include a need for prudent and deliberate planning and action. Sensitivity to the effects of possible disruption of a bountiful way of life is essential. That way of life is hard won and does not improve when overlaid with a sense of guilt for every action seeking comfort, convenience, and even a little luxury.

This sensitivity, however, cannot justify excessive forbearance of the behavior of those who, through carelessness, greed, or just plain scofflaw actions, compromise the future health and welfare of their fellow citizens. Active citizens can and must play a strong role in demanding firm and fair enforcement of laws and the regulations that are derived from them.

Chapter 4

How to Recognize a Hazardous Waste

COMPLICATIONS FROM POLITICAL DEFINITIONS

In the opening chapter, we considered how hazardous waste became a major problem, some say a crisis, for the health and environment of the people of the United States and, indeed, the world. To go forward into the study of the problem and the role of the citizen in dealing with it, it's clear that we must devote some time to defining, or at least describing, what hazardous waste is. Perhaps as important, we need to have some feel for what it isn't, too.

We begin with an only slightly facetious return to the Wonderland that Lewis Carroll created for Alice. Using this idea only as an introduction to the new and strange world of the solid and hazardous waste laws, we will attempt to clarify some of the principles. True to the complexity of these definitions, at the end of the chapter we will discuss several materials that most persons think of as "hazardous" or "waste," but which are unlikely to be either under the regulations. These are PCB's, asbestos, dioxin, and mining wastes.

Most of us can remember from our childhood Carroll's famous lines: "The time has come, the Walrus said, to speak of many things: Of ships and shoes and sealing wax, of cabbages and kings." We have seen how such "things" as the Walrus named create a long list of hazardous materials during their manufacture, use, and disposal. As Alice did, we need to suspend disbelief awhile and recognize that in the world of waste disposal, almost nothing is what it seems. We need to unravel some of the complexities so that our real options in society can be seen.

It is generally believed that Lewis Carroll wrote *Alice in Wonderland* as a political satire on British society in his time. Scholarly papers identify each character with some prominent person. Possibly, Carroll hoped to reform the excesses of *his* environment; perhaps he wanted only to entertain. At any rate, he left us a model of a world in which words meant only what the Queen decided they meant. To fail to understand brought at least the threat of instant punishment—"Off with her head!"

Whatever Carroll intended, his model has relevance today. Called to respond to a major perceived social problem, government responds in a society in which the actions of individuals are too politically sensitive to be easily regulated but in which the total output of wastes must somehow be controlled. In its hour of need, Congress basically made Hazardous Waste (note capitals) a political idea, owing little to science, logic, or other expressions of reality. On the banks of the Potomac today, as on the banks of the Thames in that earlier time, reality is redefined to suit political necessity. Political necessity arises from the reality of many people's fears and the hope of salvation from that fear by some official action.

No one can doubt the need for personal and official concern when a Love Canal is discovered. The press joins in and creates simplistic models, catch names like Love Canal or Valley of the Drums. Fearsome phrases like Toxic Waste, Hazardous Waste, and Chemical Waste are used interchangeably and often without any real attempt to inform the public of their significance. When rumors and the comments of both qualified and unqualified people are lumped together on the evening news for days at a time, the public reacts as it always has to a threat of unknown proportions that it believes was imposed by persons seeking their own profit.

The public turns in anger and frustration on the responsible parties, suspects chicanery or worse, and cries out to elected officials for relief. When the public is in an uproar about the possible appearance of a fearsome dragon in their midst, who wants to read a technical manual on dragons? The temperature of the fiery breath and the exact ability of a dragon to lift people it wants to carry away are of little interest. By St. George, make it go away!

At about the time of the revelations of Love Canal and other notorious sites, some members of Congress were already thinking about a problem with the environmental laws passed during the Environmental Decade of the 1970s. The original conception of the problem arose because the Clean Air Act of 1969 and the Clean Water Act of 1971 required removal of large amounts of waste materials from discharges to air and water. These single-minded (and single-medium) laws were difficult and expensive to comply with. The simplest technological solutions were based on the collection of the offending pollutants in solid form in end-of-pipe treatment systems. Following the practices of the time, the recovered pollutants could be buried in holes and forgotten.

It soon became obvious that collection from air and water simply to create new pollution of the ground was no solution. The old Solid Waste Disposal Act didn't really help much, and new laws were needed to "plug the gap" between the laws that generated solid wastes from air and water and the safe disposal of those wastes now suddenly appearing in this form.

Responding, Congress developed and passed the Resource Conservation and Recovery Act (RCRA) in 1976 and amended it in 1980 and 1984. Chapter 12 of this book is devoted to that extraordinary law.

The very title of the Act is a reflection of Washington-style reality. The title said nothing about waste, solid waste, or hazardous waste. It was under consideration in Congress during the early phases of the great energy crisis brought on by the oil embargo and was intended to cater to the public demand for conservation and recycling. It has, in fact, some sections on recycling which have proven to be of little effect. The realization that recycling is an economic activity with as yet a very limited public response to an ideological appeal is a fairly recent one.

We begin our journey to understanding with the language of the Act, Section 1004(5):

> The term 'hazardous waste' means a solid waste, or combination of solid wastes, which because of its quantity, concentration, or physical, chemical or infectious characteristics may—
>
> (A) cause, or significantly contribute to an increase in mortality or an increase in serious unreversible [*sic*], or incapacitating reversible illness; or
>
> (B) pose a substantial present or potential hazard to human health or the environment when improperly treated, stored, transported, or disposed of, or otherwise managed.

This seems clear enough. Congress presumes that only those wastes which are "improperly" managed will cause harm. The problems begin to arise when regulators must decide what a significant contribution to "an increase in mortality" or an "increase in serious irreversible illness" means. What is a "substantial present or potential hazard"? Many of us will believe that we could answer these questions in some commonsense way, and maybe that could be done for each of us or our families in individual cases. The regulatory authorities, however, must attempt to find answers to such questions that will withstand intensive review by courts led by judges who are bound to invoke standards of equal treatment of everyone and of due process. Complexity is the inevitable result of this ever-looming possibility of court review.

To make the identification of hazardous waste more difficult, Congress went on to exempt a long list of specific waste materials from regulation under the Act. Furthermore, to make clear that solid wastes are not necessarily solid, the Act goes on, Section 1002(27):

> The term 'solid waste' means any garbage, refuse, sludge, from a waste treatment plant, water supply treatment plant, or any pollution control facility and other discarded material, including solid, liquid, semisolid or contained gaseous material resulting from industrial, commercial, mining, and agricultural operations, and from community activities, but does not include solid or dissolved material in domestic sewage, or solid or dissolved materials in irrigation return flows or industrial discharges which are point sources subject to permits under Section 402 of the Federal Water Pollution Control Act, as amended (86 Stat 880), or source, special nuclear, or byproduct

material as defined by the Atomic Energy Act of 1954, as Amended (68 Stat. 923).

Putting these together, we find that hazardous waste is solid waste, but may be liquid, semisolid, or even gas in a container. Hazardous materials in waste waters are not hazardous waste; they are "pollutants." Similarly, air pollutants are not gases in containers and are outside the definition when actually suspended or dissolved.

This provision is the source of the EPA's frustrations when dealing with radioactive wastes—the most dangerous ones must be managed outside the framework of the hazardous waste system and by the Nuclear Regulatory Administration because of that agency's jurisdiction over the Atomic Energy Act. Generally, the public is not very tolerant of this divided jurisdiction.

This way of dealing with nuclear materials is not so unusual. Many substances, whether new or discarded, are regulated under other laws: PCB's and asbestos are regulated under the Toxic Substances Control Act, for example, and used oils under the Used Oil Recycling Act.

To compound confusion, the long definitions given apply to new or future wastes and are only one factor in defining the substances controlled as old waste and other contaminants of the soil. The latter materials—found at old landfills, spill sites, and other past incidents of disposal—are controlled by the Comprehensive Environmental Response, Compensation, and Liability Act, usually and mercifully called "Superfund" or "CERCLA" and discussed at length in Chapter 13 of this book.

The Superfund law refers primarily to "hazardous substances" and has much broader reach than does RCRA. Section 101(14) says:

> 'Hazardous substance' means—
>
> (A) Any substance designated pursuant to Section 311(b) (2) (A) of the Federal Water Pollution Control Act,
>
> (B) Any element, compound, mixture, solution, or substance designated pursuant to Section 102 of this Act [that is, CERCLA],
>
> (C) Any hazardous waste having the characteristics identified under or listed pursuant to Section 3001 of the Solid Waste Disposal Act (but not including any waste the regulation of which has been suspended by Act of Congress),
>
> (D) Any toxic pollutant listed under Section 307(a) of the Federal Water Pollution Control Act, and
>
> (E) Any hazardous air pollutant listed under Section 307(a) of the Federal Water Pollution Control Act,
>
> (F) Any imminently hazardous chemical substance or mixture with respect to which the Administrator [of the EPA] has taken action pursuant to Section 7 of the Toxic Substances Control Act. The term does not include petroleum, including crude oil or any fraction thereof which is not otherwise specifically listed or designated as a hazardous substance under subparagraphs (A) through (F) of this paragraph, and the term does not include natural gas, natural gas liquids, liquefied natural gas, or synthetic gas suitable for fuel (or mixtures of natural gas and such synthetic gas).

Even in this broader context, oil and gas are not ordinarily classed as hazardous wastes, and spills and other releases are controlled under special provisions of the Clean Water, Superfund, and Marine Sanctuaries acts. We will discuss much of this in great detail in the chapters on RCRA and Superfund, but it is interesting that the petroleum product exemption goes as far as to exempt even the additives in spilled petroleum from regulation as hazardous substances distinct from oil itself.

The Federal Environmental Protection Agency has the task of converting the definitions into specifics for regulatory purposes. This is usual, since the laws leave room for much interpretation, and times do change, while legislation may not.

The most straightforward list of hazardous wastes is just that, a series of lists of substances, often bearing names that are incomprehensible to everyone except a technical specialist. A representative list as of 1986 can be found at the end of this chapter. The complete list of "chemical substances" that are regulated under the RCRA and Superfund systems is extensive. In the chapters on those statutes, references are provided for tracking them down at any moment that the information may be needed.

In addition to the "listed wastes" just described, any waste material which has any of four "characteristics" is identified and must be managed and handled as hazardous waste. The characteristics are:

1. **Ignitability**. Defined as liable to ignite when exposed to air and/or water or having a flash point below 60° C (140° F). Ignitable compressed gases and strong oxidizers are included. The flash point is the temperature at which, under specified test conditions, a substance will sustain a flame after a small flame is passed over its surface.

2. **Corrosivity**. Any water-based material which has a pH of less than 2 or more than 12.5 or any material which corrodes mild steel at a rate greater than a quarter-inch per year. The meaning of pH is discussed in Chapter 11, "Chemistry of Waste Materials."

3. **Reactivity**. This characteristic includes all kinds of materials which react violently with themselves or with water or air, particularly if the material may explode or generate dangerous fumes or gases.

4. **Toxicity**. The toxicity characteristic is measured by a test usually called the Toxicity Characteristic Leaching Procedure (TCLP). The test and its interpretation are not yet completely standardized and are subject to controversy about its applicability in many situations. Running the test is not a task for the inexperienced or untrained but must be entrusted to a qualified laboratory. Since the results are often crucial for decisions on the management of wastes, great care is needed in its application.

Any law under the American legal system that is as broad and as intentionally ambiguous as these laws are will inevitably be subject to minute dissection by lawyers and the courts. In the case of the ones we are considering here, this process is made more active by the penalties Congress decided to impose when violations occur. Severe financial penalties are probable and jail sentences possible when companies and individuals are prosecuted and convicted of civil and criminal violations of the regulations.

These tough sanctions for violations do distinguish between "knowing" violations and others. Nonetheless, any person responsible for the management of hazardous waste who has any of the required training and education for the job probably won't be able to use the "lack of knowledge" defense—you're supposed to know enough to get help.

Since this book is intended for the general reader, we have emphasized the Federal definitions and ways of identifying hazardous wastes. It is at least equally important to realize that the Federal legislation, by design, draws the States into the regulatory system. Indeed, the Federal intent in designing the program includes State responsibility for running the entire hazardous waste permitting and enforcement process.

A major result of this is that, in addition to setting up to meet their obligations under the Federal legislation, all the States have adopted some form of State regulation that can be and often is more stringent (more difficult to achieve compliance) than the Federal one. This higher stringency is allowed, not to say encouraged, by the Federal legislation.

In a brief volume like this, it is not possible even to begin to discuss the complexities of State and regional programs. In other chapters we give some suggestions as to where citizens and business people can find the information they will need.

In the long run, citizen action may be more effective at State and local levels than it can be in Washington, and businesses need to find ways of keeping informed on the actions of their State regulators. State programs do not often get the coverage in newsletters and national association publications that is most helpful to the small business.

As our little friend Alice must have felt when the Queen haughtily swept aside her problem of finding her way with "all ways are my ways," we feel that there is more to the definition of hazardous waste than a set of nit-picking legalistic rules.

After all, the ordinary person's working-feel says that hazardous wastes must be composed of anything which is thrown away or which has escaped control and can harm persons or environmental values which we feel are important.

This sweeping definition is so broad that it includes materials we only need use our common sense to avoid. For example, a pile of broken glass is obviously a waste material and is clearly hazardous to one who walks into it barefoot. It can be removed or we can post a sign that says that there is danger unless passersby wear boots. If the sign (a regulation!) is observed and complied with, that glass is not dangerous to those who comply. What about those who can't read? If we remove the glass, how completely must we do so? Are barefooted small children to be driven across the site to test whether their feet will be injured after cleanup?

Equally wild examples can be found everywhere. Abandoned refrigerators are waste products which need to be treated (door removed, locks installed), and dams made of mine tailings have occasionally collapsed with consequent loss of life and property. The possible list is endless.

A last example is found in open dumps. Even those which contain only household waste and garbage offend our aesthetic senses and may be dangerous because of secondary effects like rodent damage and even hidden impacts on groundwater. In some places they have proved to be major ecological hazards; for example, to grizzly bears in Yellowstone National Park and even to humans through attracting birds who cause jet aircraft crashes at airports by clogging the engines with their bodies.

The common characteristic of these hazards (except for groundwater contamination) is that they tend to be immediate and are often relatively simple physical situations that can be avoided by prudent people. Grizzlies and rats are not so obvious, but even here direct observation can provide a basis for an appraisal of the danger.

Perhaps the simplest workable way of knowing a hazardous waste when we see one is to return to the basic definition:

> A hazardous waste is a discarded or spilled or lost material that, in sufficient quantity, considering its physical, chemical, or infectious nature *may*:
>
> 1. Cause or contribute to mortality or serious injury or illness, or
>
> 2. Be mismanaged in a way that endangers human health or the environment.

Clearly, this is a simplified treatment of the definitions in the laws. However, like the dragon's technical characteristics, it makes little difference to the citizen whether he or she deals with past, present, or future disposals, and some of this can be left for later consideration.

Most important, our fears and concerns almost all arise from the chemical nature of wastes and the largely unknown effects of these materials on our bodies and those of our families and neighbors. Most citizens can deal with a pile of broken glass or an abandoned refrigerator, once these dangers are found and known to them. How do we deal with threats to health through the air we breathe or the water we drink?

As John Quarles (former Deputy Administrator and Acting Administrator of the USEPA) has pointed out, basing the definitions of hazardous wastes on the assumption that they are hazardous only when mismanaged means that a very broad range of things and materials are, when mismanaged, hazardous. In turn, this results in a very broad regulatory program. Wastes that have very different potentials for causing harm are covered by the same rules. This broad reach results in much of the complexity that baffles those who must deal with these laws and their implementation.

From all this follows the conclusion that an evaluation of hazard from waste has to take into account two distinctly different things:

1. The inherent hazard or danger of the waste and its constituents, and

2. Management methods, appropriate and inappropriate, and the probability of exposure to the wastes.

The outcome of a combination of these two factors is an assessment of risk for a particular material in a particular management setup. We shall see that generalizing these matters is sometimes necessary but may lead to seriously misleading answers. Judging risk/exposure in specific situations is always simpler and more accurate than in generally defined ones.

SOME HAZARDOUS WASTES MAY NOT BE EITHER. DOES IT MATTER?

Laws, Like Shoes, Don't Always Fit

As we have discussed earlier in this chapter, the application of the word "hazardous" and the word "waste" to any material is a fairly arbitrary choice of both words and concept. The drafters of the solid waste statute (RCRA) and the regulatory officials at the Environmental Protection Agency have had their differences over the years, and citizens' groups are often at odds with both about the need for regulation. As pointed out, the various laws and regulatory programs overlap and have gaps between them that even the courts have difficulty sorting out at times.

The leadership at the EPA attempts to respond to public and congressional concern by using any authority they can find under any of the complex laws given to them to administer and enforce. Their task is often complicated by Congress's giving of authority over whole areas of public policy to other agencies, like the Occupational Safety and Health Administration, The Nuclear Regulatory Commission (formerly the Atomic Energy Commission), and the mining regulators of the Department of the Interior. Nobody really knows how many agencies of the Federal government have been assigned some responsibility for "protecting" the environment, but their number is in the hundreds for sure. Some cooperate with the lead agency, the EPA; others see it as their mission to obstruct—a challenge to which any good bureaucrat can rise with enthusiasm and skill.

For some or all of these reasons, a variety of substances and materials that may well be hazardous and less often constitute waste, or conversely are undoubtedly waste but may or not be hazardous in the language of the regulations, need to be at least described. Some of them have been the object of control efforts under various authorities; others are still looking for policy decisions on the need for and nature of public attention.

Polychlorinated Biphenyls (PCB's)

The term "polychlorinated biphenyls" or PCB's is applied to a large group of organic compounds containing large amounts of chlorine added to their structures. The original materials were and are derivatives of a coal-tar based substance, biphenyl, which has been used for many purposes in industry for a long time.

Sixty or seventy years ago, the expansion of industry resulted in the location of large electrical transformers and other gear in heavily populated areas. Consequently, the problem of fires and explosions in electrical equipment became of greater concern. All large equipment of the type was (and some still is) filled with special petroleum oils which circulate within the system to help remove the heat which is inevitably generated by inefficiencies in the transformers, circuit breakers, and capacitors which are essential to powering industry and large commercial centers. That is the purpose of the pipes and fans often seen on these large units.

Seeking to minimize the danger to the public from transformer fires and explosions, the electrical and utility industries found products called "askarel" and a few other names, which were actually PCB. These substances were substantially nonflammable and had a bonus: they have ideal electrical properties for use in the equipment, being nonconductive and possessed of a high dielectric constant, which is desirable to avoid additional losses of electrical energy in the equipment. This seemed an ideal solution to a safety problem, and one that appeared to have no adverse side. As in so many decisions, the knowledge needed to assess the impact of the product did not exist then and did not until many years later.

Not until the 1960s and 1970s did concern about the use of PCB begin to surface because of perceptions that all such heavily chlorinated substances were inimical to the survival of predatory birds and a few other species, in ways related to the problems raised by Rachel Carson in her ground-breaking book *Silent Spring*. The occurrence of a tragic case in Japan (when PCB was "inadvertently" mixed with cooking oil and used by a large number of people with some deaths and many cases of illness and even reported cases of cancer resulting) was noted as part of the argument for regulation of PCB.

At this point, it must be emphasized that the hazardous waste laws gave no one authority to deal with PCB until it became a component of the solid waste stream. The Superfund law was not passed until 1980, but it was not designed to deal with the manufacture and productive use of any chemical substance either.

When the Toxic Substances Control Act (TOSCA) was passed in 1980, Congress inserted a subsection devoted to control measures for PCB. This is still the only example of the naming of a specific substance to be banned from manufacture, use, and disposal in any environmental law.

The new law, Section 6(e) of TOSCA, ordered the EPA to eliminate progressively all manufacture, importation, and use of PCB according to a detailed timetable. Politically, it was manna from heaven for the lawmakers. The product had been successfully labeled "dangerous" and "cancer-causing" in the public mind, the only U.S. producer planned to abandon the product for public relations reasons, and the utilities were going to be able to continue using the material for the "useful life" of their equipment—it looked like an ideal measure to support about a month before the elections and was passed with enthusiasm, if not appreciation of what the action meant.

The EPA attempted, in the spirit of the original action and while under the gun, to begin the awesome task of setting up the regulatory scheme of TOSCA and proposed regulations with fairly practical timetables, requiring identification and labeling of all stocks of PCB, both in storage and in use, and substantially exempting oils containing small amounts (under 50 parts per million) of PCB from regulation.

The regulations engendered a major uproar, with environmental activists, represented by the Environmental Defense Fund, objecting to the lack of stringency as they saw it. The chemical industry, on the other hand, found that many processes previously unsuspected of a potential for making PCB as a by-product did in fact do so. Nobody, least of all the EPA, had a clue as to how to deal with this complexity. Shutting down a major part of the chemical industry to attempt to eliminate production of small unintended amounts of the banned substance sure didn't seem like an acceptable way to go.

Finally, out of all this, the EPA issued the three rules which remain the primary controls on manufacture and use of PCB, although the agency, the users, and the courts are still working out such details as: what is PCB in the first place? A decision along these lines has already said that the prefix poly- has no meaning since even biphenyl with a single chlorine atom added was found subject to the rules. Similarly, the meaning of the words "totally enclosed" when applied to regulatory permission to continue to use PCB in equipment which is totally enclosed is still being worked out. This one will take years to develop fully, during which the material will undoubtedly be eliminated from use and destroyed in the new incinerators made for the purpose.

For the present, there are complex prescriptive rules on the labeling, inspection, repair, and disposal of electrical equipment containing PCB. Some waste oils containing small amounts of PCB can be burned in utility boilers in some states but not all. Some states require PCB to be included in a management program for hazardous wastes; many do not and rely on the Federal program.

Some points are clear. Use of PCB in electrical gear in and about food and feed processing and manufacturing operations is highly restricted and will be prohibited where it is not already. Use of transformers and other equipment in vaults and other enclosed spaces, particularly in residential buildings and underground distribution systems, is coming to an end. Experience gained when fires broke out in these confined spaces shows that even though PCB does not burn, it can break down chemically and generate at least some chlorinated dioxins—which are feared more than PCB.

If you have any PCB on hand for any reason, it had better be labeled and stored, used, and inspected in accord with the regulations, because some of the largest fines the EPA has ever imposed for violation of the environmental rules have been imposed in this area of regulation.

Finally, it is unlawful to manufacture PCB or import it into the United States in any amount for any purpose unless the operation is exempted from the rules. The whole picture on PCB indicates that a prudent citizen and / or business operator will find out and carefully obey the rules as they exist and are applied in his own political jurisdiction. All State environmental control agencies provide detailed and understandable written materials for your use. Don't get caught as the small town of Gunnison, Colorado, was recently, when mishandling of PCB-containing equipment by the town-owned electric utility suddenly made it subject to fines and cleanup costs that were far beyond the town's ability to pay with its limited income.

Thus, while PCB may not be subject to the hazardous waste laws, its handling is definitely hazardous to the well-being of those who must or can't avoid dealing with it.

Asbestos

If you watch the Saturday morning cartoons and really old short subjects from the early days of the movies, from time to time you'll see a theater at the end of a performance when the curtain falls, showing the word "asbestos" painted in large letters across the fabric. This is a graphic example of the origin of our current concern, maybe obsession, with asbestos.

Before and even after the beginning of the electrical age, theaters were subject to frequent fires that began backstage and quickly engulfed the audience, who were often trapped by the subsequent panic. An asbestos curtain was accepted by the public as a nearly miraculous device for their protection. It seems unlikely today that any use of asbestos would be so openly promoted, much less touted as a major contribution to public safety.

With the kind of billing that goes with a prominent position on the curtain of the local theater and with a well-earned reputation for being strong, stable, and nearly impervious to fire, heat, and chemical attack, asbestos became widely used in nearly all facets of the modern way of life that emerged in the late nineteenth and early twentieth centuries. The introduction of steam heating and central heating in general created a need for pipe insulation that was not subject to fire and rot as cork was, and the boilers in buildings, railroad locomotives, and industrial processes were moving to ever-higher steam pressures and temperatures and had to be insulated, also. Because of its fibrous nature, asbestos can even be spun into thread and woven into fabrics like theater curtains or gaskets.

With the invention of the steel-framed, curtain-walled skyscraper in Chicago, the application of asbestos that has led to its greatest condemnation as a hazard arose naturally out of the circumstances. Since the great Chicago fire in 1871, public officials of that city have been leaders in attempts to prevent fires, particularly through requirements for building design and construction. The appearance of the new steel-framed buildings in Chicago around 1889 was bound to ensure that their performance would be closely observed in the inevitable fires that arose from the inadequate lighting systems and primitive central heat.

It soon became plain that the bare steel frames of the early experiments were extremely vulnerable to fire, which may not have been obvious to the designers—after all, steel is not flammable. The problem arises from the fact that at fairly moderate temperatures (around 700° F), ordinary steel loses much of its strength and bends and warps under the usual design load; so the building often collapses or, at best, is difficult or impossible to repair after a fire.

Chicago and other cities, along with building owners and architects, sought ways of protecting structural steel from fire, at least long enough to control the fire before expensive damage and collapse occurred. Many measures were tried, and the best solution turned out to require casting concrete around the steel to slow down the rate at which heat penetrated into the steel so that its temperature could stay in a safe range for an hour or two while the fire could be controlled.

Many large buildings were and still are built with concrete-protected steel frames, particularly industrial buildings in which the potential size and intensity of fires are great. Reacting to the unrelenting pressure of cost inflation and owners' need for containing construction costs, designers turned to cheaper ways to insulate structural steel—to make it "fire-proof." The best-performing replacement for concrete cladding turned out to involve mixing asbestos fibers with a binder, typically sodium silicate, and troweling or spraying the mixture on the steel to a thickness calculated to prevent the steel from getting too hot for one or two hours, allowing the fire protection system to work before catastrophic damage occurs.

For all these reasons and more, the Environmental Protection Agency believes that at least 30 million tons of asbestos were put in place in the United States between 1900 and 1980. Much of it is still there, and from this arises the real and perceived hazards of asbestos in the environment.

Asbestos is a generic term, often used rather carelessly, which applies to at least six different minerals. The names given to these substances are chrysotile, crocidolite, amosite, anthophyllite, actinolite, and tremolite; there are alternate names for some of them. By far the largest volume of asbestos in commerce has been chrysotile.

These minerals are usually found in massive solid forms, for example, as the hard stone called serpentine. The unique property of asbestos is its ability to be split into very small fibers instead of breaking across grain boundaries and producing granular materials, as the massive forms of the same silicate minerals do. Called asbestos under the regulations, the fibers must be at least 5 microns long and have a length-to-diameter ratio of 3 to 1 or greater. A micron is one-millionth of a meter; so the dimensions of these fibers are extremely small, which has all kinds of consequences. Asbestos mines are found in many places, particularly in Canada and California in North America.

The small size of asbestos fibers makes it difficult even to know when you are dealing with asbestos itself as opposed to fine fibers made from melted minerals in the mineral wool or fiberglass processes. Examination of carefully prepared samples under a microscope is needed to make the identification, and then only with a special technique called phase-contrast microscopy or an electron microscope. The same small size allows the fibers to be easily spread over large areas when suspended in air or water and makes them nearly impossible to capture from air or water except with very special filters. A common vacuum cleaner will do a very efficient job of distributing fibers all over the place rather than capturing them for safe disposal.

Some health personnel were aware of damage to the lungs of asbestos workers by the late 1920s. It was later found that the adverse health effects of asbestos exposure become evident only after a long time passes. This time period is controversial because it is the subject of many lawsuits. Nonetheless, adverse health impacts on workers exposed to asbestos in regular occupational settings were observed many years ago as the use of asbestos increased. Very early in the twentieth century a disease called "pneumoconiosis" was associated with exposure to asbestos—we now call this condition "asbestosis." Since some other fibrous insulating materials have fallen under suspicion for causing illnesses related to those caused by asbestos, a new name might have to be invented.

Clearly, in some individuals asbestosis leads to several forms of cancer: mesothelioma, both in the lungs and in the abdominal lining; lung cancer; and other less common cancers. These effects are all associated with breathing in airborne asbestos (or "asbestiform" fibers that may not strictly meet the definition). Cause-and-effect relationships between asbestos and disease are hard to establish because of the long time that usually elapses between exposure and clinical illness. In asbestos, this delay time or "latency period" appears to be as much as 30 years and possibly more, a subject being debated in many lawsuits.

It is also well established that tobacco smoking by persons exposed to airborne asbestos increases the health risks manyfold. The effect is so marked that some defendants in lawsuits seeking damages for asbestos exposure have attempted to shift the blame to the tobacco exclusively. So far, this defense has not been taken very seriously.

Less clear, indeed controversial, is whether ingestion of asbestos through the mouth or other exposures can cause asbestosis or result in cancer. Many streams flowing over large expanses of igneous rock, such as is found on the Canadian Shield geologic province, are found to carry large numbers of very short fibers having many of the physical and optical properties common to asbestos. While steps have been taken to remove at least some of these materials from public water supplies, there appears to be no consensus as to the danger, if any, which they cause. On the other hand, the Occupational Health and Safety Administration requires employees handling asbestos to be protected from ingestion through contamination of their hands or their food or water at the worksite.

The regulation of asbestos as a hazardous waste is possible, but the EPA has chosen, for many sound reasons, to regulate it under the authorities of the Toxic Substances Control Act (TOSCA), with other authority under the Clean Air Act, the solid and hazardous waste laws, and the special congressional mandates created under the Asbestos Hazard Emergency Response Act. This particular law affects the thousands of school buildings across the country, requiring at least some action to identify and/or remove asbestos that can reach students. It is thought by some to be a precursor of eventual legislation affecting just about all buildings of public resort or commerce.

Because asbestos is unequivocally known to be hazardous when inhaled, the regulations focus on controls which affect uses and applications which carry with them a possibility that the fibers can become airborne in ordinary use or as a result of some likely event like damage to insulation during maintenance or demolition. Many uses are still permitted under EPA rules, which apply to specified uses only; the material has not been banned from manufacturing and importation entirely. Some of these uses include use as a reinforcing agent in concrete pipe and sheeting, special gaskets and sealing materials where the product is not exposed, and other similar and important applications for which safe and effective substitutes have not yet been developed. Acceptable replacements for asbestos have been found for virtually all thermal insulation applications. It must be considered inevitable that all uses of asbestos in the United States will stop reasonably soon. Indeed, this is the announced objective of the EPA.

This complicated pattern of uses and prohibitions brings asbestos under regulation by at least six Federal agencies: Environmental Protection Agency, Occupational Safety and Health Administration, Consumer Product Safety Administration, Food and Drug Administration, Department of Transportation, and Mine Enforcement and Safety Administration, which develops safety and health procedures for the still-active mining operations in the United States.

This brief discussion cannot possibly cover the immensely detailed regulations that affect the use, handling, and removal of asbestos. The regulations are, however, directed almost entirely toward the avoidance and control of airborne asbestos. This is complicated by the nature of the fibers, which refuse to stay where they fall when released. Turned loose when insulation is damaged or disturbed for some reason, the tiny fibers will settle on surfaces which can be anywhere in the area, since they are light and remain airborne for long periods. Worse, they are easily reentrained by normal activity or even by almost undetectable air currents and made available for inhalation again and again. Reentrainment is thought to be responsible for most actual exposures.

Two basic approaches to the control of asbestos are already in place in buildings: (1) removal, (2) sealing (or encapsulation) where it is. Removal is the usual choice where a permanent solution is wanted, such as in a school building. Encapsulation may be best where the resulting coating can effectively be protected after it is applied. Either method is very expensive. Removal costs ranging from $6 to $50 per square foot of coated surface have been reported, and in many cases the area of coated surface exceeds the floor area of the building by a large factor. Reported costs for encapsulation vary even more widely, and no real generalizations seem possible. Part of the problem in estimating costs is that buildings, or at least large sections of them, must be completely taken out of service and often gutted to the bare walls and structural ceilings for the work to be accomplished safely. Some reported costs may include the costs for replacement space, others not.

As we have said, removal is the only sure and permanent way to eliminate asbestos hazard in any space. The removal industry has become highly regulated for the protection of all those with a stake in the work: the owner, the contractor and his employees, and the public who might be exposed to the material during the course of the work. Generally, the regulations are enforced by a State equivalent of the Federal OSHA, with special requirements imposed by the environmental control agencies. Most states require that only licensed contractors may do the work and impose heavy training and even testing standards on the contractor's supervisors and workers.

Methods of removal basically require handwork, scraping the asbestos from surfaces while everything is kept wet and the area is maintained in an enclosure that isolates it from the general environment. A contractor was granted a patent on the general system for removal of wet asbestos in an enclosure maintained at a pressure lower than the general environment, and he charges a fee to anyone who wishes to use this rather obvious methodology. Anyone intending to remove asbestos even in his own facilities must comply with all these rules to avoid prosecution under any of the regulatory schemes or even patent infringement risks that may apply.

Encapsulation has not really taken hold as a corrective method in spite of some powerful economic incentives coming from less-expensive materials and the potential for applications that do not require long-term evacuation of buildings. The reasons apparently focus on the fact that even in the space above false ceilings and in work spaces physically separated from the functional spaces, maintenance workers and outside contractors like telephone and electrical workers damage the coatings in the course of their work. This damage not only exposes the worker causing the damage but can result in secondary exposure of everyone in the building because of the reentrainment problem.

Finally, training and education in asbestos abatement and management has become a significant industry in itself. A number of reputable universities and technical institutes offer a wide variety of courses and hands-on training. Many private training organizations are also active. Before choosing any of these, check with the asbestos program people at your State or local environmental control agency to ensure that the program you are considering is certified or approved. If it is not, try another to be sure your training will be acceptable.

Generally, when wet asbestos is packaged in plastic bags in approved containers, it can be discarded in sanitary landfills. Some localities may require shipment to a hazardous waste landfill, but this is rare. In a landfill, asbestos never degrades, migrates little if at all, and is generally not considered to cause a hazard unless it somehow is exposed on the surface where it can become dry and thus airborne on the wind. Since asbestos cannot be burned and is proof against nearly all chemical treatment, landfilling seems the only reasonable choice for disposal.

Dioxin

In the bitter aftermath of the war in Vietnam, many former military personnel who had been exposed to the herbicide 2, 4, 5 T (2,4,5-trichlorophenoxy acetic acid) claimed aftereffects ranging from disabling injuries to birth defects in their children. In the resulting political-social-technical battle, which has lasted nearly twenty years and shows no sign of final resolution, the existence of a contaminant compound in the herbicide named (bear with us!) tetrachlorodibenzo dioxin was established. On testing, this obscure compound was found to be highly toxic and even to cause cancer in test animals. Adverse genetic effects were also predicted by tests with microorganisms and some larger animals. By 1985 the Annual Report on Carcinogens of the Public Health Service classified the compound to be one on which "sufficient evidence" existed to name it as a carcinogen in experimental animals. The evidence for cancer-causing activity in human exposure is still regarded as inadequate or marginal.

In the chapter "Chemistry of Waste Materials" we attempt to describe the way in which the structures of organic compounds like dioxin are formed. In this case, the dioxin framework is constructed of two benzene rings joined at two points through a single oxygen atom for each connection. It appears that only dioxin that has had chlorine added to its frame is responsible for the extreme levels of toxicity that have been found. Most attention falls on a substance named 2,3,7,8-tetrachlorodibenzo-p-dioxin. This substance has the reputation of being the "most toxic compound in existence" or, at best, one of a small group of the most toxic.

We use the phrase "has the reputation" to describe dioxin's status because there is little scientific data to support such a statement, which may well be true. In the real world, because of political factors developed out of the war and its effects on those who participated, dioxin stands condemned to the status of "most toxic" and facts have little to do with it. Any controlled scientific study that produced results in conflict with condemnation would probably not be accepted for publication in a peer-reviewed journal; thus, no aspiring or established scientist will waste time doing work which is likely to be unacceptable to the journals, no matter how well it is done.

In the process of arriving at this consensus on dioxin, the bad actor, we may have created a monster for other desirable environmental programs. We now know that chlorinated dioxins (there are 75 or so) are produced in nearly all combustion processes, raising questions about incineration of otherwise innocuous materials like wood, paper, and municipal waste. The status of dioxins and their associated chemical cousins, the dibenzofurans, is such that it is easily possible to panic a public audience with the possibility that these materials may be discharged either in the ash or the gaseous emissions from an incinerator. Whether or not these discharges—at concentrations ranging from a few parts per million to perhaps a million times less—are dangerous, they get people's attention when coupled with the "most toxic substance on earth" description.

Our purpose here is not to try to define a policy on management of dioxins and furans. Rather, we want to describe the issues and to provide some background for use by those who may wonder where these strange-sounding things came from and why they are the source of much controversy. In line with the main subject matter of this book, we should point out that the EPA has established a category of hazardous wastes called "dioxin wastes," which appear under appropriate numbers in the lists issued under RCRA.

With some irony, the "reportable quantity" of dioxin or dioxin wastes under the Superfund law is one pound—the National Response Center must be notified if a spill equals or exceeds that amount. According to estimates, the actual quantity of dioxin produced in all ways in the United States could be around 80 pounds per year, but no one really knows.

To sum all this up, Dr. Paul Cherimisinoff, Engineering Editor of *Pollution Engineering, The Magazine of Environmental Control*™, made a scan of the published literature on dioxins to get a sampling of the widely varying opinions:

- No deaths (of humans) from acute toxicity of dioxins have been reported.

- Dramatic effects have been seen on a guinea pig when the dose lethal to 50% of the animals is given. That dose is 0.6 micrograms per kilogram of body weight of the animal.

- Major effects may be very long-term carcinogenicity from undefined low levels (of exposure).

- The picture is unclear, but there is a growing feeling among scientists that the extent of the hazard may be overstated.

- Agent Orange used in Vietnam as a defoliant was contaminated with an average of 2 parts per million of 2,3,7,8-tetrachloro-dibenzo-p-dioxin.

- Dioxin is highly toxic to some species of animals but less so to others.

- So far, the only human illness proven to occur from dioxin exposure is chloracne (a skin rash of varying severity).

- Active dioxin poisoning can be toxic to the liver and kidneys.

- There is no conclusive evidence that dioxin causes human cancers, spontaneous abortion, or birth defects.

- Follow-up studies on accidental exposure to high levels of dioxin indicate no long-term adverse health effects.

- Some studies yet to be confirmed suggest exposure to dioxin increases risk of soft-tissue sarcomas in heavily exposed industrial workers.

Dr. Cherimisinoff listed these observations in an article published in *Pollution Engineering* for February 1989. (Used with permission.) Some minor editorial changes to clarify the technical language have been made here.

To summarize, the situation on dioxins and their close relatives is confused, as it is with many of the chemicals of concern in the environment. This is the day of the "political chemical"—one that has been convincingly presented to the public as a dangerous one, often for reasons having little to do with the actual situations in which it may be encountered in environmental exposures reasonably likely to occur in our daily lives. Yet any discussion of the substance is so highly charged with fear and emotion that politicians quickly decide that they cannot be associated with it and will, on the basis of "pressure from the public," avoid supporting any action which can, however tenuously, be tied to possible exposure.

In this kind of argument, facts are of little value, and the realist will conclude that public officials will probably be reluctant to be involved with, for example, the construction of municipal waste incinerators, no matter how badly needed, unless they can find a way to avoid being tarred with the brush of dioxin. An informed public may, on the other hand, decide that the balance of risks between exposure to infinitesimal amounts of an air pollutant and similarly small exposure to contaminated groundwater from garbage landfills comes down in favor of one or the other. No rational decisions can be made without understanding, and our representatives need public input to improve their chances of making good decisions.

Large-Volume Wastes

From the beginning of its attempts to develop a comprehensive program for the control of solid wastes, Congress has been trying to deal with a category of materials which are acknowledged to be wastes but which are generated in such huge amounts that a different approach is obviously called for in their management. In the original RCRA statute and in the 1980 amendments, most of these kinds of wastes were exempted from controls as hazardous wastes, while the EPA was directed to prepare reports and recommendations for future management approaches.

When Congress delays or declines action on a regulatory program, it often temporizes by ordering someone to do a study of the problem. Usually, these orders include deadlines, some allowing unreasonably short times for the completion of the study. In the original Resource Conservation and Recovery Act, Congress showed its difficulty with the "special" or "large-volume" wastes by even omitting the deadline from some categories. As a result, the Environmental Protection Agency, having other things to do that it regarded as more pressing, still had not completed the work after as many as six years, and other studies were far behind schedule.

The wastes that are included in this category are generally described as those which appear in huge quantities and are inherently low in hazardous effects. They are listed in the RCRA statute, both original and amended, as:

- Drilling fluids, produced waters, and other wastes associated with the exploration, development, or production of oil, natural gas or geothermal energy,

- Fly ash waste, bottom ash waste, slag waste, and flue gas emission control waste generated primarily from the combustion of coal or other fossil fuels,

- Solid waste from the extraction, beneficiation, and processing of ores and minerals, including phosphate rock and overburden from the mining of uranium ore,

- Cement kiln dust waste.

If regulated as hazardous wastes, some of these materials can become so expensive to handle and dispose of that other desirable programs, like the reduction in sulfur dioxide discharge from boiler plants, would be crippled. Congress had already been forced to accept the "political reality" that it could not legislate the use of low-sulfur coal as means of solving this problem—the then-Majority Leader of the Senate represented the coal miners of West Virginia and would not accept such a solution. Opponents also said that around 9,000 coal miners in the area would be totally deprived of their livelihood if low-sulfur coal from the strip mines of the West were substituted for the high-sulfur coal of the East. As usual, more is at stake than the health of the environment.

In a similar way, the presence of large piles of mine wastes, mostly but by no means all in remote parts of the West, signalled to politicians and regulators alike that an aggressive program to bring such materials under the hazardous waste umbrella was going to cause a lot of trouble for little return. For example, classification of copper-mining wastes as hazardous under RCRA would automatically make the sites subject to remediation under the Superfund program, which has enough trouble already. Besides, what are we to do with the vast amounts of these materials? Many of them have been lying where they are for a century or more, and few members of the public perceive any threat from them, even when the threat actually does exist. Coal-mining wastes have long been addressed in special Federal legislation which has established a number of regulatory programs and agencies which have no other function. The categories being discussed here are in the "nonfuel mining" category and the wastes associated with combustion of coal.

We must point out that solid wastes from activities incidental to, but not uniquely required by, the nature of a mining, milling, or smelting operation are still subject to the regulations which affect everyone. In this class are pesticides, waste oils, waste solvents, etc.

In its 1985 Report to Congress—"Wastes from the Extraction and Beneficiation of Metallic Ores, Phosphate Rock, Asbestos, Overburden from Uranium Mining, and Oil Shale"—the EPA estimated that 400 to 500 metal mines and around 12,000 nonmetal mines were operating in the United States. No one knows how many abandoned sites are to be found, but any casual visitor to the arid and mountainous West knows there are many thousands, usually hard to reach and challenging anyone to devise an alternate place or kind of storage for the wastes. To see the equivalent sites in states farther East is not so easy. Missouri harbors many abandoned lead mines, and so does Wisconsin, dating back to the desperate days of the Civil War. Elsewhere, Georgia had its mining boom in gold before the '49ers headed West, and Tennessee's copper and phosphate rock mines competed with a large and still-productive phosphate industry in Florida. Michigan was a major supplier of copper until very recently. In short, the mining-waste problem is not simply one involving a few Western states with a lot of Federal lands but is truly national.

The quantities of materials handled in these operations is phenomenal. No one will be surprised to learn that at least six iron-ore mines handled 10,000,000 tons or more of materials in 1982; but in the same year, seven copper mines handled similar volumes of material, nearly all waste of one kind or another. Several copper mines are billed as "the largest hole in the world"; when we realize that rarely do the ores contain more than a fraction of a percent of copper, then it comes home to us that most of the former contents of these "holes," miles across and hundreds or even thousands of feet deep, became waste. New technology is allowing many of the older piles to be reclassified as ore since the remaining "values" are recoverable by new processes. Still, 99% will again be discarded.

The eventual fate of the millions of tons of solid wastes generated by systems designed to remove sulfur from the discharge of coal-fired utility and industrial boilers has not really been addressed except by default. The wet scrubbers that were the first systems used for this purpose make a lot of sludge that is a nightmare to dispose of. A typical 1,000 megawatt power plant burning coal with 3.5% sulfur (not unusual for eastern U.S. coal) will produce as much as 250,000 tons of sludge each year, which can only rarely be processed to recover any useful products. Utilities are creating disposal facilities that look a whole lot like miners' tailings ponds where this material can be stored forever. The hope is that the newer methods of controlling sulfur discharges will also produce wastes which are less troublesome, but the reality is that wastes will be produced and must be managed.

The development and implementation of national policy for the identification and management of the large-volume special wastes will be a continuing story. Priorities in the use of resources for the development and management of the eventual program are hard to develop in an atmosphere of great conflict over air and water quality issues and the large expenditures which are the inevitable price of compliance with Superfund.

This is an area of government policy in which citizens should watch what's happening that affects them and provide guidance to legislators and government officials trying to determine what's necessary and possible. The very fact that no strong program has yet emerged from the public ferment over all things environmental indicates either a lack of a strong constituency for regulatory action in this area or a lack of perception of a significant hazard. If either were to develop, we can expect some member of Congress, in search of an issue, to force action. Whether that action will be reasonable will depend on preventing the development of a crisis atmosphere by prior deliberate evaluation of the risks in a calmer time.

Examples of Hazardous Wastes from Specific Sources

Waste ID number	Hazardous waste
Wood Preservation	
K001	Bottom sediment sludge from the treatment of wastewaters from wood preserving processes that use creosote and/or pentachlorophenol
Inorganic Pigments	
K002	Wastewater treatment sludge from the production of chrome yellow and orange pigments
K003	Wastewater treatment sludge from production of molybdate orange pigments
K004	Wastewater treatment sludge from production of zinc yellow pigments
K005	Wastewater treatment sludge from the production of chrome green pigments
K006	Wastewater treatment sludge from production of chrome oxide green pigments (anhydrous and hydrated)
K007	Wastewater treatment sludge from the production of iron blue pigments
K008	Oven residue from the production of chrome oxide green pigments
Organic Chemicals	
K009	Distillation bottoms from the production of acetaldehyde from ethylene
K010	Distillation side cuts from the production of acetaldehyde from ethylene
K011	Bottom stream from the wastewater stripper in the production of acrylonitrile
K013	Bottom stream from the acetonitrile column in the production of acrylonitrile
K014	Bottoms from the acetonitrile purification column in the production of acrylonitrile
K015	Stillbottoms from distillation of benzyl chloride
K016	Heavy ends or distillation residues from the production of carbon tetrachloride
K017	Heavy ends (stillbottoms) from the purification column in the production of epichlorohydrin
K018	Heavy ends from the fractionation column in ethyl chloride production
K019	Heavy ends from the distillation of ethylene dichloride in ethylene dichloride production
K020	Heavy ends from the distillation of vinyl chloride in vinyl chloride monomer production
K021	Aqueous spent antimony catalyst waste from fluoromethanes production

Examples of Hazardous Wastes from Non-Specific Sources

Waste ID number	Hazardous waste
F001	The following spent halogenated solvents used in degreasing: tetrachloroethylene, trichloroethylene, methylene chloride, 1,1,1—trichloroethane, carbon tetrachloride, and chlorinated fluorocarbons; and sludges from the recovery of these solvents in degreasing operations
F002	The following spent halogenated solvents: tetrachloroethylene, methylene chloride, trichloroethylene, 1,1,1-trichloroethane, chlorobenzene, 1,1,2-trichloro-1,2,2-trifluoroethane, ortho-dichloorobenzene, and trichlorofluoromethane; and the stillbottoms from the recovery of these solvents
F003	The following spent non-halogenated solvents: xylene, acetone, ethyl acetate, ethyl benzene, ethyl ether, methyl isobutyl ketone, *n*-butyl alcohol, cyclohexanone, and methanol; and the stillbottoms from the recovery of these solvents
F004	The following spent non-halogenated solvents: cresols and cresylic acid and nitrobenzene; and the stillbottoms from the recovery of these solvents
F005	The following spent non-halogenated solvents: toluene, methyl ethyl ketone, carbon disulfide, isobutanol, and pyridine; and the stillbottoms from the recovery of these solvents
F006	Wastewater treatment sludges from electroplating operations except from the following processes: (1) sulfuric acid anodizing of aluminum; (2) tin plating on carbon steel; (3) zinc plating (segregated basis) on carbon steel; (4) aluminum or zinc-aluminum plating on carbon steel; (5) cleaning/stripping associated with tin, zinc, and aluminum plating on carbon steel; and (6) chemical etching and milling of aluminum
F007	Spent cyanide plating bath solutions from electroplating operations (except for precious metals electroplating spent cyanide plating bath solutions)
F008	Plating bath sludges from the bottom of plating baths from electroplating operations where cyanides are used in the process (except for precious metals electroplating plating bath sludges)
F009	Spent stripping and cleaning bath solutions from electroplating operations where cyanides are used in the process (except for precious metals electroplating spent stripping and cleaning bath solutions)
F010	Quenching bath sludge from oil baths from metal heat treating operations where cyanides are used in the process (except for precious metals heat-treating quenching bath sludges)
F011	Spent cyanide solutions from salt bath pot cleaning from metal heat treating operations (except for precious metals heat treating spent cyanide solutions from salt bath pot cleaning)
F012	Quenching wastewater treatment sludges from metal heat treating operations where cyanides are used in the process (except for precious metals heat treating quenching wastewater treatment sludges)
F019	Wastewater treatment sludges from the chemical conversion coating of aluminum
F024	Wastes, including but not limited to distillation residues, heavy ends, tars, and reactor cleanup wastes from the production of chlorinated aliphatic hydrocarbons, having carbon content from one to five, utilizing free radical catalyzed processes. [This listing does not include light ends, spent filters and filter aids, spent dessicants, wastewater, wastewater treatment sludges, spent catalysts, and wastes listed in ss261.32]

Chapter 5

TOXICOLOGY OF WASTES

In the chapter "How to Recognize a Hazardous Waste," we explored the concepts on which the definitions of hazardous wastes are based. The theme that runs through all of the definitions is that wastes, to be hazardous, must: (1) be mismanaged, and (2) if mismanaged, must be capable of causing "harm" to human health or the environment. While the harmful effects can be many, as the "characteristic wastes" indicate, clearly the one characteristic that is of most concern is "toxicity."

TOXICITY, EXPOSURE, AND EFFECTS

The terms "toxic" and "toxicity" are in that large class of words for which everybody knows the meaning but no one can give a simple definition. Generally in official usage, the Federal government seems to abide by an effects-based definition of toxicity as the property of a substance to cause *any* adverse physiological effects, presumably to humans or at least to animals. Often, the word is equated to the quality of being *poisonous* to any form of life.

Common experience teaches that toxic substances (poisons) can affect an organism only if it is exposed to them. The public may be exposed to wastes in a variety of ways, but two are most common. Very rarely, people may come into direct physical contact with a waste material at a waste disposal or handling site or through a spill or other accidental release. Such an incident is first handled as a medical emergency, environmental concerns being dealt with later. Of more concern to most of us is the problem of exposure to hazardous waste that might migrate from its disposal site to affect people or the environment.

Completing this group are transportation-related spills of hazardous materials that did not become "waste" until they were spilled. Barges, trains, trucks, and pipelines carry many dangerous materials. When an accident occurs, wastes (often hazardous ones) are created instantly. Usually the spilled material, the contaminated soil and equipment, and all the cleanup materials become hazardous waste. Dramatic spills are expected to cause serious problems, as when a train derails in or near a town and forces an evacuation. Less obvious incidents may be more dangerous; for example, a spill of seed grain treated with a powerful fungicide may look harmless, but the trained person recognizes the threat of carcinogen exposure to a large population if the spill is not controlled.

The fear of toxic materials from waste that may act over long distances and long time periods is hardest for the citizen to bear. The study of these materials and their impact on humans and other living beings is usually lumped under the term "toxicology." Like any scientific field, a lifetime of study may be needed to understand even a small part of this important and still-emerging discipline. Still, the general principles are understandable to nonexperts and need to be described as an essential element of an informed citizenry's participation in decision making on matters that may affect their lives in important ways.

The task of dealing with toxic effects of waste materials is complicated by the careless use by the popular press, and by others who ought to know better, of language that equates hazardous waste with toxic waste or toxic chemicals or chemical waste or some similar phrases. As we shall see, the language used for discussion of toxicology and hazard control is often confusing but vital to a serious attempt at understanding what is being said about very real issues.

UNREASONABLE RISKS

The often-unspoken and unwritten qualifier to the definition of a toxic substance is the word "unreasonable." That is, not only must a substance do harm but the harm must be "unreasonable" under the circumstances of use in order for the substance to be considered toxic.

The difficulty is that one person's reasonable level of effects may be the next person's unreasonable damage. Who can say what's unreasonable with respect to another person, much less to the unseen animal, insect, and even microbial life that may be affected in the general environment?

Because humans do the research and the decision making, humans tend to think we can make valid choices as to what's reasonable when we or our fellows are exposed to harmful substances. These choices, however, are made within a context of all the values and prejudices of the decision maker. Congress uses words like "unreasonable" when it wants to signal that the regulators are not to be so literal as to shut down the economy but are to be "realistic" and choose some level of exposure and risk that will be socially acceptable in the society in which we live. This value-laden choice of "acceptable" risk is the source of much confusion about the use of scientific data to measure factors like toxicity that are essential to an effective regulatory scheme that will protect the public from unreasonable risk of harm.

This application of scientifically obtained information is also a key factor in the confusion that the nontechnical public experiences when trying to fathom the meaning of advice from experts or regulators. An unreasonable source of injury to any one person may be perfectly reasonable to another—the word "reasonable" is not a scientific term.

In another chapter we discuss the concepts of Risk Assessment and Risk Management that are used to make formal judgments about these highly subjective decisions in a "scientific" way.

SCIENCE AND THE MEASUREMENT OF TOXICITY

Science can obtain data on the effects of chemicals on living beings, may even find ways to measure the effects with great delicacy and precision, but "reasonable" will always remain a matter of opinion which no amount of scientific work will define. This necessary juxtaposition of words and terms that are on the one hand presumably "objective" because they are "scientific" and on the other hand "subjective" because they reflect the commentator's values, creates much of the emotional and even moralistic tone of many public and private debates about issues affecting our physical environment.

More importantly, this clash creates a problem for many lay people in that the opinions of scientists often are confused with statements of "scientific fact"—and they are clearly not the same. In principle, many thoughtful scientists will carefully distinguish between value judgments and observational data when addressing these issues. In practice, good data on effects of toxic substances on humans is scarce, while effects-data on nonhuman target species are even more rare and often unreliable. Under the pressure of the need to make important decisions, an informed guess about health and environmental impacts is often necessary. Qualified experts will often disagree about these guesses, but reservations and doubts tend to disappear in the heat of discussion.

In our Western philosophy, the scientific method has become the basis of all study of the world in which we live. The basis of the method is that all data and observations must be repeatable by another trained experimenter or observer and that this repeatability confirms the soundness of the information. Further, in this formal system the only "facts" are the observational data which can be duplicated and described in quantitative ways.

Once valid data are in hand, they can be interpreted in the light of "hypotheses" or theories which are fashioned in an attempt to detect the laws of nature and to understand the data well enough to predict the results of future experiments. Theories and interpretations come and go, but in true scientific work they are always provisional—that is, they are subject to replacement by better interpretations or new data. Many theories are not directly provable, but they become accepted as a basis for organizing data and further work because they are found true over and over again until any argument against them becomes trivial. Others so perfectly explain a large amount of varied observations that they become primary intellectual frameworks for generations of scientists. An example of this latter kind is the Darwinian Theory of Natural Selection—not provable beyond any exception and not acceptable to some persons as an organizing principle for the enormous mass of data which has accumulated since modern approaches to the study of the earth and its past have been developed.

The interpretation of data by an established and respected worker in a particular field usually deserves more credence than that of an amateur or nonexperienced person. The history of science is, however, littered with the wreckage of once-accepted hypotheses: the earth-centered universe, the flat earth, the theory of spontaneous generation of organisms, even Lamarck's theory of the inheritance of acquired characteristics—once offered as a primary alternative to Darwin's Natural Selection. Human hearts beat and pumped blood for millions of years before Dr. William Harvey ever described circulation. This "reasonable" description encountered resistance when it was first published.

How then can we go about determining what "substances" do "unreasonable harm" to humans, animals, plants, etc.? Sensibly enough, we first have to define "harm" and find ways of measuring it. Once we have measures, then the criterion of "reasonableness" may be applied, even if reasonable people may not agree on what's reasonable!

The measurement of toxicity must somehow deal with the determination of harmful effects on living organisms. Our ethical system does not allow us to run experiments on humans, at least not on purpose, so we must rely on two kinds of information: (1) the effects of chemicals on people who have been exposed by accident or by choice to specific materials or a particular kind of environment, and (2) the effects of chemicals on experimental animals and/or plants or even microorganisms when purposely exposed in controlled experiments. Even in this kind of work, experimental designs are restricted by public insistence on using only some kinds of animals as experimental subjects.

Without debating here the validity of the moral judgment of the animal rights activists, we must say that without sacrificing *some* species in testing for toxic effects, we will be left with only the human as an experimental subject. Accidental exposure and epidemiologic data have their limits as sources of adequate information on human safety. Here, in the choice of test subjects, whether they be *escherichia coli* (a common sewage bacterium used for chromosome-damage assessment) or an abandoned pet, lies another of those choices which are a continuing thread in this book.

Common sense tells us, too, that toxic agents may be acutely poisonous, like the ones common in detective stories or those marked with the skull and crossbones. Even worse, some may be deadly in small doses over a long time. Since most of our reasons for concern about wastes arise from the chemical nature of the stuff wherever found, it's necessary to dig into the chemistry of wastes so that we may have some basis for evaluating their biological effects, once we understand what the dangerous components of hazardous wastes really are.

To begin, we must realize that everything on earth, living or not, is made up of chemicals. Despite the apparent simplicity of this trite beginning, it is a source of much misunderstanding. Every waste is chemical waste. Only the nature of the chemicals and their relative abundance in any particular case make them hazardous (or dangerous or toxic or "chemical waste").

Chemists know and can identify many hundreds of thousands of chemicals—particular kinds of molecules or assemblies of atoms that can be identified. The number of individual chemical substances in commercial production and use in the world is probably between 30,000 and 50,000—a staggering number that might discourage any attempt at understanding. A closer look, however, shows that our real sources of concern are a few hundred substances that are toxic in concentrations likely to be encountered in the environment and a smaller number that are known to or anticipated to cause cancer in exposed individuals. Significant numbers of these are complex mixtures, sludges, and concentrated impurities from the manufacture of pure chemicals.

While it may be true that every chemical may be toxic in some amount (common table salt, for example), it does not appear to be true that everything causes cancer in humans—at least we cannot prove that is true! Exposures to acutely poisonous amounts of chemicals through the environment are very rare. Occasional major accidents like the incidents at Bhopal in India or at Soveso in Italy may expose substantial populations to acute toxics when the energy of an explosion disperses the toxics over large areas. On a more local scale, truck upsets or train derailments can create such conditions; so they cannot be taken lightly, and elaborate contingency plans are often in place to deal with them. Historically, the exposure of employees in manufacturing and using these materials has caused extensive studies of their effects in single and repeated acute exposures. These data are available in a number of standard compilations which can be found at most public libraries. A few examples for substances of interest are given in Table 3.

While the data on short-term acute effects may be expressed in a number of ways, two reported quantities are usually of major importance in assessing hazards of these kinds of exposures. The first of these expresses the potency of any poisonous effects of a chemical or mixture by describing the dose of that substance which, when administered to a particular animal species, results in the death of half of the animals in the test group in a specific and short period of time. This lethal dose to half the mice, rats, hamsters, or other test animals or plants is called the LD_{50} dose. The risk of poisoning by contact with the material is obviously greater when the reported LD_{50} dose decreases. The dose is usually reported as milligrams of the test material per kilogram of body weight of the test animals.

Industrial hygienists work with toxicologists to interpret these figures. From comparative experience and data on short-term acute and long-term exposures in workplaces, maximum allowable concentrations in air, for example, for a regular work schedule (8 hours a day, 50 weeks per year is typical) are set for enforcement by the Occupational Safety and Health Administration and by various state OSHA units, Industrial Commissions, or Health Departments. These so-called Threshold Limit Values (TLV's) are useful guides to the relative danger of materials that can be compared. They may have little relevance to risks to the general public through environmental pathways of exposure, and like all such information, must be used carefully in decision making. Table 3 shows values of lethal doses and Threshold Limit Concentrations for some common materials used in industry.

Long-term (or chronic) toxicity testing begins at a concentration less than the minimum lethal dose and may be extended to very low levels of exposure, often limited only by analytical sensitivity to the compound being tested, but more often by the difficulty of detecting pathologies in the test subjects. Progress on analytical sensitivity and the difficulties that may arise from the very low concentrations now measurable is discussed in Chapter 11, "Chemistry of Waste Materials."

Living organisms are tough and have grown and evolved in a world that has many toxic substances floating around at low concentrations. For example, the seawater in Tampa Bay is said to contain enough arsenic naturally to kill every person on earth. Many natural foods contain traces of extremely poisonous chemicals. Perhaps the tomato was thought to be poisonous for a long time because it belongs to a genus of plants (Solanaceae) that has many poisonous members. Even the friendly Irish potato contains a toxic chemical that, if concentrated, can be lethal to humans!

Table 3

LD_{50} and TLC Values for Some Substances of Interest

Chemical Substance	**Lethal Dose, 50% (LD_{50})**	**Threshold Limit Concentration, 8-hour workplace exposure**
Acrylonitrile	35 mg/kg	0
Benzene	ca 2000 mg/kg	1 ppm
DDT	400 mg/kg	1 mg/M^3
Ethyl Alcohol	300–400 ml/hr	1,000 ppm
Methyl Alcohol	60–250 ml (one time)	200 ppm
Perchloroethylene	4,000 mg/kg	50 ppm
Trichloroethylene	580 mg/kg	10 ppm
Arsenic (oxide)	120 mg (one time)	0.002 mg/M^3
Chromium Salts (valence 2–3)	ca 5,000 mg/kg	0.50 mg/M^3
Lead (adult)	500 mg total absorbed	0.15 mg/M^3
Cocaine	30 mg/kg	——
Cyanide Salts (inorganic)	2 mg/kg	——

Note: mg/kg=milligrams dose per kilogram of body weight
ppm=parts per million in air
mg/M^3=milligrams per cubic meter of air

Source: Robert H. Dreisbach and William O. Robertson, *Handbook of Poisoning,* 12th ed. (Norwalk, CT: Appleton & Lange, 1987). Used with permission.

Since we know from experience that organisms grow and thrive in this "soup" of toxics, there must be natural resistance or an ability to handle most toxics in small amounts or concentrations. Toxicologists find that for most toxics a low level of exposure can be determined that causes no detectable effects on living things. This level is specific both to the chemical substance and to the particular plant or animal that is exposed. In testing toxics, then, the experimenter seeks the dose level at which there is "no observable effect," called the NOEL in the inevitable language of acronyms favored by regulators and other bureaucrats. This no-observable-effect level is a precise scientific piece of data, measurable in principle for any test species and any substance. However, since different species and even different individuals vary greatly in their reaction to a toxic substance, it is hard to know from animal tests how humans, for example, might react.

THE JUDGMENT FACTOR

The use of these test results is best described as a process of arriving at a best guess as to what levels of exposure might be safe for humans. A panel of experts is usually convened to review all the available information on animal tests of a chemical or substance. Using their judgment and whatever experience may be available, this group arrives at some concentration or dose in milligrams per kilogram of body weight that is thought to result in no observable effect. Next, to take account of the uncertainty and poor data quality, a safety factor is arrived at by consensus. The safety factor is applied to the NOEL and the result is called the Reference Dose—thought to be a level at which humans (or, rarely, other organisms) can be safely exposed to the chemical throughout their life.

There is little science and a lot of opinion involved in setting these levels of Reference Dose. Until recently, the amounts were called the Acceptable Daily Intake, which they really are. Some people who believe that only zero exposure is acceptable have objected to the implications of the term "Acceptable"; so the name has been changed—but nothing else.

The idea that there is a "threshold" rate of exposure to chemicals below which there is no adverse impact is regarded by some as highly controversial. Certainly it is hard to prove, since the testing methods are limited in their ability to measure small changes over a lifetime. More important, it is generally believed that cancer-causing action, together with the damage that causes birth defects and adverse mutations in offspring, result from actions within cells that are different from those observed in direct toxicity. If this is true, and it seems to be logical, then there may be no threshold concentration or dose below which none of these oncogenic, teratogenic, or mutagenic effects will happen.

The experimental toxicologist seeks to find and to measure two very different kinds of effects of chemicals in experimental animals. The first of these is indicated by changes in the general health or behavior of the animals when compared to an identical group (the controls) not being dosed. Of course, frankly expressed injury at sites of administration or other ways must be observed also. The test animals may die sooner than the controls, they may behave differently, becoming extremely active or torpid and inactive, or a multitude of other things lumped under the headings of "Mortality and Morbidity."

At the end of the experimental period, some or all of the dosed mice or rats or whatever will usually be killed and dissected to try to determine what, if any, organs were affected by the test chemical.

The test must be conducted on many animals at more than one dose level, and this takes time and money. Test animals and their feeding and housing are expensive, as is the time-consuming work of the highly skilled pathologists needed to evaluate the effects. Even a short-time evaluation of highly toxic material is a major effort, requiring extreme care in its execution. Table 4 shows some 1989 estimates of costs for toxicological tests of interest.

Test animals must be carefully chosen. Wild mice, rats, etc., are so tough and accustomed to contact with all kinds of substances in their ordinary lives that they cannot be relied on to show reproducible effects when exposed to a single substance. (Remember that the scientific method insists that we be able to repeat the results.) Test animals are specially bred to be as alike as possible, sometimes to a point that the untreated controls, being highly inbred and sensitive, will live only a little longer than the treated ones.

When preparing to test for long-term toxic effect, it is usually necessary to do short-term tests first. There is obviously no sense in long-term testing at exposure levels that kill the test animals quickly. Long-term toxicological evaluation of a substance is a complex and expensive research program which must be designed for each particular situation by experts, carried out under rigorous control to avoid effects from outside the experiment itself, and conducted by highly trained and experienced persons.

Mice and rats are usually chosen for test subjects out of necessity. Hundreds of animals are often needed to draw any meaningful conclusions, and larger animals are prohibitively expensive to buy, house, feed, and generally maintain for periods of months and even years. Larger animals must sometimes be used in special test situations. The potential for irritation to skin or even the absorption of certain toxics through the skin is often tested with rabbits. The guinea pig is the most sensitive small animal to many toxics but is larger and slower-breeding than the

Table 4

Cost of Toxicological Testing

Kind of Test	**Approximate Cost 1989**
Ames Test (Screening for chromosome damage)	$1,500
Determination of LD_{50} dose in rodents (oral administration)	$4,000
Determination of LD_{50} dose in rodents (skin application)	$6,000
Screening test for skin irritation	$1,000
Screening test for eye irritation	$1,000
Oral toxicity study—90 days	$200,000 to $400,000
Oral toxicity study—2 years	$1,000,000 to $1,500,000

Note: These cost figures are approximations which are subject to major variation depending on the numbers of compounds to be tested, difficulties of administration resulting from chemical or physical characteristics of the test substance, market factors resulting from laboratory workloads, and many other considerations.

Source: Private communication, Dr. Donald E. Stevenson, Shell Oil Co. Used with permission.

rat or the mouse. Larger animals than these are, by cost factors, restricted to testing of drugs and extremely high-value formulations.

Tests usually expose the animals on a regular basis in feed or by some other method of administration for the lifetime of the animal. Exposure to vapors or mists, painting on the skin, or injection by needles are among other common methods. Some kinds of tests, as we shall see, may require continuing the test through more than one generation of animals—which makes mice especially attractive as test subjects. Testing seeks knowledge about a variety of effects: (1) Excess mortality—do the test animals die more quickly than the controls? (2) Morbidity—are the test animals less healthy in some way? (3) Oncogenicity—do the test animals develop tumors or cancers more frequently than the controls? (4) Teratogenicity—do the offspring of the test animals have more birth defects than the offspring of the control group? (5) Mutagenicity—do the offspring have birth defects which are inherited by the third generation?

Aside from the intimidating words, the tests are designed to tell the experimenter whether the long-term exposure makes animals sick, the level below which there is no observable effect, and whether there is an increase in the incidence of cancer and birth defects in the first generation and in their offspring.

An important piece of information that is sought in these tests is called the dose-response relationship. Unless it can be shown that a larger dose of toxic chemical produces more effect (mortality, morbidity, etc.), then there is doubt that the effect being seen is actually related to the chemical being tested. While this may seem obvious to common-sense thinking, it is not always accepted as a requirement in finding a material hazardous.

We have discussed the problem of wild animals as test subjects. This problem arises with even greater impact in long-term testing. To gain sensitivity, to increase the likelihood that any effect is observed, very special inbred mice and rats are used as subjects. In some cases, substantial numbers of the test animals (30% or 40%) will develop tumors even if they are not in the treated groups. This complicates the interpretation of the experimental results, introducing the need for sophisticated statistical analysis to be applied and often making the result intelligible only to experts who are comfortable in the arcane language of the statistician. In this world, simple questions have complicated answers.

This situation is not the fault of the scientist or the statisticians, who are often the same people. It is the result of the fact of observation: no two individuals of any species are truly identical; therefore, some will be more or less resistant to toxic effects than others. Statistical techniques are the only recourse in attempts to measure changes in a variable population.

As shown in Table 4, such testing can easily cost half a million dollars just to get provisional answers for one compound or substance. Naturally, repeating the work for ensuring scientific validity will at least double the cost. Testing for long-term effects like carcinogenicity and potential for genetic damage resulting in birth defects is far more expensive, and the results are still more difficult to interpret as they may be applied to humans.

Here, in the difficulty and expense of testing, lies the greatest problem in understanding and, for that matter, enforcing the rules and laws dealing with toxic materials in our environment.

There are thousands of chemicals, hundreds in our wastes, but only a few can be tested at any one time. The test results can be interpreted in different ways by equally qualified experts. Finally, we have no common basis for judging whether something is "unreasonably" dangerous.

If we refer to our basic definition of hazardous waste—something discarded, spilled, or lost which can do harm to people or the environment—a little analysis can throw some light on risks we may encounter. Since management of the materials is an essential part of the picture, some practices need to be reviewed at the outset.

First, having already found that nearly every activity generates waste, some of which may be hazardous, we need to understand where hazardous wastes come from so we can examine how people are exposed. There is no hazard without exposure and no way to apply toxicological data without an estimate of the amount of toxic exposure. In an excellent review of hazardous waste regulations, John Quarles, formerly General Counsel and Acting Administrator of the U.S. Environmental Protection Agency, has suggested broad categories of activities that generate hazardous waste:

1. Manufacturing processes
2. End-use activities
3. Finished products that have become unusable, unneeded, or unwanted
4. Spills during transportation of hazardous nonwaste materials

It is easy to see how manufacturing processes may be substantial producers of hazardous waste. Our earlier discussions on chemical equilibrium and the inevitability of waste production are enough to establish this as a source. The EPA has estimated that as much as 88% of the hazardous waste generated in 1975 came from four kinds of manufacturing and that 60% of the total hazardous waste was generated from chemicals and allied products production.

End-use activities are those in which we the consumers are often implicated. Building construction, agriculture, and the daily life of American consumers result in used pesticide containers, spills of potentially hazardous products, broken or nonfunctioning products like used motor oil, solvents, dry-cleaning fluids, old electrical equipment which may contain PCB's, and even asbestos from construction waste or discarded insulation. One end-use category often overlooked includes ash and sludge from antipollution equipment—even though much of such material is not "officially" hazardous.

The third category of unusable, unneeded, or unwanted products is a source of much concern and frustration both to the public and small business and to the regulatory and waste management people who are seeking ways to handle them safely. Garages and gardeners' sheds all over the country have millions of full and partial packages of stuff sitting around. Periodic attempts by waste managers to get the public to turn these in have repeatedly shown that many toxic products, even many that have been banned from the market for years, are to be found in consumers' hands, often without a clue as to the potential hazard involved.

Among such products are DDT and other chlorinated pesticides, most banned; cleaning fluids containing carbon tetrachloride and related chemicals, paint removers, automotive fluids like motor oil, brake and transmission fluids, banned or restricted herbicides like 2,4, D, or similar products.

Small businesses have the same kinds of materials on their hands, usually in larger amounts. Finding legitimate means of getting rid of these things, even if unused, is difficult and expensive. Often these materials will simply be left where they are in the vague hope that some opportunity to use them will arise. Checklists of dangerous materials likely to be found in the home or business are available. In the appropriate chapters, we include some helpful lists for improving the safety of homes and businesses and of the environment.

In our daily lives, under the constant prodding of press and politicians, it is hard to gain a sense of perspective about toxic materials, particularly carcinogens, which are identified or cleared by the testing methods we have been describing. Usually, the word "carcinogen" is applied to chemicals which cause tumors, although it properly can include those exerting adverse effects on genetic materials. The terminology is not reliable, since most writers in the popular press are not well-informed in this field and do not understand the terminology well enough to explain it to their readers.

At any rate, the U.S. Department of Health and Human Services in the Fifth Annual Report on Carcinogens for 1989, lists 20 substances or groups of substances which are "known carcinogens"; that is, they are known to cause cancer or related effects in humans at some level of exposure. One hundred forty-one others are listed because they "may reasonably be anticipated to be carcinogens." The use of our old friend "reasonable" indicates that this is a judgment call by some expert group.

No one really believes that all possible chemicals capable of causing cancer are in these lists. What does matter is that the number of such substances is relatively small when compared to the enormous number of candidate chemicals to which humans are exposed.

It is particularly interesting that of the listed materials:

- 2 are natural substances occurring in food raw materials
- 13 are pesticides, all either banned or restricted in use
- at least 31 are drugs, many used in cancer therapy
- about 100 are commercial chemicals, including 7 metals in various forms

Toxic wastes, then, are those wastes that, given enough exposure, will cause someone to become ill or disabled. Cancer is only one possible result—other effects are real and possible, provided a person takes in enough of the poison to affect him. In principle, the same statements can be made about all living things, but the truth is that we humans, in our self-centered way, have spent relatively little effort or resources on determining the effects of toxic substances on nonhuman targets; so it is difficult or impossible to draw conclusions about the effects on these little-known organisms.

How much is enough exposure? More important, how much is too much? This is the crux of the problem we face in understanding our risks from waste disposal in our environment—understanding that is essential to managing those risks to achieve safety with reasonable achievement of other goals in society.

Many people who have devoted much thought and concern to the issue of safe disposal and handling of wastes, as well as to the exposure of persons in our industrial society to chemicals which are not waste, have concluded that there is no safe level of exposure, at least not when dealing with carcinogenic substances. Zero exposure was a common objective in the public debates of the 1960s and early 1970s. Zero proved an elusive concept in the light of advances in analytical chemistry that were coming about at the time.

ANALYTICAL PROGRESS

In the chemical laboratory of the 1940s and even the 1950s, most quantitative analyses were accomplished by separation of the target material from the sample and weighing it. Very sophisticated balances were developed, but even the most advanced were limited in sensitivity to a few micrograms. A microgram is a millionth of a gram; so for a one-gram sample, it was possible to detect a few micrograms—a few parts per million.

Wartime research brought about rapid introduction of new methods of analysis based on spectroscopic techniques which did not require precise weighing and depended only on the ratios between amounts of constituents in a sample. During the 1950s we were measuring concentrations of a fraction of a part per million, depending on the compounds. By the 1970s it became literally possible to count the molecules and to detect organic constituents at one one-thousandth of the 1950s levels—or a few parts per billion. In some research laboratories, sensitivities in the parts per trillion or even per quadrillion are claimed.

One part per trillion is one gram in 1,000,000,000,000 grams (for mathematicians 1×10^{-12}). How can we visualize such extremely low concentrations? Let's try this: A drop of water weighs about 30 milligrams. A liter of water weighs about 1 kilogram, or 1,000 grams, or 1,000,000 milligrams. One drop, then, in a liter (a little more than a quart) is roughly 30 parts per million (ppm).

If your bathtub at home holds about 35 gallons of water, then one drop amounts to one-half part per million. To push this calculation to the limit, one drop of contaminant in 500 bathtubs amounts to one part per billion (ppb); in 500,000 bathtubs, one part per trillion (ppt).

Now, it's hard to believe that such small quantities of anything could be harmful. Remember, we live in a world of chemicals and have evolved as animals in the presence of many of these things over millions of years. The real question is: what does it mean when some official or some activist announces that a few parts per billion of some suspected toxic has been found in our drinking water or air? No one really knows. Does it constitute an "unreasonable" threat to health and/or the environment? There are no standards by which most such questions can be judged, and it is doubtful that such standards can be developed in a meaningful way for many years.

This gap between knowledge and understanding is hard to fill. A report on testing a chemical for cancer-causing power will go something like this (in part):

Carcinogenicity

- There is sufficient evidence that benzene is carcinogenic in experimental animals ...
- There is sufficient evidence that benzene is carcinogenic to man.

Production

- Aggregate for the U.S. more than one billion pounds

Exposure

- In 1976 an estimated 1 to 3 billion pounds of benzene were released in the atmosphere from 132 million stationary and mobile sources ...

- Ambient monitoring data indicate that levels of benzene (in the air) range from 1 ppb to 100 ppb ...

- Additionally, benzene has been identified in drinking water and in subsurface water at concentrations to 10 parts per million.

- Benzene occurs naturally in fruit, fish, vegetables, nuts, dairy products, beverages and eggs.

- The National Cancer Institute estimated that an individual may ingest up to 250 micrograms per day. Benzene is reported to occur in cigarette smoke at between 50 and 150 micrograms per cigarette.

Experts and even the courts hearing the experts often disagree as to what this kind of report means. The workplace allowable level of benzene in the air is 10 parts per million for 8 hours of exposure per day, presumably for a lifetime. The Supreme Court, on procedural grounds, stopped an attempt to reduce the level to 1 part per million. The Court specifically avoided a ruling on the reasonability of the standard itself.

CHRONIC TOXICITY AND CARCINOGENS

Most scientists who work in the field of cancer-causing chemicals believe that the basic effect of a carcinogen occurs within the cells of a living organism. It is possible, but may be impossible to prove, that a chemical has to affect only one cell in some way so that that cell goes "wild" and begins to multiply rapidly. Most of these "wild" cells may be caught and destroyed by the organism's own protective system, but some may survive to become tumors, even cancers.

In this so-called "one-hit" model, the likelihood that a cancer will be started is related to the concentration of chemical molecules capable of triggering the reaction which arrive in the body of the animal or plant. While this probability will decrease as concentration decreases, it never will become zero unless the amount of carcinogenic chemical is reduced to zero. In this way of thinking, then, and unlike simple poisons, carcinogens cannot be said to become ineffective at some low or threshold concentration.

Research continues to show that the life of cells and therefore of their parent organisms is controlled completely by their genetic material: the genes, chromosomes, and their accompanying constituents. This leads many scientists to believe that the potential for cancer-causing action can be measured by exposing cells to chemicals and finding out whether the genetic material is affected in some way. It is, however, not yet possible to know whether breakage of chromosomes in a cell is related to any specific change in the cell or the whole body—the code is too complex. Nonetheless, much screening of substances for carcinogenic potential is conducted using this technique, applied to simple organisms like bacteria. While such methods are useful in identifying candidates for more sophisticated testing, the data generated in the Ames and related tests must be used with caution.

Indeed, if there is a conclusion to be reached at this relatively early stage in the development of the science of experimental toxicology, that lesson is that the data must be used with caution. Even experts will disagree about the nature of the data and about methods of interpretation. Citizens should beware of the self-identified expert whose qualifications are a general interest or a general education in biology, when attempting to assess their own position on these matters. Better the common sense of the informed lay person than the well-intended but often erroneous judgments of the half-informed.

Some of the finest minds and laboratory facilities in the world are devoted to improving our understanding of the ways in which tumors, cancers, and other conditions resulting from our life-styles and exposures can be understood and controlled. This section, by a nonspecialist, will be of little use to the specialist because it is too general.

Readers who wish to pursue this topic can do so at almost any level of specialization they wish to achieve. Begin with the popular journals, *Science, Scientific American,* and their companions. Beyond those, the scientific libraries are full of the international journals devoted to this field, but they often require a Ph.D. to comprehend them properly.

Chapter 6

RISK ASSESSMENT AND MANAGEMENT

THE SCIENCE AND ART OF RISK ASSESSMENT

A traveler in the United States today, at least one who reads the local papers and watches the evening news, is struck by frequent references to hazardous wastes or toxic waste or industrial waste sites and their threat to the community. Graphic artists for the papers and TV photographers try to communicate some sense of the issues in pictures of evil-looking piles of junk or streams of liquids. The stories nearly always end with a statement that something in the junk will escape and is capable of causing cancer or some other dread disease. Fairly extensive travel in the other developed countries of the world has not encountered this same level of publicity and, presumably, of public concern.

Recently, the "Green Parties" in Europe have been gaining in political strength by calling attention to the problems of water and air pollution. The relative freedom of European governments to foster low-cost exports by accepting very high levels of pollution may be ending, led, as in the United States, by an aroused and worried public.

American journalists coin catchy names, easily fitting into the headlines. Love Canal, Stringfellow Acid Pits, the Twin Cities, Rocky Mountain Arsenal, or a hundred others form a basis for stories. It is hard to imagine a more foreboding place name than the "Valley of the Drums"! Several sites have been branded the "worst in the nation" or first on the EPA's National Priority List of sites identified as needing cleanup under Superfund. For several years, one Minneapolis television station used the same clip of the same backhoe puncturing the same drums of goo almost as if it were a main title for any story about hazardous waste.

Practically all these stories deal with old disposal sites where wastes of some kind (usually chemicals) were dumped. Most stories are illustrated with pictures and quotations from people in the neighborhood expressing feelings which range from concern to outright claims of serious illness or death among their families or close neighbors.

The publicity is very effective. No one doubts the importance of the subject in the public perception of things. Few doubt the reality in any objective way. The question for these citizens is: how can we be safe? What risks do we run from possible exposure to toxic and hazardous wastes, and how can we live in a world in which these things are already dumped, much less in one where new quantities have to be dumped, used, treated, or otherwise disposed of every day?

Someone in society has to decide what can and cannot happen when wastes finally have to be managed. After all, the regulatory laws are based on the idea that *wastes are dangerous only when mismanaged*. To use another favorite term in the laws, mismanagement occurs when an unreasonable risk to human health or the environment results. The elusive concept of "unreasonable risk" is discussed in Chapter 5, "Toxicology of Wastes."

Decision makers, then, look for ways of managing waste. Research in this field is highly technical and expensive, but the options for final disposal are few: on the land, in the air, or in the water. According to the Law of Conservation of Matter, we can change the chemical or physical nature of waste materials, but we cannot ever destroy their basic building blocks, the elements out of which our universe is built. Treatment methods are always being sought to detoxify wastes or to convert them into inert and inactive materials. It is likely, however, that we will always have some waste to dispose of that is still toxic to something at some level. The reasoning behind this will be found in other chapters of this book. At any rate this viewpoint, that we can never completely stop producing wastes that have some toxic or dangerous properties, while controversial in some ways, guides the decisions being made on management of all wastes—not just toxic ones—in regulatory action by governments.

Anxious to ensure that wastes are treated or otherwise made less dangerous before disposal, in 1984 Congress simply prohibited disposal of some kinds of waste on or in the land until they had been "treated." The Congressional Budget Office, in a 1984 report, estimated that this action would more than double the cost of waste management to American industry to a new range of $8 to $11 billion per year. On the other hand, in the same report, costs of returning to a site where improper disposal has occurred and taking action to remedy the damage are thought to be from 10 to 100 times the original cost of proper disposal.

While important, costs are not by themselves likely to dictate a decision on how to handle wastes. After all, simple burial or dumping on the ground was the traditional low-cost method. Feeling society's concern, Congress has now decided this form of dumping is too risky to human health and the environment unless carefully managed. Costs are clearly secondary in this public policy, except that the policy in part relies on the high cost of getting caught violating the new laws to encourage compliance. If costs are not taken into account and the inevitability of waste generation is accepted, then again, the "someone must decide" issue arises.

We Americans cling tenaciously to the Jeffersonian ideal that ours is a government of laws, not of men. In our context, that means that few citizens are willing to accept the unsupported opinion of an expert as to the safety of waste management, and it means, also, that government officers and bureaucrats need a decision-making tool which has two characteristics: it must appear to be "scientific" because it must appear to be "objective." Ideally, this kind of decision maker would love to have a formula into which one puts numbers, runs a calculation, and obtains another number which, by its magnitude, tells whether the proposed way of handling a particular waste is "safe" or "not safe."

At once, there is a problem—what in the world does the word "safe" mean to the citizen who is supposed to be protected by this regulation? After all, the *Titanic* was the safest ship in the world except for one small misjudgment. Even from the perspective of today, she may rank as one of the safest ships—no one else will ever suffer because of her! No, real safety is found somewhere between expert judgment and the finality of the grave where nothing good or bad can ever happen.

When William Ruckelshaus began his second stint as Administrator of the U.S. Environmental Protection Agency, he reviewed before the National Press Club the difference in environmental problems which had developed while he was away from the agency:

> Dealing with toxics, either as products, emissions, or leakage from waste dumps, puts us in a very different sort of business. Instead of being able to speak of allowable doses and adequate margins of safety, we now must speak in terms of risk.

How do we do this?

Grasping at an idea like drowning sailors grasping at floating straws, the regulators found Risk Assessment and Risk Management to be the kind of technique that answers their need. Fortunately, the ideas are basically correct. If we can find, understand, and measure all the factors needed, we can indeed produce a risk assessment formula that is, in principle, capable of predicting some risks, at least in a comparative way.

If the assessments are correct and we can interpret them in real terms, then Risk Management is a possible tool for decision making. As an alternative to near-clandestine consultations among small groups of experts or, even worse, the self-serving declarations of politicians seeking votes as saviors of the people, it may be our best hope.

As an idea, Risk Assessment is elegant and looks simple on the surface. A general outline of the approach is shown in Figure 2. One begins by imagining some practice for managing a hazardous waste. A landfill is an example, and the landfill can be constructed in several ways and located in many different kinds of places. At any rate, it's possible in principle to select, for the particular situation, some particular toxic substance (chemical) that is in the landfill.

Having made these choices, it is possible to decide what pathways the particular toxic chemical could follow if it escaped the landfill, and in principle to imagine who or what will be most exposed to the chemical. If we know an individual's patterns of sensitivity to harm from any exposure to the selected chemical substance, we can decide whether that person or another organism is at risk. Difficulties arise, however, when one tries to apply the usually scarce and uncertain information available to an actual assessment.

For example, for a leaking landfill, pathways for pollutants are many:

1. Into the groundwater
2. Into the air through vaporization
3. Into surface water running off the top of the landfill
4. Through groundwater that may reach the surface through
    a. Wells for drinking water
    b. Springs or seeps
5. Through dirt blowing off the top of the landfill
6. Through other, often less obvious pathways

Finding how much of a given toxic substance may travel each possible path requires a great deal of knowledge about the properties of the material itself, how it vaporizes or diffuses into the air, and many other properties. Usually, these properties are fitted into some mathematical "model" that takes all the known factors into account. Such a "model" is a mathematical equation which is developed from analysis of all the factors involved.

The assessment then requires some judgment about who or what is exposed to the toxic substance traveling any given pathway. It's easy to imagine a person drinking well water; it may not be so easy to visualize cows drinking contaminated well water or grazing on fields where dust has settled and passing the substance of concern on to drinkers of milk and eaters of cheese and butter and beef. These models can be simple, involving only a few factors, but simplicity is achieved at the cost of leaving out factors which are judged to be unimportant. This is the point at which formal mathematical risk assessment often begins to unravel.

Figure 2

Elements of Risk Assessment and Risk Management

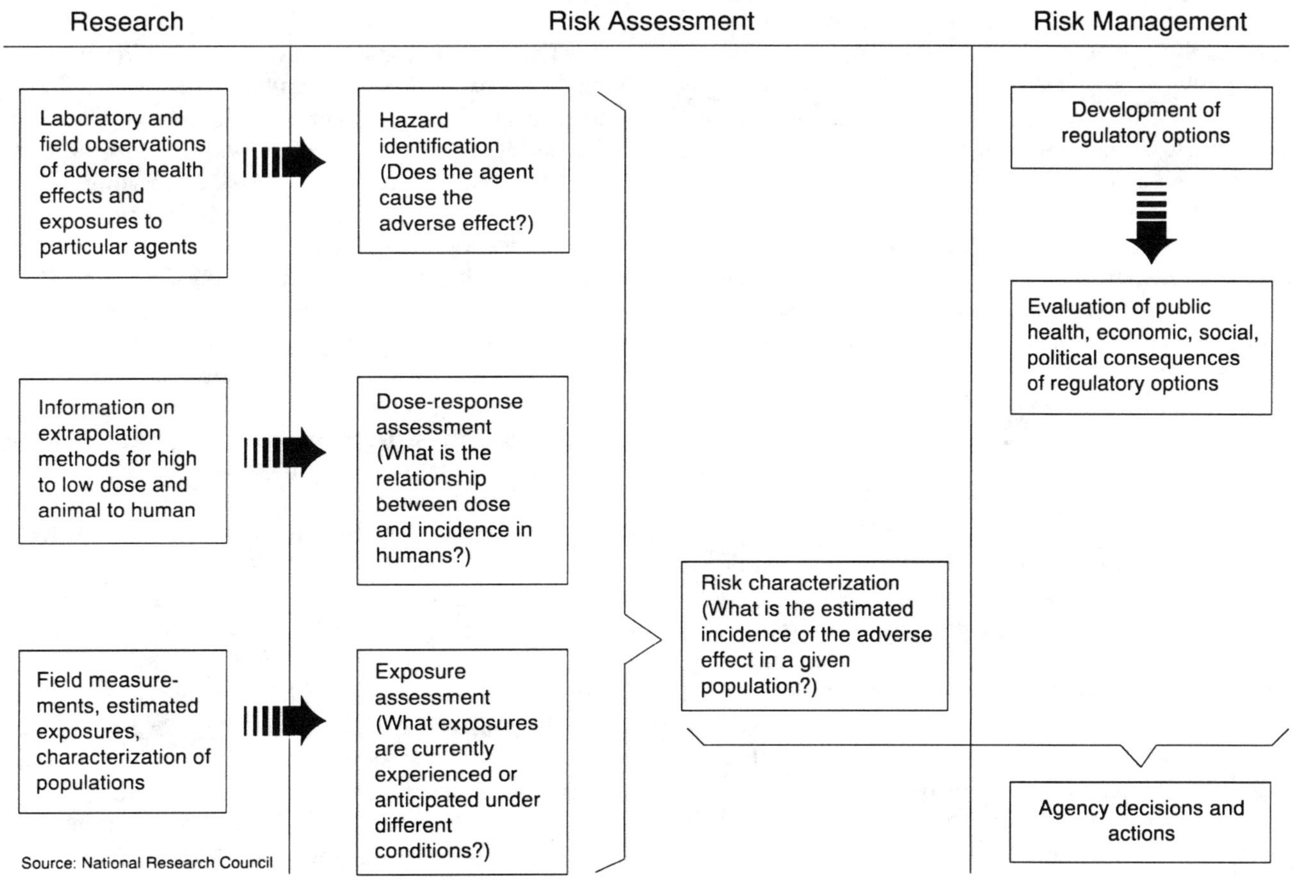

How do we decide who or what the most-exposed individual is? Is it a cow, a human, an earthworm, some fish or shellfish that live in nearby waters invisibly contaminated by seepage of groundwater? Even if we can identify that individual, do we know enough about health effects to decide what the risk is after we have calculated the exposure?

Naturally, being human, we tend to rank human health as the most important criterion of acceptability of an exposure from just about any source. This anthropomorphic slant infuriates and depresses the ecologists, who are also concerned about the health of plants and animals for reasons defined by their own professional insights. In the end, those insights lead us to believe that the loss of any organism is likely sometime, somewhere to lessen the viability of man's existence on the planet.

A typical output of one of these model studies is expressed in ranges of concentration or even in probabilities that the concentrations will be at certain levels. The hard facts are few, the ranges tend to be broad, with concentration ranges possible over two or three orders of magnitude—one hundred- or one-thousand-fold uncertainties.

To complicate life even further, the health-effects data we must use to estimate the likelihood of harm are themselves very uncertain. Risk-reference doses—the amount of a toxic which an organism can tolerate without observable effects—are guessed at for humans based (usually) on effects on animals. To protect against wrong guesses, the doses often have been reduced by 10-, 100-, or 1,000-fold before they are considered usable for assessments of health risks.

The decision as to whether a risk is acceptable or even if one exists is vital to this whole process. In the simplest way, Reference Dose, Acceptable Daily Intake, Virtual Safe Dose, whatever it may be called tomorrow, is chosen as a marker. If the total exposure to a particular chemical by some pathway, including exposures from other sources, such as food, is at or below these limits, the exposures are regarded as safe—no risk at all for noncarcinogens. For cancer-causing agents (carcinogens) the safe dose is often that which increases the risk of cancer in the exposed population by one case in a million persons per year, given lifetime exposure at the levels identified in the assessment. (See the discussions on this subject in Chapter 5, "Toxicology of Wastes.") Finally, if one follows the logic to its conclusion, the whole process has to be repeated for each toxic substance likely to be present and for all the pathways of possible exposure.

Even without the detail, it is easy to see that the longed-for objective of a "scientific" and an "objective" method remains a chimera that will elude our grasp until we have a great deal more information about our world and the creatures therein. Not only do we lack information on how organisms interact with chemicals, we know little about how the chemicals themselves behave in the environment. Do they interact to create new and different ones? Do they react with soil minerals and microorganisms? Obviously, but in what ways and with what effects?

All this negative comment, which is most often made by people who are actively trying to develop and use the methods of Risk Assessment, should not blind us to a reality: though the information is incomplete, merely thinking through the issues in the detail we have outlined does produce a better understanding of the working of the system that is analyzed. What is required is responsible use of the mechanism so that decision making in the next step is done in the light of the assumptions and uncertainties of the assessment process.

Wisdom is fostered when analysis shows that consistently assuming the worst case in each step will result in paralysis. Humankind has not evolved in a risk-free world; indeed, it has evolved in a world full of danger and threats which must be overcome to maintain a decent life. The next step, in which wisdom and judgment are at a premium, is Risk Management.

THE ART AND SCIENCE OF RISK MANAGEMENT

This distinction between the "scientific and technical" emphasis of risk assessment and the "wisdom and judgment" of risk management is by no means intended to denigrate the many value judgments that have to be made to carry out a Risk Assessment. Separating decisions of management from those on assessing risk merely recognizes the differing points of view that must be brought to this next set of decisions.

In Risk Assessment, the primary input and judgments are those of a wide variety of scientific and technical persons. Health scientists, mathematicians, engineers, earth scientists, even oceanographers and atmospheric physicists have much to contribute, along with other specialists and generalists. Value judgments are made on the validity of approximations, on the relative importance of pathways, on the validity of health-effects information and its applicability to real cases, and on many ways to bridge the gaps left by lack of knowledge.

In Risk Management, on the other hand, societal values, relative costs, acceptable levels of risk within the values of the society as reflected in its laws and ethics—all must somehow be taken into account. By comparison with the technical arguments that occur in Risk Assessment, the decisions on management of risk are largely political, using "political" in its broader meaning. Politics, in this context, means making politic decisions, which Webster defines as "having practical wisdom, prudent, shrewd, diplomatic" and even as "prudently or artfully contrived: expedient, as a plan, action, or remark, etc."

Thus, when Risk Assessment indicates that there is a low probability of harm from landfill disposal of certain wastes, wisdom may well suggest that regulators permit that disposal. If on the other hand, whether for good reason or for bad, the Risk Manager finds that the people or their surrogates, the Congress, won't stand for the action, then expediency will prevail, and the practice will be banned. This is the course followed for recent legislation banning many materials from land disposal. Having this option foreclosed, the Risk Manager must then return to Assessment for another set of options.

Those options may be very numerous or very restricted. In some cases, ways can be found to avoid producing the dangerous waste. The costs imposed by restricting the cheaper method of disposal in a landfill may do more to accomplish this than any direct regulation. Certainly, there is incentive to treat wastes so as to allow disposal in an economical way that is allowed by regulation. This in a single phrase is the notion of technology-forcing legislation that has become so dear to the Congress.

The risk manager has at his disposal many tools of the political economy to enable action on the disposal options that are available. One tricky issue arises: should a particular technology be specified for treatment and disposal, or should the desired result be specified and reliance placed on the economic incentives of our system to achieve acceptable results at minimum cost?

One of the few things clear about the mixed capitalistic economy is that individuals are the only members of society that have money to spend and to pay taxes. All corporate costs will in the end be paid by consumers of the corporate output. Stockholders, financial institutions, and others with a stake in the corporation are made up of consumers, also.

Risk management decisions cannot be made in a political vacuum or without sensitivity to what actions cost in the total economy and what they achieve in the well-being of the citizens. Hazardous waste management is an area in which fears that may be groundless or overblown provide fertile soil for the politician seeking a reputation for wisdom by the most expedient means over the short term. The location and permitting of waste disposal sites is one of the outstanding examples of this weakness in the system today.

Under the original version of the Resource Conservation and Recovery Act (RCRA) which sets up a regulatory scheme for newly created hazardous wastes, many of the operating disposal sites were allowed to continue operating under "interim status," meaning basically that they had to register their presence and scale of activity and to begin to get ready to meet the new standards of operation that would be issued sometime later by the Environmental Protection Agency. Some 4,000 of these facilities for treatment, storage, or disposal of regulated wastes gained interim status. In the process, they made their presence known to anyone who wanted to know.

When the regulations for a permitted landfill were finally issued, many of those operators decided to close rather than meet the new requirements. This decision to close, by the way, did not release them from future liability. More importantly, the generators of waste already placed in those facilities still are liable for certain things under the laws for at least 30 years. At any rate, when the final date for the implementation of the full permit requirements came along, at least 1,200 of the "interim" sites closed down. The status of many others remains uncertain even now.

In the meantime, only three new fully permitted sites had been brought into being. A number of states have no licensed (permitted) disposal sites, at least any for use by anyone other than the owner of the site. It is not uncommon for regulated wastes to be shipped several hundred miles through several other states for disposal, with the attendant risks of transportation accidents.

There have been few proven technologies around for reducing the amount of waste or for treatment to render the wastes nonhazardous. The Congressional Budget Office Report of 1985 guessed that by 1990 waste reduction/minimization efforts might reduce the total waste stream by as much as 14% compared to the 1983 amount. At the same time, the total volume of wastes could increase at 1% per year. Thus the need for final disposal capacity remains high, even further pressed by the congressional restrictions on landfill disposal of many, indeed most, hazardous wastes that have not been treated to "reduce their toxicity."

Finally, the 1984 changes in the RCRA law (see the chapter regarding RCRA) bring all generators of 100 kilograms or more of certain wastes each month under the obligation to use only permitted hazardous waste disposal sites. The original requirement was set at 1,000 kilograms/month, and the expansion is believed to have added a new population to the regulated community that expands its numbers by a factor of ten, adding some 400,000 small-volume generators.

So we find ourselves in a classic contrast of hope over reality. The legislators hope that by cracking down on generators and prohibiting some kinds of material from landfills, new technology or something will take care of the waste. The reality is that little research is in progress on this subject, although recent legislation to mandate "treatment or detoxification" before landfilling is stirring interest in the field.

This concept, often called "technology-forcing" legislation, was first seen in the Clean Air Act and in the Clean Water Act. The regulators were told to set the requirements at those levels which the very best technology could achieve and force those with less effective treatment systems to catch up to the best. In some cases, the allowed levels were arbitrarily set well below those achievable by any proven technology on the assumption that the polluters would somehow figure it out. The approach almost worked in the older problems. After all, wastewater treatment has over a century of history and developing knowledge behind it. Air pollution control is nearly as old, having been essential to safety in the flour mills of the 1880s which used the then-new Hungarian milling system.

In the case of solid wastes, including hazardous wastes, little if any technical activity had been supported over the years before the 1970s. Even into the late 1980s not much has been done in this field. Some fairly sophisticated systems have been devised, like the deep-well injection of extremely dangerous waste from nerve gas plants. In the 1950s some difficult wastes were treated to avoid conflict with communities, even if not thought dangerous.

It's perhaps understandable that the "best" solution quickly available was to seal the landfill off from the groundwater, the air, and the surface runoff. The "secure" landfill was invented, but convinced few of its long-term "security." Research in other technologies began but with little enthusiasm or visible support from the large generators.

Overhanging all this technology and regulatory philosophy is the public's attitude. Remember the NIMBY syndrome? Do you want a hazardous waste dump in your backyard? Never mind that today's best examples of hazardous waste management facilities are as far from a dump as a Rolls is from a Ford—people turned out in droves for the required public input meetings under the permitting procedures for new disposal sites. Attending some of these meetings and public hearings, one is struck with the lack of trust shown by the ordinary citizen when hearing information provided by industry or government. Time after time, municipal officials echo their citizens' concerns and are no better informed than those individual citizens.

It takes a strong stomach to advocate a waste disposal facility. In some communities it may take physical bravery, too! Examples abound, many showing how public officials hunt for cover when faced with these siting decisions. In Minnesota, home of some of the most stringent state regulations, the State Waste Management Board repeatedly failed in its assignment to find even one site within the state and has now been substantially eliminated without any replacement. The Minnesota regulations basically deal with hazardous wastes by requiring careful control of the material while it gets sent to some other state for disposal. A few private facilities are permitted, but most business has no place to go within the State.

This situation is not unique to Minnesota, but it does raise serious issues of fairness and competition among states and indeed the world. The U.S. Supreme Court has declared that hazardous wastes are materials which are protected from state restrictions on interstate commerce like other commodities.

Returning to our original subject, if Risk Assessment is a valid approach, and no one has an effective alternative except authoritarian rule, then should it not apply to all? Surely, risks in South Carolina or Idaho or Minnesota can be measured in the same way if local factors are properly taken into account. This is not the real problem. The real problem comes in the Risk Management responsibility.

Often we hear in public hearings the lack of trust in the Risk Assessment approach that is revealed by such questions as: can you assure us that we will never have any exposure to the wastes being handled or disposed of? The only honest answer to this question is that the *probability* of significant exposure is very small and that the judgment of the "experts" says that it is "safe." All judgments are relative, since we have seen that the information we need to compute risk exactly does not exist.

After all, life is a risky proposition. The general guideline for "acceptable" cancer risk is that which results from exposure which is believed, on the basis of all the evidence, to increase the number of cancers in an exposed population of one million individuals by one per year. The data rarely are available to support this so-called risk-reference dose, which is arrived at by consensus in a relatively small group of informed specialists.

There are many things we do every day which increase our chance of *dying* of cancer (not just becoming a cancer patient) by one in each million of the exposed population. Some of these are shown in Table 5, the data for which were originally published in "Analyzing the Daily Risks of Life" by Richard Wilson in *Technology Review* (Feb. 1979), a publication of the Massachusetts Institute of Technology. Other people have done estimates that are consistent with these.

Leaving out obvious life-style habits like smoking or taking illegal drugs or drinking too much alcohol for our liver to metabolize, we are still left with a lot of common activities that increase our potential for developing cancer. We are not deterred from a trip to Paris by the increased exposure to cosmic rays that we experi-

Table 5

Actions That Increase the Chance of Death by One Chance in a Million

Action	Related Cause
Smoking 1.4 cigarettes	Cancer, heart disease
Drinking 1/2 liter of wine	Cirrhosis of the liver
Spending 1 hour in a coal mine	Black lung disease
Spending 3 hours in a coal mine	Accident
Living 2 days in New York or Boston	Air pollution
Traveling 6 minutes by canoe	Accident
Traveling 10 minutes by bicycle	Accident
Traveling 300 miles by car	Accident
Flying 1,000 miles by jet	Accident
Flying 6,000 miles by jet	Cancer caused by cosmic radiation
Living 2 months in Denver while on vacation from New York	Cancer caused by cosmic radiation
Living 2 months in an average brick or stone building	Cancer caused by natural radioactivity
One chest X ray	Cancer caused by radiation
Living 2 months with a cigarette smoker	Cancer, heart disease
Eating 40 tablespoons of peanut butter	Cancer caused by aflatoxin B
Drinking Miami drinking water for one year	Cancer caused by chloroform
Drinking 30 12-oz. cans of diet soda	Cancer caused by saccharine
Living 5 years at the site boundary of a typical nuclear power plant in the open	Cancer caused by radiation
Drinking 1,000 24-oz. soft drinks from now-banned acrylic plastic bottles	Cancer from exposure to acrylonitrile
Living 150 years within 20 miles of a nuclear power plant	Cancer caused by radiation
Eating 100 charcoal-broiled steaks	Cancer caused by benzopyrene
Risk resulting from accidental exposure by living within 5 miles of a nuclear reactor for 50 years	Cancer caused by radiation

Source: *Technology Review* (1979), reprinted with permission.

ence during high-altitude jet-airplane travel, or from driving to Grandma's house by the possibility of a car wreck on the way. Will we give up our vinyl shower curtains or sewer pipe and return to mildewed cotton and expensively installed cast iron? Will we choose soap instead of synthetic detergents?

One way to deal with this new (to most) and fearsome (to most) group of threats to our well-being is to place it in a broader context. Every time we turn on the electric stove to heat the TV dinner or the Thanksgiving turkey, we are increasing the probability of some coal miner's dying by accident or of black lung disease. We are also creating some new acid rain which may affect our forests and lakes, and even helping generate radium release from the chimney of the power plant. (Much coal contains measurable amounts of radium.) All of this and much else occurs because someone or some group has decided that the adverse consequences are worth it for the ease and convenience of electric power, as well as for its essential role in economic activity.

Risk analysis is often compared to measuring the ripples when a stone is tossed into water—but in dealing with wastes, one must think of many stones, not one. That same life-style affects us all, and relative risk is our key to comfortable living in a technological age. For the individual citizen, it is important to remember that the health-effects data and impacts are based on an *exposed* population. Before becoming terribly concerned, find out what your exposure is and that of other people and things of interest or concern.

Having decided on exposure, if any, then the kinds of materials need to be evaluated. A statement that something is a "suspected carcinogen" is a warning sign. One needs to find out whether any authoritative source has made this claim; then decide how serious the problem can be when the judgment is based on a lifetime of exposure—few people will ever have such exposure.

Finally, we have made and will continue to make Risk Management decisions based on the risks to nonhuman target organisms. Banning DDT could not be based on human health effects, but the deeper values placed on the eagles and other birds of prey which were so seriously threatened by the pesticide prevailed. Their population recovery and greater chances of survival have proved the rightness of the decision to many.

It is regrettable that so little solid information can be found on exposures and hazards to nonhuman species, at least for the higher animals. Until this gap is filled, we cannot be real ecologists in these decisions. Too many times, environmental policy decisions are made in a confrontational atmosphere between those who oppose any action unless all the impacts on all the biota are known and those who often rightly argue that such knowledge is reserved to the divine and we poor humans must make do with imperfect knowledge. Each side qualifies its argument with one or the other introductory phrase, then presents what is usually a self-serving proposal. These discussions are seldom really helpful to the decision makers, who then turn to some form of risk analysis and risk management analysis to erect a rational structure in the face of essentially irrational argument.

Part II

Hazardous Waste Management Technology

Chapter 7

Waste Management: Land Disposal and Protection of Groundwater

WASTE DISPOSAL IN AND ON THE LAND

From the earliest traces of human existence until the middle of the current century, all kinds of wastes were routinely and without much thought disposed of by placing or leaving them on the land. If they were particularly troublesome or disgusting, waste materials might be buried. Often, they were simply carried a convenient distance away from the population and then dumped or placed in or on the land.

Early in the saga of mankind, population was thinly spread and often in constant movement. It was easy to load up the shelter poles, pack up the yurt or the tepee, and simply move on, leaving your mess behind. Better still, the mess tended to be composed of readily degraded natural materials that would be gone by the time the band returned to the spot again. People who led such migratory lives left few traces for the archaeologists to follow; so we often know little about them.

When new technologies led to the development of settled communities, however, one problem must have gradually emerged: every house had some wastes to get rid of almost every day. Human wastes are easy to visualize as a problem, but wastes from food preparation and even leftovers must have eventually become problems. We know that at least some of the earliest people paid little attention to these wastes because the caves and rock shelters which they occupied often have many feet of accumulated debris still in place. Careful study of these sites provides a great deal of insight into the food sources and other life-style characteristics of ancient peoples.

A little later, we find a different feature—the waste pile or midden, often found in association with ancient habitation sites. In the American Southwest, wastes were often thrown or pushed off the edge of the rock shelters in which cliff dwellings were built. These slopes are often good hunting for anthropologists. There is no reason to doubt that these middens were early versions of the dump and evolved into the landfills of recent times.

Solid waste disposal, like all other features of life, is affected by enormous cultural differences. Confronted by land shortages resulting from dense populations, European countries sought alternatives to open dumps sooner than we North Americans did and developed the technology of waste incineration. Underdeveloped and still-industrializing countries continue to depend heavily on open dumping, pleading lack of financial resources to do better. The huge open dumps near Mexico City and other developing country capitals are witness to the collision of old disposal practices with the new generation of wastes.

As we discussed in earlier chapters, changes in society brought and continue to bring about great changes in the waste stream of society and in society's perception of that stream.

In his book *The Third Wave*, Alvin Toffler points out that the products available to First Wave societies, that is, before the Industrial Revolution, were basically simple ones. He suggests that in First Wave societies (which endured in the West until late in the seventeenth century) products were almost all made locally from natural materials found in the same general region as their consumers. Bad transportation and worse communication prevented mass distribution of just about anything; thus, markets and manufacturers were small and local.

When industrialization began with the development of new ways of converting fuels into energy along with the means of applying that energy to work, everything began to change. Driven by the availability of cheap mechanical energy, Toffler's Second Wave began to gather force and sweep away the old world. In the modern world that was its successor, vast supplies of materials—at prices unimaginably low only a short time before—were available to begin to create the mass market that is so characteristic of modern society.

Napoleon Bonaparte demanded and got from his French savants a revolutionary way to feed his troops in the field—thus releasing them from the need to forage for food and making his army more "productive." This was not insignificant in the remaking of Europe. The development of the tin can, driven by military ambition, was made possible by the early development of the industrial (as opposed to decentralized crafts manufacture) production of metals—iron, steel, and tin. It was all right for the army, but ordinary citizens still could not take advantage of the technology. That had to wait for the invention of the blast furnace and associated methods for the large-scale production of steel. While the steel was needed for the new railroads, it became so cheap anyone could use it.

Suddenly, it seems to us now, canned and preserved foods were everywhere. Abandoned miners' cabins from the 1880s in the high country of Colorado all have piles of old rusty cans somewhere nearby. Napoleon's ambition and Andrew Carnegie's business sense had freed the miners from the need to hunt, fish, or garden. All their efforts could go into the often fruitless search for riches.

The development of canned foods and the iron, steel, and other industrial technologies behind them is a parable of the impact of the Industrial Revolution on waste and the disposal of wastes.

As Toffler points out, the Second Wave, constituting our industrial civilization, did not change only technologies. Exploitation of the new methods required the development of the factory system, with many impacts on social structure and living styles. Toffler says:

> Industrialization was more than smokestacks and assembly lines. It was a rich, many-sided social system that touched every aspect of human life and attacked every feature of the First Wave past. It produced the great Willow Run factory outside of Detroit, but it also put the tractor on the farm, the typewriter in the office, the refrigerator in the kitchen.
>
> (Alvin Toffler, *The Third Wave*, p.16)

What Toffler didn't say was that this same change produced an equally major change in both kind and quantity of the wastes of the new society as compared with the old.

Managing the Wastes of an Industrial Society

The new wave of technology swept away the decentralized population of the agricultural society. The factory system needed large numbers of people near the production centers, whether these were large integrated factories or clusters of interdependent smaller companies. In the century after 1880, the agricultural (therefore dispersed) population fell from about half the total in the United States to less than one-twentieth.

When people threw something away in the crowded industrial cities, it landed on their neighbor's doorstep—or on his head. Managers of these new cities had to cope with a rapid growth of the idea that mass production had made many things so cheap that they could be thrown away after one or a few uses. The technical advances in the paper industry, for example, made high-quality "virgin" paper so inexpensive that the reuse of newspapers as wrapping for foods in the markets vanished. It still survives in some traditional uses, like the newspaper cones in which British fish and chips are often still served. Similarly, the idea of a shopper's basket brought from home gave way to the grocery bag packed by the clerk.

The natural reaction of those managers was an extension of old practices. Open dumps grew and grew. It became more difficult to find places for dumping near the cities because of the rats and other vermin attracted by the uncovered waste. Paper and other light materials blew away and littered the countryside, sometimes for miles. These dumps were "unsanitary" and believed to spread disease. Almost no one considered them as threats to water supply or groundwater, although it was easy to see that they might contaminate a river or stream with rainwater runoff from the piles of garbage and other wastes.

From Dumps to Landfills

As we have observed many times in this book, the continuing changes in the society have been reflected in the changing character of the wastes which found their way into the dumps. In earlier times industrial wastes were often seen as indications of prosperity and were kept on factory property as a kind of scoreboard device not unrelated to the pride some communities took in the smoking chimneys and furnace stacks of the early industrial period. Also important was the growth of chemical industry and the acceptance of many of its products by individual consumers who discarded excess and used materials into the municipal waste stream. The industry itself, following age-old ideas, placed the waste products of manufacturing in pits, lagoons, and dumps. Everybody believed that these materials would be absorbed, broken down, or in some way rendered harmless by contact with the earth. The idea of migration away from the place of disposal just didn't seem to occur to many people.

Hardly anyone who reached adulthood in the United States before 1960 would have hesitated to dispose of any waste by burying it in his backyard or anywhere nearby. The rationale is easy: "It's only a little bit of stuff, and everybody knows the earth will break it down." We all learned to know better—or should have!

As population and waste generation grew, it became more and more necessary to do something about the unsanitary open dump. Those serving small towns or individual families might be used for generations without serious problems, but cities and industries needed to do something. Since dumps were unsanitary, the solution was sanitation—some way of removing the tendency of a dump to attract vermin and to litter the landscape. The answer was the sanitary landfill.

A sanitary landfill is a dump in which the wastes are covered with earth as soon as practical after they are dumped. Landfill is a term chosen to indicate that a social benefit is being attained by converting waste land into useful property that can be used at some future time. Swamps, marshes, and steep ravines that could all be raised to higher levels and made useful were the obvious candidates. The vast marshes of northern New Jersey were filled with generations of waste from New York City because they were after all only "wastelands" and no one could really understand their role as wetlands in the ecological cycle. So, too, the margins of San Francisco Bay and hundreds of other such sites that were "unbuildable" were "reclaimed" from nature for other uses. Sanitary landfills offer the only practical solution to the waste disposal needs of many communities. Properly sited, designed, and operated, a landfill can serve well without contamination of drinking water or creating unsanitary conditions.

The reason landfills have been under attack in recent times arises from some basic characteristics which have to be taken into account. The method of operation, requiring covering of wastes within a few hours at most after they are deposited, does a fine job of hiding any problems and dressing up the top of the refuse pile. It does nothing for the bottom of the waste. At the bottom, soaked by rain or by water in the ground, the wastes are attacked by bacteria and molds and undergo chemical reactions among their components to produce a liquid called leachate which, in the simplest cases, drains out the bottom of the landfill. When ravines, etc., are filled, it is common to see a small stream of obviously contaminated liquid draining out the bottom, often into a stream. The same thing happens in flat-land locations, only the leachate mingles with the water already in the ground under the landfill. Groundwater usually moves, and the mixture of leachate and groundwater may surface in a spring or seep or, more ominously, may travel invisibly underground until it reaches a well. If the leachate makes the water unsuitable for its intended use, like drinking water for people or livestock, trouble may ensue.

THE LAND DISPOSAL–GROUNDWATER SYSTEM

Up to this point, our focus has been on landfills as a primary means of disposal of wastes in contact with the land. There are other forms of land disposal which we will describe later in the book; landspreading, landfarming, and deep-well injection are the primary ones. Disposal in pits, ponds, and lagoons, and even by storing waste in piles on the ground surface are probably best thought of as variations on landfills. Each method has its particular technical base and reason for being, which we will take up later in this chapter.

Nonetheless, all methods of waste disposal that involve bringing wastes into contact with the land in any way must deal with the inseparable relationship between what's placed in or on the land and the quality of the water found under and within that land or in another area to which the water may migrate.

Groundwater is the most important source of drinking water in the United States, with more than half the population getting their water supply from around twelve million wells. Overall, about a quarter of the total domestic, agricultural, and industrial water needs of the country are supplied by exploiting water from the ground. While many believe that this source is primarily a rural and small-town supply, the Environmental Protection Agency points out that thirty-four of the nation's largest cities obtain at least a part of their drinking water from underground sources.

The magnitude of the groundwater resources of the world is emphasized by the statistics of the distribution of water: nearly all the earth's supply of uncombined water (that is, water that is not chemically tied up in crystalline minerals) is not available for human use because it is either loaded with dissolved materials—mostly salt—or is in the form of ice. Only 1% of all this "free" water is fresh and liquid most of the time. All the rivers, lakes, and other surface water features are formed from an incredibly small fraction of the "free" water, about 2%. Our hidden reserves of underground water are roughly fifty times as large as those surface waters that we can see.

Contamination of surface water is relatively easy to see and detect, at least compared to the problem of detecting the same contamination of underground supplies. Yet it was not until recent times that any real steps to protect and improve the quality of rivers, streams, and lakes were taken. The need for expensive and inconvenient ways of protecting the salty oceans and other nonpotable (undrinkable by man) waters is still not by any means widely accepted even today.

It seems perfectly obvious today that anything dumped on the land or buried within it is very likely to work its way through the soil until it appears in the water that is held by that soil. Why, then, did it take so long to recognize the connection between waste disposal on the land and contamination of groundwater? The answers to this question lie in both the scientific and the humanistic worlds.

From a scientific point of view, the connection between soil contamination and groundwater quality is not always absolutely true or proportional. There are several mechanisms working in the soil which can break down some contaminants so that no dangerous chemicals are likely to reach the groundwater. Microorganisms in soil, too numerous even to be identified, can bring about biodegradation by utilizing the contaminants in their metabolic systems and producing wastes from that process which may be less troublesome. Some soils contain minerals which can assist in (catalyze) chemical reactions with soil constituents or with the groundwater itself to detoxify the waste materials. When small amounts of wastes are discarded over huge areas of land, the contamination may not be detectable or significant for the uses to which the groundwater is put. When wastes are injected deep in the earth into groundwaters which are so loaded with minerals and even petroleum and natural gas that they have no conceivable use on the surface, it is not likely we will ever have to deal with them again. Therefore, they are effectively removed from the biosphere in which we live.

The bad news is that, while all of these mechanisms exist and are truly effective in some cases, our past indiscriminate dependence on the powers of Mother Earth to decontaminate just about anything we dumped has caught up with us. The kinds and amounts of wastes we need to manage force us to decide intentionally what can and cannot be handled by land disposal. The developing knowledge about the results of land-disposal techniques can guide us in choosing which, if any, of the methods are safe. It seems likely that there are some materials which simply cannot be disposed of in or on the land and for which some other means of disposal must be developed.

The Origin and General Characteristics of Groundwater

According to the Federal Environmental Protection Agency, some 4.2 *trillion gallons* of rain fall every day on the United States. More than half of this evaporates to fall again as rain or snow elsewhere—about 2.8 trillion gallons—while 1.2 billion gallons flow through the rivers and streams to Canada, Mexico, and the surrounding oceans.

About 61 *billion* gallons soak into the ground, where they are stored in soil and rock formations called *aquifers*, from the Greek for *water bearers*. An aquifer is simply an underground layer which contains water which can be removed by simple mechanical means—pumps and the like.

Every day, the United States is believed to withdraw up to 90 billion gallons from underground sources. Twelve billion gallons of this are for public water supply, and yes, we are using more than the rainfall is replacing. The huge amount of water in underground storage allows this unbalance to continue—but not forever. The issue of the quantity of water needed and used in this country is a very large and controversial one. A whole book would be required to examine that problem. For our purposes in this book, we can see that an approaching shortage of high-quality water adds force to our need to protect the available supply.

Being geological features, aquifers are mostly large scale and regional in nature. The Ogallala Aquifer, as an example, is a connected series of water-bearing gravels and sands which extend from South Dakota to northwest Texas and cover an east and west area nearly as wide as the state of Oklahoma. Thousands of farms and ranches and dozens of communities depend on the groundwater from the Ogallala for their existence.

Other aquifers cover smaller areas of land and may lie much deeper under the surface. Some are thought to contain "fossil water," trapped in rock formations for geologic ages and unavailable for use until the technology of well drilling developed.

Aquifers exist when a more-or-less porous rock, sand, or gravel layer lies over a nonporous layer of rock or clay that *confines* the water so that it does not continue to sink further into the earth. The first layer of water encountered when one digs or drills into the ground is often called the *water table* and represents usually the accumulated local rain and snow water that has been stopped from further infiltration by a nonporous layer of bedrock or perhaps clay left by a glacier or ancient stream.

Often, below this bedrock layer and on top of another *confining layer*, we find another water-filled layer of earth and/or rock, constituting a confined aquifer. Such aquifers often have their water sources, their *recharge zones*, at some distance from the site of a well drilled into them. A recharge zone is a place where rainfall or meltwater can soak into an aquifer, usually where the aquifer rock or sand reaches to the surface, or "outcrops." If the pressure is high enough from the distant source, the water may rise to the surface or even spout into the air, giving an *artesian well*.

Many areas have several aquifers stacked one over the other in complex systems that may be partially interconnected but with each different layer in the array having water of different quality from the others. There is really no consistent pattern; sometimes deeper waters are of higher quality. At the bottom of all aquifer systems is either dry monolithic rock or, frequently, salt-water-bearing formations which may even be sources of special chemicals like the bromine-containing brines of Michigan, but which have no utility for human or animal drinking supplies.

It is characteristic of all aquifers that the water in them is moving from one place to another. Only rarely is the image of an underground river applicable, because the movement is usually very slow. The EPA often uses an "average" rate of movement of 30 *feet* per year when estimating the contamination potential of certain chemicals.

Like all resources, groundwater is of limited usefulness to man until some way is found to make it available in convenient form. Springs and seeps are places where groundwater naturally emerges at the surface. Since ancient times, springs have controlled the location of human habitation and even of cities. The ancient Hebrews dug a long, twisting tunnel from the top of the Temple Rock in Jerusalem to the natural spring waters of the Pool of Siloam—making the city much more lasting and perhaps altering our history. Other ancient cities captured rainwater for storage for the same purpose.

It was surely only a straightforward exercise in logic which resulted in the first dug wells—those made by hand excavation from the surface. Perhaps the first examples were intended to be cisterns for the storage of rainwater but which reached down to the water table and provided their own supply. Today's technology of borehole wells has come a long way from those primitive times but introduces a new hazard to groundwater.

The ability to drill deep wells, which may and often do pass through several aquifers seeking one that will produce both the quantity and quality of water desired by the user, provides the possibility for leakage to occur between aquifers which are otherwise separated by impermeable zones. There are many cases in which deep groundwaters have been contaminated by intrusion of materials from the surface or down an old well. To help prevent this, most States have rigorous requirements for the plugging and abandonment of old and unused wells. Unfortunately, these standards are often ignored.

A new kind of concern arises when sampling and monitoring wells are drilled to evaluate groundwater contamination. Careless placement or poor sealing of the points where the casings of such wells penetrate the presumably contaminated aquifer can result in spreading the undesired materials and making a real cleanup impossible.

The Hydrologic Cycle

All of the elements of the groundwater-atmosphere-soil-human impacts relationship are summarized in the endless natural processes that make up the Hydrologic Cycle. A pictorial summary of this large-scale system is depicted in Figure 3, originally published in the *EPA Journal*, a monthly magazine published by the Federal Environmental Protection Agency.

A description can begin anywhere in the cycle, but it is easy to start by imagining the huge amounts of water which evaporate from the soil and the surface of all bodies of water on the planet. This evaporation requires an enormous amount of heat to accomplish the conversion of liquid water and even ice into water vapor, the invisible gas which makes up a great deal of our atmosphere. We all experience this need for heat to be supplied when we feel the cooling effect of dry air on a nice day when moisture on our skin evaporates. Humid days, however, do not cool so well because our skin moisture evaporates only slowly into wet air and does not remove so much heat.

On the global scale, the sun supplies the heat needed to start the cycle. You may remember that about 60 trillion gallons of water evaporate every day from the surface of the United States alone. Only the sun itself provides enough energy to keep such a gigantic system in operation.

Once water has entered the atmosphere as a gas, it is carried widely by winds and convection currents caused by temperature and other density variations in the atmosphere. Everyone who has watched a hot air balloon in action knows that warm air is lighter than cold; not everyone appreciates the fact that moist air is lighter than dry air. The almost constant presence of clouds formed by microscopic droplets of water which form at the top of columns of relatively warm moist air, or where layers of air of differing moisture-holding ability come into contact, gives testimony to this tendency to rise.

Under the right conditions, the clouds release precipitation as rain, snow, or ice. When it reaches the ground,

Figure 3

The Hydrologic Cycle and Sources of Groundwater Contamination

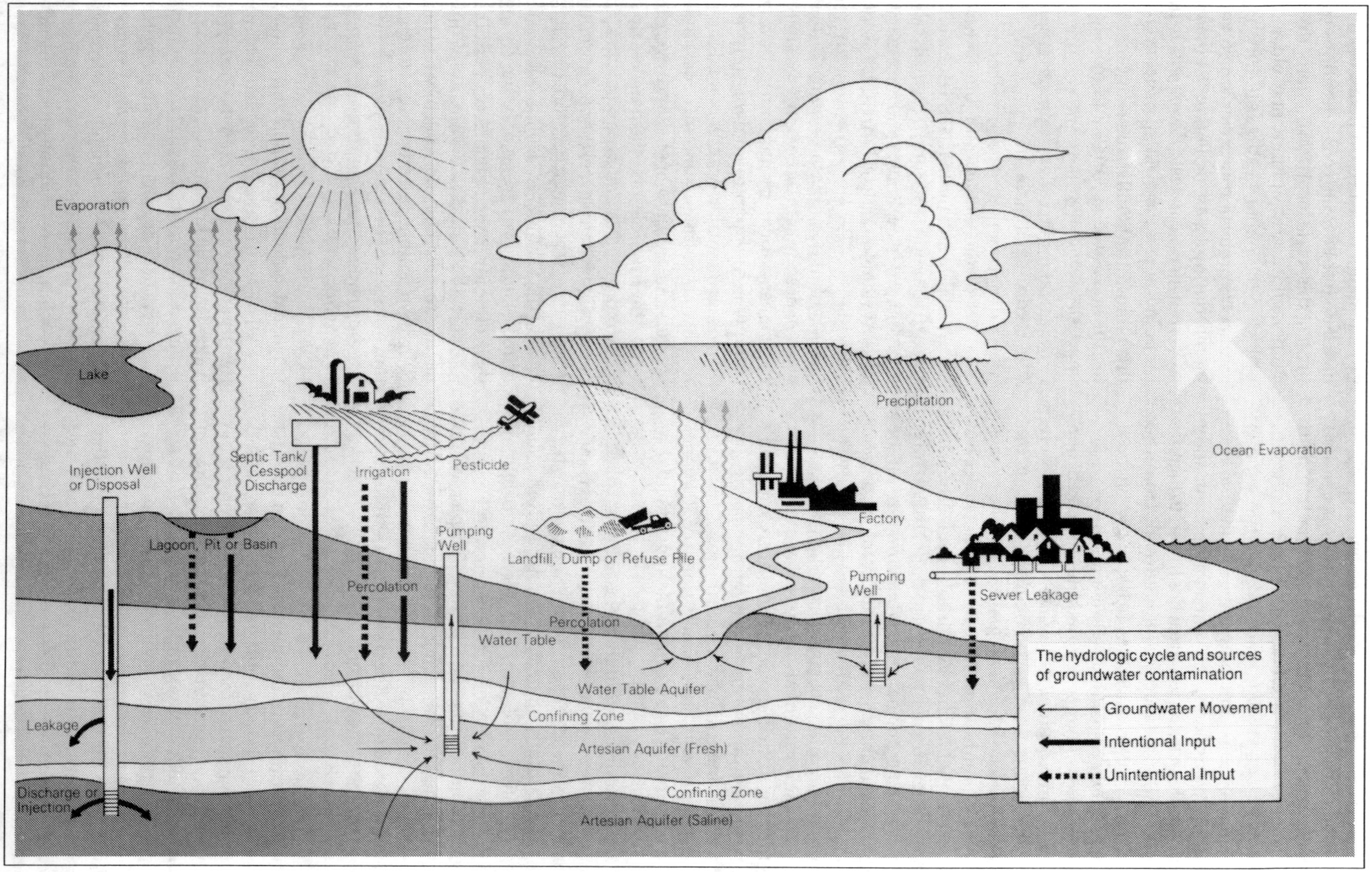

Source: *EPA Journal* (September 1986).

the division occurs between that water which runs off into streams and eventually into rivers and the oceans and the small amount that penetrates into the ground and becomes groundwater. In Figure 3, we see that in its journey into the groundwater, the new supply can carry many forms of contamination along and bring them into the aquifers.

Landfills are a significant source of contamination of groundwater but by no means the only one. When the demand for homes on cheap land outside the reach of sewer systems developed in the 1950s, many of those homes depended on shallow wells in the surficial aquifer—the one between the ground surface and first confining layer—for their drinking water, while septic tanks were installed for sanitary sewage disposal. Serious illness in some small infants in such areas was traced to excessive nitrates in the groundwater. The nitrates could have come from septic systems, from which nitrates in human wastes would be readily released. It was also traced to the pride people took in their lawns—they applied a lot of nitrate fertilizers to make them green.

Today, in farming areas, the excess nitrate fertilizers and pesticides which have been applied to the land in increasing amounts in recent years are appearing in the groundwaters. In some porous or sandy soil areas, the effect can be dramatic in its sudden appearance. In most areas, the entrance of agricultural chemicals into the groundwater may take many years but must eventually be taken into account.

Leachate from landfills and from other kinds of hazardous disposal sites can move in the same ways. The hydrologic cycle, essential to the existence of life on earth, also behaves as an endless conveyor of harmful materials into and through the land. This is the basis for all groundwater-protection action, given the knowledge that at least some of the components of wastes, and even of agricultural chemicals, can do harm to vulnerable members of the human and nonhuman residents of the earth.

INSTITUTIONAL AND HISTORIC ISSUES IN THE PROTECTION OF GROUNDWATER

Fifty and more years ago, public health authorities conducted campaigns to alert the rural population to the dangers of contaminated well and rainwater cisterns, which were then fairly common as sources of drinking water for isolated homes. The problem was the use of privies for human wastes, which allowed disease-causing microorganisms (pathogens) to enter the earth directly. Sometimes, these germs and parasites traveled through the soil into shallow wells nearby and caused illness. However, it has been proved that soil, even sand, is a pretty effective filter for most sickness-causing bacteria and parasites. It is not so effective as protection against viruses, and we are still learning about soil and chemicals.

The first measures to protect groundwater were based on public health considerations from this cause. In most states and local jurisdictions, the public responsibility for protection of groundwater was and still is placed in the public health authority, not in the environmental protection agencies which were established largely during the 1970s. For technical advice on design and construction of wells for potable water supply and for local sewage-disposal methods like septic systems, it is almost universally true that one must turn to the health departments or their local equivalents. Many States license well drillers and require permits for the withdrawal (often called "appropriation") of groundwater. In most of them, the permitting authority is the health department or its equivalent.

While the public health regulation of groundwater protection and use worked pretty well with regard to isolated threats of acute disease, the development of serious cases of groundwater contamination by chemicals overwhelmed that system. It is interesting, however, that in many States the health departments still operate the only public laboratories capable of analyzing water samples for either microorganic or chemical contamination.

It quickly became evident in the face of the Love Canal and many other disclosures about large amounts of presumably toxic chemicals contaminating the ground and groundwater, that if regulation were to be undertaken, the public health agencies were not, in general, able to handle the job.

Once the groundwater problem became apparent, the development of a national strategy to counter it began. The Federal Environmental Protection Agency, under pressure as usual, developed a Groundwater Protection Strategy. That strategy is still being reviewed and further attuned to the reality of the issues. Many technical and scientific problems arise in any attempt to develop a strategy for this purpose. Issues of jurisdiction of various government agencies and various levels of government also complicate the problem.

The fundamental difficulty arises from the highly local nature of the sources of contamination and of the use of groundwater. These features are in conflict with any desire to develop any national program, more conflict than was encountered in attempts to control air and surface water pollution. The root of all the conflicts is the issue of land-use regulation.

Land-use regulation is a much-cherished area of authority for local governments. Even at that level, it is not universally agreed that the assignment of land uses to private property is an acceptable use of government power. It is almost an article of faith among those who have a stake in the decisions on land development and uses that the Federal government can be allowed no role in the matter. Whatever historic reasons may lie behind this attitude, there is little question that it exists and is a major obstacle to the development of a national program to protect groundwater.

Still, the nature of the hydrologic cycle ties the uses of the land inseparably to the risks of and the nature of any contamination of the water lying under it—and often to water many miles away.

THE PHYSICAL NATURE AND BEHAVIOR OF GROUNDWATER

Groundwater carries its share of the burden of impurities washed from the air by rain. It also contains many substances which are dissolved in its passage over and through the soil. We know this process takes many forms. A visit to a cave or even just a tunnel through rock shows the once-dissolved materials—the stalactites hanging like warm icicles from the roof and any cracks in the walls and roof.

This ability of water in the ground to pass through and transport materials creates many interesting artifacts, like the spectacular limestone caves found in many parts of the country and the "petrified" wood of many places where water has deposited silica, the principal ingredient of sand, in the structure of ancient wood. It also demonstrates that water is a powerful agent of change in the earth.

Unlike the power of water in motion that carves spectacular eroded landscapes on the surface, the movement of groundwater is invisible and can seem mysterious and even threatening, especially if it is believed to carry dangerous contaminants. Moreover, its ability to erode big caverns and to deposit layers of rock prove that it can also transport contaminants which enter the ground as a result of human activity.

Hydrogeologists study the occurrence and movement of water underground by learning to interpret the information they obtain from observations of wells. Most well-drilling permits require the driller to maintain a careful log, or record, of the various soil and rock formations through which the well passes. The chemical quality of any significant amounts of water found in a formation is also usually recorded. A study of the logs of many wells over a substantial area will gradually reveal the structures of the underground soils and rocks and the nature of the water associated with each formation. This background data is essential to any understanding of the availability, quality, and behavior of groundwater in any region. With this information in hand, engineers can begin to predict the impact of land disposal, past or future, on the groundwater and its users.

Figure 4 shows a simplified cross section of shallow layers of the earth where water may be found. The drawing also indicates how the layers of soils and rocks form separate aquifers. These are separated by water-resistant layers called aquacludes because they exclude water.

Figure 4

Simplified Cross Section of Land with Aquifers

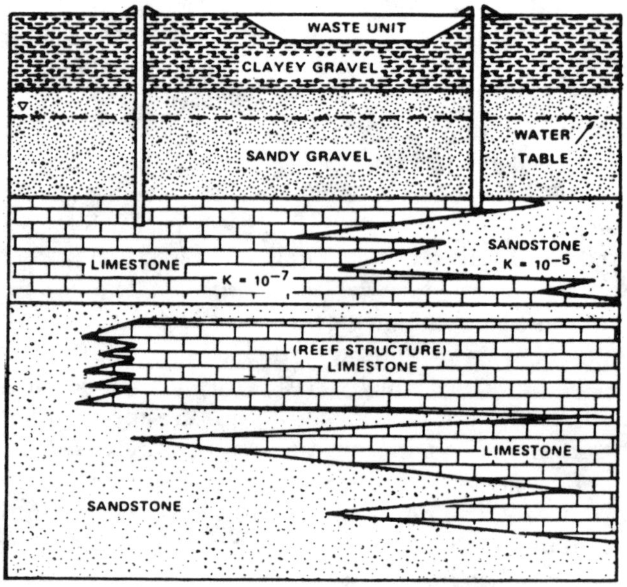

Source: USEPA

As one digs below the surface, the soils typically become more and more wet until finally a level is reached at which the bottom of the hole or well fills with water seeping from the ground. The level of this water is called the "water table" and is a very significant one in all engineering and regulation of groundwater use and protection. Water found in this water table of the "surficial" aquifer is the most vulnerable of all to contamination from any contact of toxic materials with the land, whether on the surface through spills or dumping or from disposal facilities. This aquifer frequently provides the water supply for local springs, lakes, and ponds or seepage marshes and swamps.

Most places also have one or more aquifers lying deeper than the surface aquifer and more or less isolated from it by an aquaclude which prevents or restricts flow of water between the surface aquifer and the deeper one(s). Since these deeper supplies are more protected from contamination, they are often the sources of drinking water sought in well drilling for either individual or community supplies.

This simple layer-cake structure does not always apply, of course, and structures of endless complexity are encountered. This is the reason that not all wells produce useful water supplies and the reason that not all measures designed to protect groundwater are effective. Since this is not a treatise on groundwater, we will not go further into these complications.

The flow or migration of water within the underground structures is controlled by the pressure available to push the water through and by the resistance of the formation through which the migration occurs. The pressure is generated by the higher levels of water built up near the sources of fresh water entering the formation from the surface—from rain or infiltration from surface water sources—and is determined by measuring the level (the "head") to which the water rises in a well open only to the particular formation chosen for measurement. The levels are always measured relative to some fixed level—mean sea level or, more often, some local "datum," which is merely a point chosen by local engineers for convenience and accessibility.

The resistance of the formations to the flow of water through them is not easily measured or predicted. The most reliable method measures flow-time between wells by injecting dye or radioactive tracers into the water, but this is expensive and, to say the least, not welcomed by users of the water. Others involve laboratory testing of samples of the formation and have larger ranges of results. The laboratory methods are usually capable of good precision and reproducibility, but problems of obtaining representative samples from underground are difficult to overcome.

The *direction* of flow of groundwater is usually more important to know than is the speed. Measurement of the pressure head in the aquifer at three points, requiring three test wells, is theoretically enough to determine the flow direction. In real formations, irregularities are encountered which may make it necessary to drill many wells to improve the prediction of flow directions.

Working with the information just described, it becomes clear that groundwater usually moves very slowly by comparison to the speed we are accustomed to in streams and even in the circulation of lakes and ponds. The fabled underground rivers do exist in a few rare places, but most groundwater flow rates are just a few feet a year or less. The EPA has suggested a likely rate of about 30 feet per year in homogeneous consolidated formations. The actual range is very broad, from a few feet *per century* in clay or hard rock to several miles in a few days in porous or cavernous rocks and earth.

The highly variable patterns of groundwater flow are responsible for much of the concern about contamination of groundwater. It is not easily possible to determine or predict the behavior of wastes from general information about an area; it is always possible that contamination entering an aquifer at any point may unpredictably show up anywhere else.

Groundwater protection programs all begin with requirements to investigate the flow directions and quantities of water and to establish a "baseline" water quality before any potentially contaminating operation is permitted to occur on the surface or in nearby aquifers that may be individually allowed to receive wastes or other contaminants. Experience has shown that contaminants will spread in the direction of migration of the groundwater, although often at a rate of travel which may be substantially slower than the movement of the water itself.

Chemical contaminants in groundwater interact with the soil and rock in a wide variety of ways. Occasionally, chemical reactions take place which can detoxify the contaminants. Generally, the process of interaction results in a sequence of deposition of the contaminant on the surface of the soil particles followed by a stripping off of the material by later-arriving water. The result of all this is not just the delay of the passage of chemicals but also the separation of the chemicals resulting from the differing rates of deposition and stripping of different chemicals which results from their own nature and from the nature of the soil through which they move.

The complexity of the soil-groundwater interactions and of the movement of water through geological formations makes the decision to use a method of waste disposal in or on the land an equally complex and difficult one. Remember, we have already discussed the basic truth of waste management: wastes will be produced in some amount and have only three possible places for their disposal—the land, the surface water, and the atmosphere. Decisions on methods of disposal will always ultimately be decisions on *relative risk*, and there well may be conditions and wastes that demand land disposal.

METHODS OF LAND DISPOSAL FOR MINIMUM RISKS

There are four basic methods of land disposal of wastes available to us. These are secure landfill, landspreading, landfarming, and deep-well injection. In this context, for simplicity, variations like waste ponds and other impoundments, piles, and other variations are considered special cases of the other surface and shallow-formations approaches.

Secure Landfills

When we find that wastes escaping from a disposal landfill, pit, or pond are contaminating the groundwater and causing damage to adjacent property, our natural impulse, at least among engineers, is an attempt to prevent the material from escaping.

The outcome of this approach is the so-called secure landfill. This high-tech facility is about as far from a "dump" as a Rolls Royce is from a Model T Ford, as you can see from the cross-sectional drawing in Figure 5.

To compound the complexity, advancing knowledge has shown us that the clay-sealing materials, even when carefully selected, are sometimes surprisingly susceptible to penetration by some of the organic solvents often found in hazardous wastes. This observation is the prime reason for the total ban on the disposal of "free liquids" in landfills which was included in the latest revisions of the Hazardous and Solid Waste Act. This act also incorporates into law the principle that "all landfills eventually leak," which the Environmental Protection Agency has followed for a long time.

Figure 5 Cross Section of a Secure Landfill

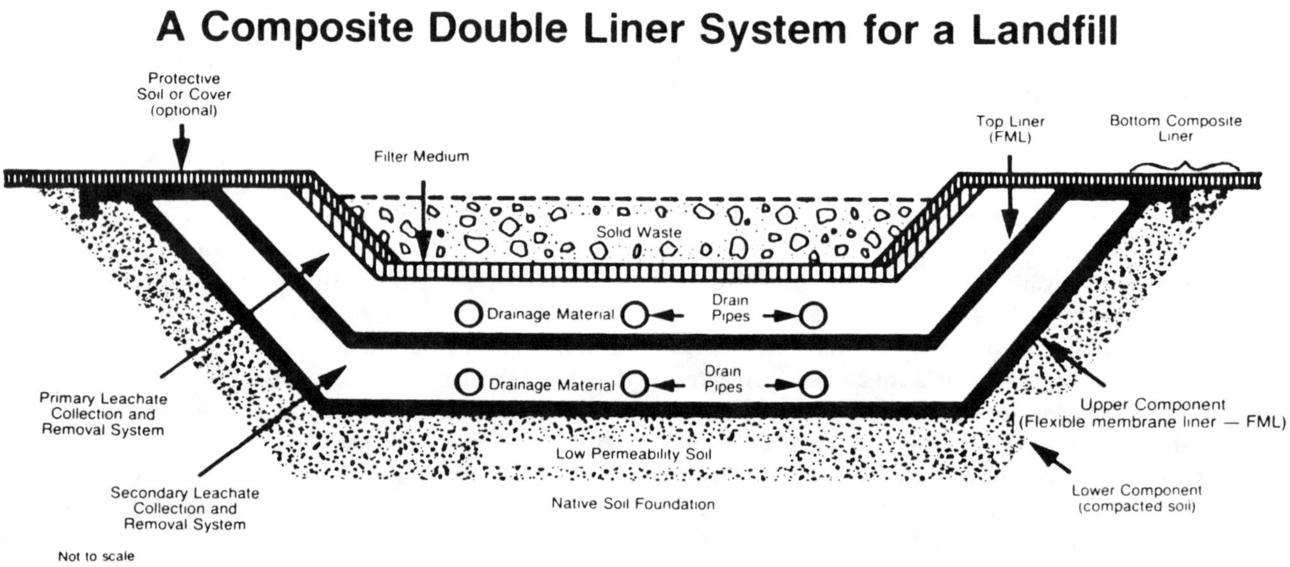

Source: the EPA's "Draft Minimum Technology Guidance on Double Liner Systems for Landfills and Surface Impoundments — Design, Construction and Operation" (May 25, 1985).

The design philosophy is based on the provision of several separate barriers to prevent escape of the contents of the landfill, with a top cover to prevent rain water from entering to create more liquid contents through leaching—the dissolving of materials from the waste by water percolating down through it. The main barrier is a thick layer of clay mineral carefully selected to resist the passage of water from above. Above the clay is a two-layer system of "liners" with a space between them in which leaks through the upper liner can be detected.

The idea is that if the upper liner leaks, the material can be caught in the lower level, pumped out, and treated to prevent damage—away from the landfill itself. There are many problems with this engineered solution. The liner materials have to be organic materials, rubbers or plastics, to be strong and flexible enough to withstand the stresses of installing them on a rough and uneven base of packed clay in a large hole in the ground, hundreds—maybe many hundreds—of feet in extent. Furthermore, the materials have to be capable of being sealed together to form an essentially seamless sheet over several acres of land.

Despite all these negative observations, well-designed landfills seem destined to play a continuing part in the management of wastes, and for a long time. To accept this, we must accept a couple of realities. One is that we will have some wastes which are not susceptible to further reduction—for example, the ash from incineration of municipal waste. The other is that no method of disposal of toxic and hazardous waste can forever guarantee zero exposure of all persons and all environmental species.

After all, the definition of a hazard is that which results from *exposure to toxic materials in toxic amounts.* Land disposal methods, including well-designed landfills, can so delay and attenuate the escape of any materials into a medium which can result in exposure that no significant hazard will exist. This principle, while hard for many to accept, is the only one on which sound policy can be developed. It is the reason that Risk Assessment (see Chapter 6 on that subject) has become the basic structure for decision making in the management of hazardous waste, indeed of all wastes.

Landspreading

Disposal by landspreading is just what its name describes. Wastes (almost always in liquid or semiliquid form) are simply spread over the surface of land. The land may be in use for agricultural or silvicultural purposes; sometimes it is wasteland which, at least for the time, has no other use.

Landspreading has an ancient basis in the traditional farmer's practice for disposal of manure, and indeed, the manure spreader is often the machine of choice for spreading hard-to-handle materials like sewage sludge even today. Landspreading as a method is most often used for disposing of dilute organic wastes like whey from cheese manufacturing, waste water from food-canning plants, and similar liquids. Often, there is a strong incentive for a farmer to allow cropland to be used in this way because of the nutrients available from the wastes. Many fruit canners maintain large woodlots and areas of brushland near their factories for the sole purpose of providing space for the spray-irrigation disposal of their dilute wastewater.

Landspreading is of limited usefulness for wastes which contain significant amounts of heavy metals which tend to be retained in or near the root-zone of crops grown in the soil. It has been shown that some metals, particularly lead, cadmium, copper, and related metals, can be picked up by food crops grown on land used for spreading wastes containing them. This naturally poses a danger of exposure of persons eating the crops and requires strict monitoring of the crops—not readily done in most agricultural situations.

There have been some large-scale applications of sewage sludge to farmlands in the never-ending search for economical ways of getting rid of this ever-growing problem material. The presence of metals and some hazardous organic pollutants in the sludge have been of real concern, although much research still needs to be done to establish finally the safety of the method.

Landfarming

Landfarming is an important variation of landspreading that is often applicable to the disposal of dilute oily wastes, like those left after a petroleum processor or refiner has removed all the recoverable oil from a wastewater stream, but the water still contains small amounts of oil. Wastewaters like these are not suitable for disposal in most sanitary waste treatment plants and are troublesome to handle because of the large volume of water relative to the oil content. There have been some applications of the method to treatment of sludges and other materials, but this is still an experimental procedure in most cases.

The landfarming operation utilizes the microorganisms (bacteria, fungi) naturally present in soil to digest and destroy the oil in the wastewater. The land is prepared by grading and by building berms or low embankments around the disposal area to prevent liquids applied to the surface from escaping across the land. In a few cases, the area has been excavated and supplied with a "liner" to restrict penetration of the applied waste to a shallow layer of soil. In most cases, this is so expensive as to rule out the method on economic grounds.

Generally, the waste liquids are applied in carefully controlled amounts so that they penetrate only to a shallow and controlled depth, and so the soil does not become flooded or saturated with water. This maintenance of a nonsaturated soil layer allows the microorganisms in the soil to have access to oxygen for their metabolism as they ingest the oil. Ideally, the organisms convert the hydrocarbons in the waste to carbon dioxide, water, and perhaps small amounts of sulfur and nitrogen compounds.

Limitations on this disposal method are imposed by any presence of heavy metals in the wastewater, which may have to be removed before disposal or at least strictly limited to avoid groundwater pollution under the site. The system requires a substantial amount of land because of the limitation of the amount of liquid which can be applied without saturation. For the same reason, its use in areas of heavy rainfall and low evaporation has to be questioned.

Properly designed and operated landfarming systems can do a very effective job of disposing of a class of troublesome, if seldom highly dangerous, wastes.

Deep-Well Injection

For a number of years, well before the current era of regulation of hazardous wastes, some producers of particularly dangerous waste materials sought permanent means of disposal. One of those was the injection of dilute solutions of the wastes deep in the earth, into formations which were so deep, so hot, and so laden with dissolved solids that it seemed impossible that their groundwaters could ever be of use on the surface.

Some of the first materials chosen for this form of disposal contained acid and cyanide mixtures that were hard to treat. Later, the method was extended to such difficult chemicals as nerve-gas by-products and pesticide residues or process wastes. Today, the underground injection technique is still the method of choice for some extremely hazardous waste materials.

Injection into the ground is by no means a simple or low-tech activity. The equipment and methods are basically derived from the experience of the petroleum exploration and production industries and depends for its success on the availability of the equipment designed for that use.

First, a suitable location for a disposal well must be found reasonably near to the point of waste generation. Hauling such wastes is generally not an acceptable risk; so pipeline transportation from the point of generation to the point of disposal is highly desirable. The underground geology is critical, of course. Rock formations which are suitable for deep-well disposal must lie well outside any zone of petroleum production, often, but not always, at a deeper level than any known oil deposits. The disposal zone must also have suitable porosity to allow the injected material to diffuse away from the injection point without requiring immense pressures in the well and at the surface to get it to flow. As we have already said, the formation must be one for which there is no other conceivable use and which has no detectable communication with any other formation or aquifer which might be of interest in the future. This set of requirements (and there are others in specific cases) severely limits the places in which deep-well injection is even a possibility.

Since the bottom of any formation system is the so-called basement rocks which make up the crustal structure of the earth itself and which generally have low permeability to any liquids, then a suitable site must be able to provide the proper conditions in shallower formations. This pretty well limits the suitable locations to those in which very deep sedimentary rock with trapped salt and other dissolved solids overlies the basement rock. Such areas have been found in a number of areas of the United States, but not all, by any means.

Once a site has been qualified as suitable, the injection well must be designed to accept the liquid wastes and to withstand any corrosive tendencies they may have, even at the high temperatures often found at depth in the earth. Bottom-hole temperatures of 700°F and higher may be encountered, and corrosion under such conditions is very difficult to overcome.

The injection well will almost always pass through aquifers of actual or potential use for drinking water or other critical applications. Therefore, the well itself must be constructed to avoid leakage into the other aquifers to which leaks are not desired or permitted by the regulatory authorities.

The result of all these needs is a complex well structure as shown in Figure 6. It is not possible to describe all the features of a successful design here—that would be the subject of a highly specialized engineering course. Nevertheless, the drawing gives some idea of the multiple well casings, leak detection systems, and formation-sealing systems that are needed for safe use of the method.

Figure 6

Deep-Injection Well

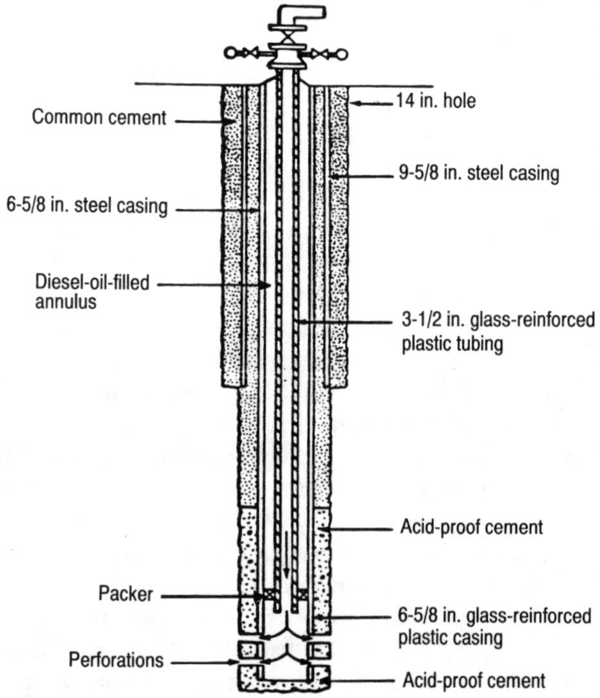

Source: *USEPA Report to Congress on Injection of Hazardous Waste* (EPA 570/9-85-003), iv–5.

Finally, the fluid for injection must, in general, be capable of filtration to a very high degree to avoid plugging of the formation into which injection is made. Any failure in the removal of *all* particulate matter will limit the life of the well, increase the pressure needed for injection, and increase the risk of casing failure, which can allow the escape of the toxic materials into useful aquifers. This need for ultrafiltration is a severe limitation on the candidate materials for this form of disposal.

LARGE-VOLUME SOLID WASTES: A MAJOR CONCERN AWAITING SOLUTION

This discussion has focused throughout on the kinds of wastes that one can transport readily to a disposal site. At that site, presumably, technical methods can be applied to reduce the amount of waste or to reduce the impact of the waste on the environment of the waste disposal site itself.

At many places in the United States and the world as a whole, enormous quantities of wastes have been dumped essentially where they were generated and left to the mercy of wind and weather. Often called simply "mining wastes," they may also have been generated by certain process industries and even, in the case of some sludges, by regulations designed to guard against air pollution.

In the American West, the mining of many metals over the past 140 years has left huge amounts of various kinds of solid waste. We call these tailings piles or ponds, mine dumps, overburden piles, mill waste, or many other names.

Most of these materials were put where we find them by people and companies long gone and unable or unwilling to do anything about them. Because of the nature of the problem, Congress substantially exempted such wastes from the laws passed in the 1970s. Little has been done, nor will much be done for a while because of the scale of the problem and because of disagreement about the risks which have resulted.

The problem was first identified in the Eastern U.S. coalfields, where abandoned mines were found to produce millions of gallons of highly acidic water which drained into the streams each year. Later, the drainage water was found to carry a heavy burden of dissolved and suspended metals like iron, manganese, lead, and others.

The mechanism of the acid formation is simple. Coal and most metallic ore bodies contain sulfur minerals, largely in the form of iron pyrite but also in more complex compounds. So long as the pyrite remains below ground and away from the air, it does little harm. When mining is done, however, either by tunneling or by strip or surface methods, the minerals are exposed to the oxygen of the air in the presence of the ground water. Such conditions quickly convert many sulfur compounds into sulfuric acid, a very powerful agent for dissolving other metals from its surroundings.

Studies of the thousands of abandoned mines and their leavings all over the West showed that the same phenomenon is at work there. Drainage from the piles of waste materials and from the old mines themselves is carrying huge amounts of metals into the streams in that arid country as well as into those of the wetter East. The Federal Environmental Protection Agency has begun to study this problem. Many of the Western States are also learning about the effects.

When dealing with mining wastes, one is dealing not with millions of pounds of toxic and potentially hazardous materials but with millions of *tons* which may need management. It is not unusual for the old dumps to contain more of the original valuable metals sought than the original miners ever recovered. So, too, do they contain the sulfur compounds and heavy metals they always did but in mobile form which can contaminate the water supply of half the continent.

A century and more ago, the gold and silver miners of our romantic past saved a lot of money and produced cheaper resources by ignoring the cost of disposal of their wastes. Now a new generation is faced with the costs. This is the ultimate parable of the externalization of costs—making someone else deal with them while living with the illusion that profits are high because those costs were "avoided." We still have to pay the costs some way and some day.

FINAL THOUGHTS ON LAND DISPOSAL AND PROTECTION OF GROUNDWATER

For two or three generations, public health professionals have understood the important relationship between human health and the protection of drinking water from underground sources from infiltration of human and animal wastes. Admonitions were frequent: keep your well away from the privy, from the barnyard. The preferred solution for the isolated home was a properly located septic tank. Many felt betrayed when the combination of small suburban lots, septic tanks, and inadequate wells produced nitrate poisoning in little kids in the tract homes of the 1950s.

In these more complicated days, when wastes are more difficult to recognize and to understand, the principles still apply. We, as a nation, have failed to exercise prudence and care in the way we locate and operate waste-disposal facilities. The results should not surprise us. Neither should those occasional incidents of real trouble discourage us.

If we are not willing to invest in the needed research and evaluation of waste-disposal sites before the decision is made to use them, we are likely to get into trouble. We know much about how to do these studies, and we know what materials to keep out of the waste to make the disposal safe within reason with relation to all the other risks in our society. The issues before our citizens are institutional and political. No one can forever guarantee that any disposal facility will operate without problems. It *is* reasonable to design and operate such facilities so that the problems, when they arise, can be controlled.

As long as we insist on maintaining a life-style that is closely related to the one we live in the latter part of the twentieth century, we will always have some residues which are best managed by burial on the land or in the depths of the sea. Self-seeking political polemics will not change this. What we need is a lower level of controversy and a higher level of pragmatism.

Chapter 8

WASTE MANAGEMENT: INCINERATION

ADVANTAGES OF INCINERATION

Wastes have been burned for a long time. Open dumps always seem to catch fire, producing noxious smoke and resisting efforts to control the smoldering mess. Indeed, one of the main reasons the public often supports the use of more expensive sanitary landfill technology is the disgust that people feel for the smoke and odors of a dump. For example, until just a few years ago, the heavy chemical and petroleum processing areas along the Gulf coast of Louisiana and Texas were marked by black plumes of smoke rising from "burning pits" used by many plants to dispose of liquid wastes and sludges that were awkward to handle and not readily recovered or recycled.

The combustion process is the only thing a modern incinerator has in common with the old burning systems. The elaborate technology for combustion and for the control of emissions and residuals marks incineration as a relatively new approach to waste management, not to be compared to the older system. Figure 7 illustrates the hazardous waste incinerator installation of the 3M Company near St. Paul, Minnesota. This was one of the first systems of its type in the United States and remains one of the best.

The heritage of public distaste for "burning garbage" remains a burden to be overcome when incineration is proposed in a community. Despite this heritage, literally hundreds of municipal waste incinerators are operating, or being built or proposed. Incineration of hazardous wastes is rapidly becoming more attractive as other alternatives become prohibited or severely restricted by law or by extreme costs.

Forces driving the growth of incineration are: the scarcity of acceptable locations for landfills and the expense of building and operating landfills which can meet the new standards. Materials processed by incineration are perceived to be destroyed once for all with no lingering liability. Landfilling and related methods often merely store the waste for an indefinite period after which it reappears and requires new decisions.

In spite of the apparent advantages of incineration in the management and permanent disposal of hazardous waste, clearly the public and even the permitting authorities are not easily convinced of the relative safety and general usefulness of the technology. This regulatory and public reluctance to accept incineration systems is best measured by the difficulty encountered in obtaining the required permits to build and operate these systems. For most of the recent history of public concern about safe waste management, incinerators for mixed municipal waste had little opposition and that was usually based on aesthetic concerns—littering, rats, and odors. Since the publicity broke about the dioxin scare at Times Beach, Missouri, public opposition has seemed to center on fear of exposure to dioxins from the burners. One effect of this shift has been a tendency for the public to consider all waste incinerators as if they were intended to handle hazardous waste—which confuses the issue for all elected officials almost beyond their ability to cope with the issue at all.

Figure 7

The 3M Company Hazardous Waste Incinerator in Cottage Grove, Minnesota

Reproduced with permission of the 3M Company and photographer Lowell Ludford.

THE PERMITTING PROCESS

The permitting process begins, as all hazardous waste management proposals do, with the Federal permit under the Resource Conservation and Recovery Act (RCRA). Part A of the application for the RCRA permit contains the basic description of the types and quantities of wastes to be handled and disposed of and of the intended means of meeting the complex requirements imposed on all facilities that handle, store, or treat hazardous waste. These requirements are described in the Code of Federal Regulations under the title *Standards for Owners and Operators of Hazardous Waste Treatment, Storage, and Disposal Facilities*. While often (and accurately) described as complex and difficult, the requirements under Part A are intended to prevent the future development of the unacceptable storage and handling facilities and practices that have led to the gigantic fires and consequent disruption of communities in a number of places in recent years. Like all such applications, these are public documents and serve to inform the affected public about the proposed facility.

Part B of the permit application contains a complex series of documents which are intended to address all the important issues which can arise out of the proposed treatment technology—in this case, the incineration of the particular kind and quantity of waste described in the Part A filing.

While the standards that apply to any treatment technology are frequently changing as knowledge grows and needs are redefined, an incinerator proposal must identify the Principal Organic Hazardous Constituents in each waste feed for which the system is to be permitted and show that the system proposed will destroy and/or remove each one with an effectiveness of at least 99.99% (often called the "four nines" standard). A simplified diagram of the process for getting the needed permits for a hazardous waste incinerator is shown in Figure 8.

Performance standards for incinerators under the Resource Conservation and Recovery Act are focused on a figure called the Destruction and Removal Efficiency (DRE) determined at the trial burn of the finished incinerator. This is the origin of the "four nines" standard that is usually mentioned in hearings and releases about new incinerators. However, the regulators have other and higher requirements available to them if they wish to use them.

The authority to regulate the special incinerators that burn polychlorinated biphenyls is found in the Toxic Substances Control Act, not in one of the solid and hazardous waste laws. The rule that has been developed sets the temperature and the time required for the chemical to remain within the combustion zone to ensure the destruction of PCB's. This and other requirements end up with a minimum acceptable destruction and removal efficiency of "six nines"—99.9999%.

In addition, particulate matter emitted by incinerators is strictly controlled, and hydrogen chloride gas (which becomes hydrochloric acid when dissolved in water) emission is limited to one percent of the chlorine entering the incinerator or to about four pounds per hour. (Special rules apply to PCB burners.) Hydrogen chloride gas is formed in an incinerator when feed materials containing chlorine are burned. Common materials which contain significant amounts of chlorine include many industrial and household solvents and most vinyl resins found in common plastics.

The development of the required information and the preparation of the Part B application for an incinerator is a highly technical and time-consuming process. Anticipating the complexity and the difficulty of analyzing the supporting documents, the Environmental

Protection Agency, alerted by the Part A filing, closely follows the generation of the application, so closely that the EPA itself makes the decision as to whether the application is complete. Only after that decision is made can the actual request for a permit be made. Responding to complaints that permitting decisions took too long, the EPA set deadlines for its own decisions, but these deadlines are often meaningless because the agency reserves for itself the decision as to whether the application is "complete"—and all the deadlines are measured from the time a "complete application" is filed.

In those States which have been given approval (delegation) to operate the Clean Air Act and Solid and Hazardous Waste programs, the formal application and permitting process is actually handled by the State agency, like any permit for a new or revised source of air pollutants. Incineration of hazardous waste, however, is still so sensitive that it is unlikely that the State will be left alone to carry out the regulatory function.

Figure 8

Simplified Diagram of the Permitting Process for a Hazardous Waste Incinerator

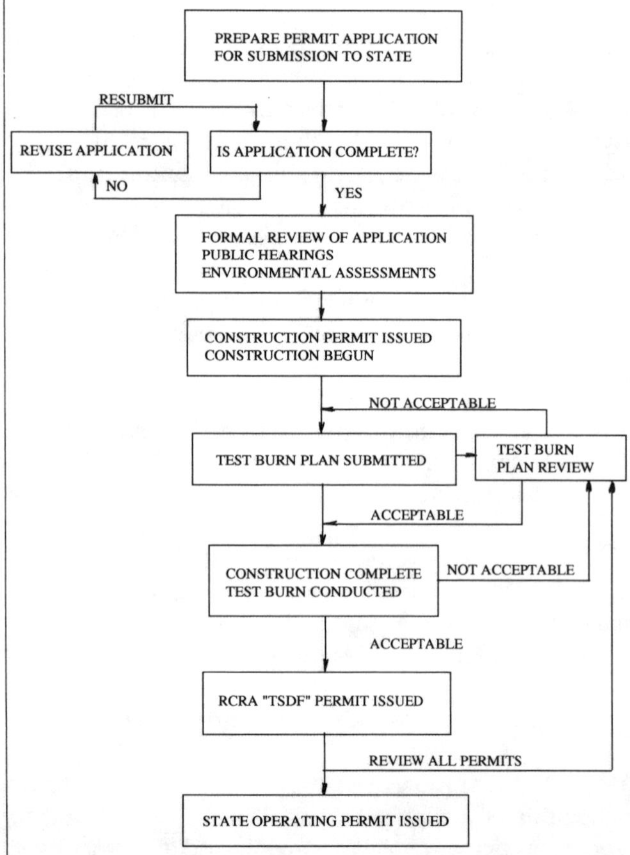

Each EPA Region is going to conduct something like a joint process with the State which is designed officially to ensure that the State lives up to its obligations. Unofficially, the EPA needs to protect itself from any mistakes if it can.

When the complete application is submitted to the State Agency and/or the EPA, the Federal process requires one or more public hearings, held reasonably near the proposed site. Considering the application, the information generated at the public hearing(s), and any investigative work by the agency staff, a decision is made either to deny the permit or to grant a temporary or draft permit. Issuance of this draft permit allows the owner to build the incinerator and to operate it under severely limited conditions for particular purposes. At this and any later stage of the process, the owner of the incinerator is proceeding with a substantial investment without any assurance that a final operating permit will ever be granted. Even if the conditions for granting of the basic Federal permit are met, that is only one of a potentially large number of government permits that will almost certainly be required.

Even if the proposed incinerator can meet all required performance standards, it will have to have an emission permit under the Clean Air Act, which will generally involve a more-or-less elaborate review of the State Implementation Plan under that program. Sometimes the process of obtaining this permit alone can take more than a year, especially if there is vocal opposition to the plan. Since wet scrubbers are often needed to clean the discharge gas from the incinerator, it is common for a permit under the National Pollution Discharge Elimination System under the Clean Water Act to be required.

Both the Clean Air Act and the Clean Water Act permitting processes can involve both State and Federal agencies and policies. Either can trigger environment assessment or environmental impact analyses of varying levels of complexity, needing time to complete. Many other kinds of government permission will be found essential, from zoning and land-use controls to waivers of road restrictions on trucks hauling certain kinds of waste. Each state, county, or city has its own requirements that must be met, so that simply defining the required permits in advance of beginning the process has become a specialized field of consulting.

OPERATING CONSIDERATIONS

Choosing Wastes for Incineration

Because of the attractiveness of incineration as a "permanent" solution for destruction of dangerous or other difficult waste, it is tempting to apply it broadly to all kinds of materials. When difficult issues arise in siting of landfills or other disposal facilities, political officials may succumb to the idea that incineration is the general solution to these kinds of problems. The facts are not so encouraging. One group of EPA technical persons suggests that in choosing wastes for possible treatment in an incinerator, the following factors need to be considered:

- Moisture content
- Inert (ash-forming) material content
- Heating value and auxiliary fuel requirements
- Physical form (liquid, solid, or gas)
- Corrosiveness
- Dioxin or substituted furan content
- Potential pollutants expected in incinerator effluents
- Potential health and environmental effects of the effluents
- Known content of carcinogens

The first five of these factors are of primary importance in the operation of the incinerator system itself. The physical form, as will be seen in a later discussion on the various forms which incinerators take, has major effects on the complexity of the feed system. If a proposed incinerator feed is so low in heat content that it refuses to burn on its own at a high temperature, then it will be necessary to fire additional fuel to get the desired breakdown of the feed. This raises costs and causes other changes in the design of the incinerator furnace that may be self-defeating and make the process unsuitable for that feed material. A highly corrosive feedstock or one that leaves large amounts of inert ash in the furnace can cause obvious operating problems in a high-temperature process. Pretreatment before burning is often needed to offset these kinds of problems.

The last four items for consideration are those which affect the community in which the facility is to be located. Incinerators which are to be licensed for treating PCB or polychlorinated furans (related compounds with somewhat similar chemical character) will be subjected to more stringent requirements on destruction and removal efficiency reaching for a so-called "six nines" level, or 99.9999%. This difficult-to-achieve performance requirement may in some cases be replaced by operating standards which define the time-temperature-oxygen regime within which these compounds are to be treated. In any case, because of great public concern about these materials, regulatory agencies may be expected to proceed slowly and with maximum public input in licensing any facility to handle them. It is worthwhile to remember that any waste oil containing less than 50 parts per million of PCB is not considered under these stringent regulations, and that oil containing up to 250 parts per million may be incinerated from a Federal standpoint, at least, in approved utility and industrial boiler furnaces as mentioned before. As is so often the case in interpreting these environmental regulations, PCB is not always PCB from a regulatory point of view.

Selection of incineration as a means of treatment/disposal will obviously have to take into account the potential pollutants expected to be found in the incinerator effluents and emissions, the potential health and environmental effects of these effluents and emissions, and in particular any known discharges of carcinogens.

All of this information is in principle available to the public during the permitting (licensing) process. At first, it must be predicted in the Part B application which will be the subject of public hearings. After the required trial burn of the system, the results of that trial are also public documents available from the EPA.

It is to be expected that the regulatory agencies will take a Risk-Assessment approach based on the effluent and emissions data. For an understanding of this approach, it is necessary to deal not only with the health-effects data but, in the case of an incinerator, with the meteorological models which will be used to predict the magnitude and nature of the exposure of any "target" population to the pollutants which may escape the control systems. This particularly arcane predictive process is often beyond understanding by anyone but the experts and may be the source of the uncertainty which can be detected in the regulatory officials who must make the final decisions on issuing permits to incinerator operators. If the public feels a need to become involved in this part of the process, it can rely only on its own set of consultants—an expensive and often frustrating process. Recent changes in the regulatory rules make it possible for the EPA to provide grant money to community groups to permit them to retain professional advisers for participation in this process.

Key Design and Operating Factors

In its simplest form, incineration of anything is a process of subjecting the feed material to a high temperature for a long enough time, usually in the presence of air, to break apart the molecular structure of the feed and react it with the oxygen in the air. If the process is effective, nearly all of the original feed material disappears, and the only products are simple nontoxic compounds: carbon dioxide, water, perhaps sulfur dioxide, and a few others. After the combustion is complete, the hot gases must be cooled, often by recovering heat for useful purposes, then purified to remove any air pollutants and suspended solids left over as ash from the burning.

An acceptable incinerator for hazardous waste

has been defined as one capable of reducing the amount of any particular toxic in the feed by 99.99%. This is the "four nines" standard already mentioned. To achieve this level of destruction, the incinerator must be equipped with effective combustion-gas cleaning equipment following a combustion furnace that reaches a time-temperature combination needed to destroy the most difficult components of the incoming waste.

While the particular design of elements of any individual incinerator is unique, all today's systems contain the basic elements shown in Figure 9. Each of the steps shown in the various boxes can be designed to fit a particular kind of waste or location, or simply to reflect the owner's preference for a particular kind or style of equipment. It is rare to find any two waste incinerators with identical arrangements of equipment or operating conditions.

Using the general scheme shown in Figure 9, we can describe some of the most commonly used approaches to each step of the process.

Figure 9

Basic Elements of an Incinerator

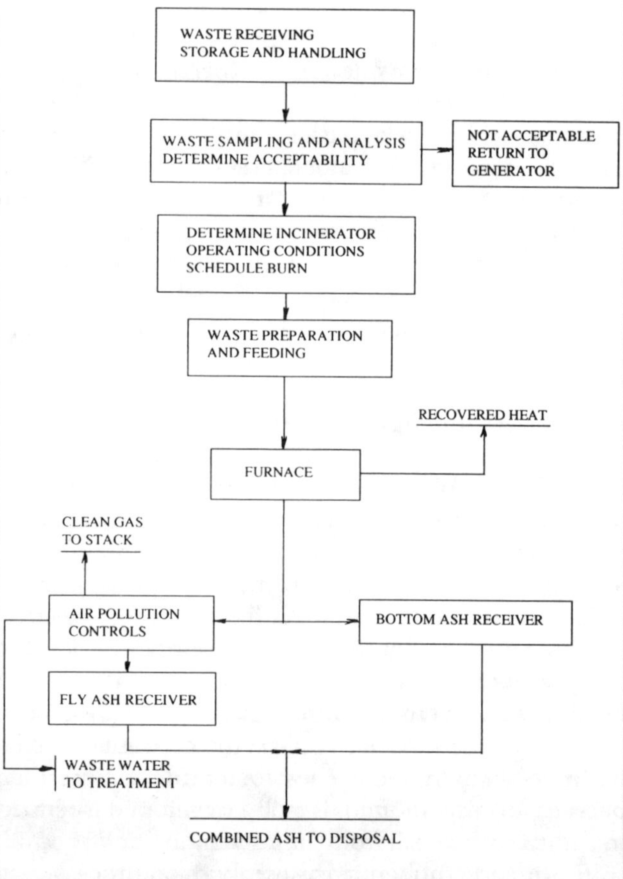

Waste Receiving, Handling, and Storage

Methods and equipment for these operations are chosen to accommodate the ways in which wastes are delivered to the plant site. If the wastes are hazardous, this system must meet design standards for hazardous waste storage. Usually needed are bulk tanks for liquids, bins or silos for solids, and areas for container storage. Storage methods must allow for sampling of the incoming wastes to ensure their suitability for the particular incinerator, specifically to ensure that the particular wastes are allowed to be treated under the facility's permit. Safety of the facility is another prime consideration. More than one municipal waste incinerator has been victimized by the arrival of an illegally dumped container of gasoline or some other flammable material. At municipal waste incinerators, trucks hauling waste have been known to unload burning substances directly onto the tipping floor or into the bin from which raw waste is fed to the incinerator itself.

Waste Preparation

Arriving waste often requires grinding or some size reduction of solid materials to fit the feed system. Wastes are often blended to maintain more uniformity in the operation of the burner, usually to even out fluctuations in the heat content so that supplemental fuel use is minimized.

Feed Metering

Liquid feeds are usually simple to feed to the incinerator, usually with a fairly simple pumping system. Solids present complex problems in maintaining a smooth and uniform flow of waste into the combustion unit. Some units are designed to allow filled but unopened metal or fiber drums to be fed into the combustion zone. Such charging systems have complex airlocks to allow these large "lumps" of feed to enter the combustion zone without leakage of hazardous materials in the surroundings. The extremely high temperatures needed to destroy hazardous chemicals will convert the steel in the drums into an inert metal oxide powder from which new iron can readily be made in a steel mill.

TYPES OF FURNACES

The heart of any incinerator is the furnace, where the combustion process actually takes place. The design and operation of the furnace has as its objective the achievement of the required degree of destruction of the incoming feed without making excessive amounts of unburned toxics, undesired by-products, or other negative results.

Many different types of furnaces are in use in waste incineration. Each has a particular type and range of wastes for which it is best suited. Each has special needs for controls, operation, and maintenance. Each has more-or-less impassioned advocates. Among the more common types are:

Mass-Burn Furnaces

These combustors are built very much like a coal-fired steam boiler. Wastes are spread on grates or hearths and burned with plenty of air at a fairly low temperature. Unburned solids fall off the hearth or through the grates, while combustion gases are usually heated by supplemental fuel injection while being delayed for a minute or two to allow resistant chemicals to complete their breakdown. The hot gases usually are cooled by water in boiler tubes that generate steam for heating or other uses. The cooled gases can be purified by washing with water or chemical solutions in a scrubber, often followed by a filtration system to remove the last traces of dust or solid particles before being exhausted through a stack. Mass-burn furnaces are often chosen for treatment of municipal solid waste, but are not ordinarily satisfactory for disposing of hazardous materials.

Liquid-Injection Furnaces

These types of furnaces are used to burn liquid of high heat content, like contaminated oils or solvents. Such furnaces are often built like oil-fired steam boilers to permit recovery of heat in the form of steam. Since many liquid wastes contain acid-forming constituents like sulfur, gas scrubbers are essential to safe use of these systems. In recent years, tests have shown that large utility boiler furnaces used for steam generation in electric power generating plants can safely be used to burn liquids contaminated with some kinds of hazardous substances. Successful use of this method of disposal has great potential for minimizing the need for special treatment of a large volume of these wastes while recovering useful amounts of energy.

Rotary Kilns

Similar to those used in the cement industry, these kilns are gently sloped cylinders that rotate slowly while being heated to very high temperatures internally. The heat is generated by burning waste in combination with other fuels—oil or gas are usually chosen, but other solid fuels can be used. The kiln system probably can handle the largest variety of kinds of wastes, since nearly anything that is not large enough to interfere with gas flow within the kiln can be fed and passed through the incineration zone. Since kilns of eight- or ten-foot diameter are common, they are the equipment of choice for handling large drums or containers of wastes that may be fed intact into the combustion zone. They can be built as long as may be needed to achieve the time of exposure to the high temperatures that is desired. Cement kilns several hundred feet in length are common.

Kilns are usually lined with refractory materials to permit very high temperatures to be maintained without hazard to the structure. Some kilns are designed and operated to maintain a layer of melted glasslike slag on the inside surface during operation to help retain and treat fine particles that are released in the combustion of the feed. Gas scrubbers and dust removal systems are readily attached. Heat recovery from a rotary kiln may be mechanically complex when compared to other systems, but it is practical when conditions are right.

Multiple-Hearth Furnaces

Originally developed for thermal processing of metal ores, like roasting of copper or zinc sulfides or the manufacture of lime by calcining of limestone, these furnaces have rotating plows that stir and move solids over a series of round hearths in a vertical stack while hot gas passes through and over the solids. Because the solids are constantly being stirred and moved around on the hearths, sticky or lumpy material is constantly exposing new surface to the hot gas or air rising through the furnace enclosure. These furnaces are mechanically complex and require a fairly steady workload to give good results. One favored application is the incineration of sewage sludge, where the agitation on the hearths overcomes the stickiness of this messy feed material and achieves complete combustion when other systems are less successful.

Solids-Contact Systems

This system generates a bed of suspended sand or pebbles (fluidized bed) that is heated to a suitable temperature by heating the air used to fluidize the bed. Wastes and additional fuel, if needed, are fed into the hot solids and are burned. In some systems, some of the hot solids in the bed are circulated through another combustion chamber to complete the consumption of the wastes and to facilitate the separation of ash or solid residues to keep the bed clean. It is also possible to make the bed of some solid that reacts with constituents of the waste to detoxify them. This system, with limestone pebbles forming the bed, is being developed for burning high-sulfur coal with minimum air pollution by sulfur dioxide formed in the combustion process.

Another variation uses molten sodium carbonate (washing soda) or other molten salts as a source of heat. Wastes are introduced into the hot pool of liquid (750° to 1000° C) and completely burned with air that is injected into the liquid simultaneously.

All these solid-contact approaches require supplemental fuel for the initial heating-up; many need continuous heating in operation. Gas cleaning and scrubbing is readily attached, although wear on the solids in the beds may create extra dust for the system to handle.

Liquid-Phase Systems

These systems are designed to oxidize wastes just as in a combustion reaction, but achieve the destruction by contacting the waste with air or pure oxygen in a water solution or suspension at high pressure and temperature. The liquid-phase oxidation process was first developed for waste disposal aboard ships, where compactness and the avoidance of open flames and very high temperatures were essential.

Today's systems obtain the high pressure and temperature by a variety of methods, but the results tend to be similar. After the reaction is completed, the mixture is allowed to expand to lower pressure and the unreacted materials are settled out for disposal. The liquid phase can be returned to the system with new wastes or sent to a wastewater treatment system.

An advantage of these systems is the near-total absence of any gaseous emissions. This unique feature may in the long run increase the attractiveness of this approach.

Pyrolitic Systems

This system destroys wastes by subjecting them to very high temperatures, usually but not always in the absence of air. If there is no air in the reaction zone, then usually the breakdown products are in gas form and are fed to a burner of some kind, followed by the usual gas cleaning and pollution control systems. Variations include heating by a pool of molten glass maintained by electric resistance heating. The glass absorbs any toxic materials and when cooled, immobilizes them for easy disposal.

In general, these systems appear to be useful in highly specialized applications but are not yet proven on a large scale. One possible application might consist of the destruction of cotton or paper protective clothing contaminated by radioactive metals. The process could produce a glass containing all the metals in a nonleaching form while destroying all the organic materials.

AIR POLLUTION CONTROLS

In many cases incineration offers permanent disposal of a waste in a manner that cannot be done when land disposal or other forms of dumping into the environment are practiced. However, incineration will create at least the possibility of polluting the air. Little is known about the health and environmental impacts of possible chemicals in the emissions from an incinerator; indeed, little is known about the long-term impacts of *any* air pollutant at the low concentrations usually encountered in the ambient air near a source like an incinerator.

The products of the combustion process in an incinerator are a bottom ash that falls out of the burning mass and settles to the bottom of the furnace, and a complex mixture of air, dust, and chemical gases that have been formed or released in the burning system. The ash is a special problem of disposal, as will be seen. The control of the air pollutants is a very difficult and expensive engineering problem that accounts for much of the size, complexity, and cost of an incinerator installation. As is often the case, lack of real knowledge of the effects of the many trace materials in the combustion products forces a severe restriction on the amounts of known pollutants which may be emitted, in the hope that the unmeasurable and unknown components will also be controlled to levels that are not likely to be harmful.

Among pollutants that are likely to be produced in the furnace are:

- Particulate matter (dust)
- Sulfur oxides
- Nitrogen oxides
- Hydrogen chloride and related gases
- Trace metals
- Products of incomplete combustion

As of this writing, only hydrogen chloride and related compounds and particulate matter are the subject of specific regulations on the amounts that may be emitted from incinerators.

An extended discussion of the technology of air pollution control is beyond the scope of this book. It is useful, however, at least to describe the general arrangement of the system, which is broadly the same in all facilities. First, the hot gases must be cooled to freeze the chemical reactions and to prevent new and undesirable compounds from forming during the time the gases are in a temperature range that allows chemical reactions to continue, and to protect the later filtration equipment from the very high temperatures reached in the furnace. This operation, called quenching, usually involves contacting the hot gas with a large amount of liquid water which evaporates and cools the gas stream very quickly to near ambient conditions. The quenching water is separated, and the steam formed is condensed so that the liquid containing many contaminants can be sent to a wastewater treatment plant—another necessary auxiliary of the incinerator.

The quenching operation may be combined with or be a preliminary to a scrubber in which the gases are more intimately contacted with a liquid which may be water or a solution of chemicals chosen to react with specific compounds in the quenched gas. Again the liquid used must be cleaned in a treatment plant of some kind, then recycled again to the scrubber.

Scrubbing usually removes almost all the gaseous chemical burden of the gases on their way to final release through the stack, but a final step is usually needed to remove the last of the suspended dust in the stream. This may involve filtering the gas through cloth in a baghouse or, rarely, cleaning in an electrostatic system.

The concentration of materials which may remain in the emitted gases from the incinerator stack is spelled out in the facility permit. The owner-operator of the incinerator is required to measure the concentrations in the stack emissions either at regular intervals or, more often, continually with appropriate instruments, and to file regular Monitoring Reports with the regulatory agency responsible for the permit. These reports are public documents, since all monitoring reports are by Federal law required to be made public.

INCINERATOR ASH

The ash from an incinerator is naturally composed of the unreactive residues from the combustion reaction. There is nothing left in it that will burn and there is little choice as to its disposal: landfill or dumped in the ocean. It will contain some oxides of metals originally in the incinerator feed, at least those metals which did not vaporize and leave the furnace in the gas stream. It may contain organic chemical compounds, either products of incomplete combustion (known as PIC's) or things like dioxin which are actually formed in the combustion process.

Whatever the facts about incinerator ash, it always is a significant problem to the public when a project is proposed. Probably the only way to evaluate any hazards from this source is to start with the results of the trial burn required before any incinerator is permitted to go into regular operation. A thorough analysis of the ash produced under the permit's operating conditions will make possible an evaluation of the ash management plan proposed in the permit. Fortunately, the ash is usually produced in small quantity in an incinerator properly designed for its input feedstock, and land disposal methods can almost always be found which will secure such material with reasonable long-term safety.

Many hazardous wastes produce little ash when incinerated, but facilities which can accept whole drums of waste or even old electrical transformers and capacitors will naturally have to dispose of the metal oxides left over from burning the steel or other metal containers. Fortunately, this kind of "ash" is easily recycled at a steel mill or other metal-processing plant, as we have already said earlier.

SUMMARY AND CONCLUSIONS

Incineration is a method for actually destroying wastes, whether hazardous or not, so that they are not merely stored in some form until they escape or are broken down by slow natural processes. This should make it a method of choice for advanced research and the development of means of offsetting or avoiding its weaknesses.

Unfortunately, well-meaning opponents of incineration have so confused the situation that political leaders are forced into a no-win situation. Landfills are full or nearly so, and there is no place to put new ones. As we have seen earlier in the book, wastes keep being produced and must be handled somehow.

If nothing else, incineration can so reduce the waste's physical volume and so improve its physical resistance to settling that the life of landfills can be vastly extended.

The air and water pollution potential of properly designed and operated incinerators can be controlled—our European counterparts accept this conclusion fully, because they encountered the land shortage earlier than we did in the United States. At the very least, adequate research funds should be provided and a sound program should be maintained to find ways of assuring the public of safe incineration service.

As we shall see in later chapters of this book, the universal alternatives to waste management sought in recycling and source reduction are still utopian dreams, while public authorities need to manage waste today.

Chapter 9

WASTE MANAGEMENT: TREATMENT OF HAZARDOUS WASTE

THE SEARCH FOR PERMANENT DISPOSAL

As the realization spread that land disposal of hazardous waste, indeed all wastes, was at best a method that could return to haunt those who threw the material away, a modest interest in methods for treatment of hazardous materials before disposal began to develop. Some large companies had, for their own reasons, used treatment techniques to get rid of difficult or embarrassing materials for some time. Even as early as the 1940s and 1950s, examples can be found. In Texas City, Texas, Monsanto incinerated quench water containing acetylene black, a particularly difficult form of carbon black, to avoid dumping the highly visible material into the waters of Galveston Bay.

Yes, it is possible to incinerate water, at least when natural gas could be had for around 10 cents per thousand cubic feet. Other similar kinds of preventive steps were adopted by other companies, as was revealed by the frequent sight of plumes of black smoke rising from the "burning pits" associated with just about all large chemical complexes of that period. In those pits many sludges, liquid wastes, and even off-specification products were often burned to avoid having to develop any other disposal methods and the hauling of wastes any distance unless it had real economic value.

A brisk business in collecting and "regenerating" sulfuric acid sludges resulting from acid treatment of lubricating and other specialty oils developed because it was messy to get rid of these materials and it was proven economical to recover the sulfur in them for reconversion to useful acid. In a similar vein, but with a completely different approach, hydrochloric and sulfuric acids used for "pickling" steel (to remove the oxide coatings remaining after processing and shaping in the mills) were the subject of many attempts to utilize "waste pickle liquor" for all kinds of uses from additions to water purification systems to recovery of iron sulfates for agricultural use in correcting excessive alkalinity of soil.

Few of these approaches to treatment of wastes before or as an auxiliary to disposal became widespread, except, of course, the burning of materials in open pits or even in incinerators or flares. Burning was and is a permanent solution to waste disposal, and in those times in which air pollution was not such a sensitive issue, an attractive one. It was no accident that the nights were bright with the light of many flares wherever large chemical plants and refineries were concentrated, and for that matter, still are in some places.

When air pollution became an issue (beginning in the 1950s) or the sheer volume of waste and its dilute form became problems, the wastes were frequently just stored in a pit, pond, or lagoon in the hope they would go away or something. Many of these were still around when the solid waste control programs were developed in the 1970s and are, indeed, the subject of some of the most difficult Superfund cleanups now and recently under way.

Through all this, one constant factor remained: it was always less expensive and less complicated to dump wastes into a landfill than to do anything else with them, assuming that their economic value as recyclable materials was so low that this route was not attractive. As long as this relationship remained true, change was unlikely.

New factors emerged to alter the assumptions and the economic realities surrounding solid wastes in general and hazardous wastes in particular. The volume of material needing disposal grew enormously with increasing population, increasing affluence, and increasing public preference for synthetic fabrics, plastics, pesticides, herbicides, and the other chemical trappings of a modern consumer society. Not so obviously, the new laws demanding more wastewater treatment and higher levels of air pollution control encouraged methods of treatment which transferred the water and air pollutants into materials that entered the solid waste stream and had to be handled there, where the technology and capacity did not exist.

As we have said before and will again, nothing ever goes away in this universe. We can change its form, even its chemical structure in combination with other materials, but we cannot destroy matter. What comes out of the water or the air must go somewhere. Treatment methods can at best reduce the toxicity, volume (landfill space needed), or ability of any waste which still presents hazards to migrate from its chosen disposal site in a way that endangers human health or environmental values.

Today, we are told by the EPA and the Congress that, even if ways were found to increase the available capacity of land disposal facilities, there are some wastes which are simply too dangerous to place in the soil over the long term. This "land-ban" has been achieved by prohibiting the disposal of many types of waste "unless treated" such that, if they are placed in landfills they will, in the judgment of the Administrator of the U.S. Environmental Protection Agency, not present any threat to human health or the environment. Finding treatment methods to meet this standard is a tall order. When the tendency of Congress to second-guess the Administrator is taken into account, the task is much harder.

CHANGES IN WASTE MANAGEMENT THROUGH INCENTIVES AND DISINCENTIVES

This flat prohibition of certain kinds of disposal is an expression of Congress's frustration with the difficulty of designing a regulatory program that is politically acceptable to its many constituencies, legally acceptable under Constitutional law, and economically possible in the capitalistic system. Only rarely has Congress been strongly influenced by the sheer technical possibility of accomplishing its directives, and the ever-tightening restrictions on waste disposal have paid little heed to this factor.

Since Congress began to consider seriously the national program for management of solid wastes, it has turned to ever-stronger legal sanctions in an attempt to compel industry and the public to change behavior in the handling and disposal of hazardous wastes. Any reading of the chapters of this book dealing with the Resource Conservation and Recovery Act and the Comprehensive Environmental Response, Compensation, and Liability Act will quickly convince you that the structure is exclusively built on a foundation of coercion: "You must do the following" and "You must not do the following" or you'll be fined or thrown into jail. These are powerful incentives, as the thought is caught by the sage, "It wonderfully concentrates the mind to know that one is to be hanged tomorrow."

Many environmentalists have looked for other, basically economic, incentives for more desirable ways of managing hazardous waste. Recycling, in particular, is often cited in public meetings and political speeches as *the* answer to all waste management issues. Regrettably, basic chemistry often prevents the use of this supposedly benign solution, since molecules are much tougher to break apart than they are to form in the first place, and undesired secondary reaction products may be impossible to change to the desired ones. The reasons for these problems can be further explored in Chapter 11, "Chemistry of Waste Materials."

Nonetheless, the coercive laws have had the desired effect in many ways. We can argue whether the outcome is efficient, whether it is indeed needed at all, but not with the fact that for the first time on a large scale, the search is on for methods of reducing the amount of hazardous waste produced, and for treating hazardous waste to make it less hazardous and simpler to dispose of. In this search, the objective is more and more to find treatment systems that also constitute permanent disposal so that generators can eliminate the possibility that their wastes will return from the "grave" to cause more trouble, hazard, and expense.

It is the nature of business managers dealing with problems that they seek solutions which are permanent to "put it behind" them. Thus the conditions all begin to fit together. The large investments needed to build a modern hazardous waste landfill have changed the cost relationships, and landfilling is not necessarily the cheapest method anymore. Further, if one accepts the current premise that all landfills will leak and, if so, will presumably contaminate groundwater which regulations will require to be cleaned up, then it is long term madness to use landfills for disposal at all. It is this understanding that has led Congress and many State legislatures to prohibit or at least strongly inhibit the disposal of many wastes from landfill disposal in their original or untreated state.

In legislating the "land-ban" Congress has imposed on everyone involved in the management of hazardous waste a set of broad prohibitions of materials from disposal in landfill and land-treatment systems. These prohibitions apply unless the EPA can certify that such disposal will *not ever* result in migration of toxic and hazardous substances which will endanger human health or the environment. *Not ever* is a long time, and few scientists will commit to a conclusion that anything is completely impossible in a very long time span. Furthermore, to make its intent clear, Congress prohibited most "free liquids" (even those in containers) from land disposal and required the EPA to review all the lists of hazardous wastes by 1991, and to determine by rule whether any of them could be placed in landfills under the broad restriction on migration and harm already mentioned. Congress has concluded that landfills are to be phased out completely at some time.

This sweeping conclusion unfortunately cannot be implemented today and may never be. As we shall see, even most treatment methods result in the production of *some* residue for which the only final disposal means is in a landfill. The idea is to make that residue as small as possible and to reduce its hazard potential to an absolute minimum to allow the landfill technology to work as it is intended.

To ensure that the economic lesson is well learned in the economy, Congress has taken steps to so restrict the use of landfills that, regardless of the cost incentives, every generator must seek disposal methods which can meet the criteria spelled out in the declaration that one must make when signing a Uniform Hazardous Waste Manifest:

> I also certify that I have a program in place to reduce the volume and toxicity of waste generated to the degree I have determined to be economically practicable and I have selected the method of treatment, storage, or disposal currently available to me which minimizes the present and future threat to human health and the environment.

Such a declaration, made under penalties of perjury, should provide enough incentive for any responsible individual or business to evaluate properly the alternatives to landfill disposal.

CHOOSING TREATMENT METHODS

To paraphrase the words of the Federal declaration, any method of handling and treating wastes should at a minimum, be designed to reduce the volume and toxicity within the limits of economic feasibility and should minimize the threat that waste will ever cause to "human health and the environment." In order to deal with this large assignment, we need to break it down into workable parts:

1. The obvious way to avoid waste disposal problems is the avoidance of making the waste in the first place.

2. If there's no way to eliminate the production of hazardous waste, then be sure only an absolute minimum is produced.

3. Determine whether the remaining irreducible waste is disposable through licensed transporters and treatment, storage, and disposal operations which are reasonably available to you.

4. Find out the cost of disposal through the legitimate system, if it is available.

5. Do the research on methods of treatment and/or destruction of wastes which are appropriate for your materials, including the costs of these methods for comparison with the costs already established for contract disposal service. Proposals from outside disposal services may be of help in this phase, but remember that disposal-site operators will naturally propose to use the system or systems they have in their facility—whether or not it is the optimum one for your wastes. No reputable TSDF operator will suggest an unworkable or unsafe method, but the responsibility is yours to select the method and to ensure you receive value for your expenditure.

This work, some highly technical and demanding that you have qualified staff to evaluate the possibilities, will determine whether in-plant treatment offers advantages to your operation. After evaluation of all the relevant factors, a decision can be made.

Once again, as pointed out before, such treatment systems are best when made an integral part of the manufacturing process, so that the issue of becoming a storage and disposal facility *per se* does not arise. In-process waste treatment amounts to waste avoidance or at least minimization in the best possible sense.

Waste Avoidance and Minimization

The surest way to accomplish all the goals of a treatment program lies in finding ways to avoid making the waste in the first place or in being sure that the quantity of waste subject to regulation is kept to a minimum. Chapter 10, entitled "Waste Management: Minimization and Recycling," should be consulted for more detail. Most of an effort to reduce and eliminate waste, at least hazardous waste, is a matter of common sense analysis linked with the necessary technical support for making changes in processes and operations.

Waste minimization work begins with a waste inventory and a search for opportunities to segregate hazardous materials from ordinary waste as soon as possible in the manufacturing cycle to ensure that it is not mixed to create a far larger quantity of hazardous waste unnecessarily. It goes on to more sophisticated process analysis and change that won't be discussed here, but is considered in some detail in the chapter on waste minimization.

Waste avoidance implies that changes are made in operations and processes that eliminate the use of hazardous substances that appear in the waste stream because they are essential to a process. The automobile manufacturers reduced their contribution to both air pollution and land pollution by changing their paints from enamels and lacquers to water-based enamels. The development costs in this transition were enormous but were largely borne by paint suppliers who wanted to participate in this very large market. The change not only eliminated the solvents in the paints themselves—significant contributors to ozone smog—but greatly reduced the amounts of other, more difficult cleanup solvents used.

An open-minded evaluation of waste avoidance possibilities will nearly always pay off in better operations producing less regulated waste. The 3M Company has achieved savings in the hundreds of millions by the application of these principles in the program they call the 3P Program—"Pollution Prevention Pays." The State of South Carolina has an industry assistance program under the same name.

Treatment Choices for Practical Objectives

The reasons that an individual enterprise considers going into treatment of its wastes may be manifold. Saving money is an important one, but it is often seen as a result of decisions made for operational reasons or simply to avoid the complications and perceived trouble of managing even small amounts of hazardous waste through off-site services.

Remember, the object of treatment is first to avoid producing a hazardous waste at all, to reduce the volume of waste to be shipped to disposal, and to reduce its toxicity and/or its tendency to migrate from any location where it may finally be placed for disposal so that it will not return to require remedial action in the future.

The decision to do treatment in your own facility or through a contract treatment, storage, and disposal operator is not always an easy one. The choice of method will often force the choice between on- and off-site treatment. Few businesses can justify the very large investment needed to build and license a hazardous waste incinerator. Many can afford the installation of cyanide-destruction and heavy metal recovery systems needed to support a metal-finishing plant. The kinds of "treatment" steps that can and should be incorporated into the manufacturing process itself may be so advantageous that the decision is made for you.

An example of this kind of thinking comes from a process which uses a large amount of a chlorinated solvent which, in "spent" or used form, is a listed waste, that is, subject to regulation as hazardous just because it is on the list, not because of any hazardous properties resulting from its use in the process itself. It was economical for many years to recover such used solvents, accumulate them in a spent solvent tank, and either distill or otherwise purify them when a sufficient batch amount had accumulated, or to ship the accumulated truck- or carload to a reprocessor who would return fresh solvent to specification.

Today, depending on many factors, including how long the spent solvent is allowed to accumulate, this approach can require obtaining an RCRA Permit for a Treatment, Storage, and Disposal Facility, a most difficult and expensive thing to do.

Why not incorporate solvent recovery into the process unit itself? Often, a small solvent-recovery unit that would have been too expensive as an addition to a process when recovery costs were cheaper can be a very economical means of reducing this hazardous waste stream to no more than the small residue from purification or distillation—from perhaps hundreds of pounds per month to a few pounds of residue that can be managed under the simpler requirements which apply to small quantity generators under the regulations. Possibly, the residue can be a solid material that is easier to handle and ship or even to incinerate than is the liquid or tarry stillbottom material.

We could continue with other examples of the best approach, but the principle we want to express is clear: treatment of hazardous wastes is good public policy and can often be good practice for a generator. Recent changes in the regulatory world have altered the economic balance forever in ways that may make in-process and on-site treatment pay off in lowered indirect costs of compliance with the regulations when their direct payout may not be immediately attractive.

"Delisting" of Wastes

Many wastes appear on lists of hazardous wastes which are published in Federal and State regulations.

Often the listing is based entirely on the origin of the waste:

Waste F019 Wastewater treatment sludges from the chemical conversion coating of aluminum

Waste K062 Spent pickle liquor from steel finishing operations

The regulations provide for a generator's right to file a petition to have a particular waste stream at a particular facility excluded from the definition of "hazardous waste" even when that waste is on the list of such wastes under RCRA. Recent pressure from Congress and the public-interest groups has made the EPA less interested in pursuing this process. To be successful, the petitioner must somehow convince the EPA that his waste does not fit any of the criteria which were used in deciding to add the material to the list in the first place. As is true in so many cases, the unambiguous proof of a negative conclusion is very difficult or impossible.

Early in the implementation of RCRA, a significant number of delisting petitions were successful, but under today's conditions, this approach to dealing with listed wastes is not likely to work. Some waste-treatment process vendors still offer the possibility, and it may be that some will indeed be useful in this way. It is increasingly hard to prove that there is not some amount of the original toxic component present after treatment, and the regulators have a hard time approving a delisting petition which has any ambiguity on this point.

Characteristic or Unlisted Wastes

Waste materials which are classified as hazardous because of their characteristics of ignitability, reactivity, corrosivity, or extraction-procedure toxicity are in some ways better subjects for treatment than are the listed wastes. If a method can be found to eliminate the characteristic which causes the material to be classified hazardous, then the waste is ordinarily no longer required to be managed as a hazardous waste. Adequate and continued testing to ensure that the hazardous characteristic(s) do not return because of some failure of the process is still needed for such materials, but the method of disposal is limited only by the local situation on disposal of solid wastes—which may still present some substantial limitations.

In thinking about the management of these wastes by treatment, it is essential that we keep in mind the fact that "solid waste" under the hazardous waste rules includes liquid and even gaseous wastes, as discussed in some detail in Chapter 4, "How to Recognize a Hazardous Waste." Since this opens up the possibility of using wastewater treatment methods for some aqueous liquids and even some gases which might be captured by a water scrubber, etc., it is important to evaluate all waste streams in a plant or operation, whether they are obviously solid hazardous waste or not.

Characteristic wastes also present the possibility of a plant generating a class of wastes which John Sabatino, a consultant in this field, calls "nonobvious hazardous wastes." Usually, he says, these wastes are produced as by-products of a manufacturing process or from the handling or even from treating of hazardous wastes. Examples are found in the fact that heavy metals like lead and cadmium are often present in steel alloys and turn up in levels defined as toxic in cutting oils, lubricating oils, or even stillbottoms from recovery of degreasing solvents. Often these can be treated to eliminate the hazardous characteristic, but not unless they are identified through testing or through a careful review of all the materials which are processed or handled at the plant. The Manufacturer's Safety Data Sheets (MSDS)—required by the Occupational Safety and Health Administration (OSHA)—are the best source of data for these reviews.

TREATMENT METHODS AND PROCESS TECHNOLOGIES

Given today's interest in treatment technologies, there are dozens, perhaps hundreds of processes offered for the purpose by many different engineering firms, equipment suppliers, and technology-marketing organizations. It is not possible here to cover this rapidly changing field in detail. Many of the processes are proprietary or require specialized equipment which is available from a limited number of suppliers or at only a few locations, often only one. Some have been "approved" by the regulatory agencies in the sense that one or more systems at particular locations for treating specific wastes have been licensed. In general, while the granting of a permit at one location for one or more wastes is a help in obtaining permits for other applications, it is not a substitute for the often-difficult and always-expensive process of obtaining necessary TSDF permits, which, particularly for incinerators and combustion processes, can take years to accomplish, and then only after the commitment of the investment.

We are seeing the beginnings of a policy for the EPA to grant permits for certain mobile treatment systems for use all over the country or in a region. However, the States may or may not be willing to accept this shortcut; so claims should be carefully evaluated.

To try to bring some order into a discussion of this complex set of processes, we will discuss the general principles of waste treatment beginning with separation; then, destructive systems based mostly on thermal processing; then, methods of applying chemical change to reduce or eliminate toxicity; and finally, methods of reducing or preventing migration from disposal sites. Nearly all existing practical treatment methods can be encompassed in these broad categories. Many processes apply more than one of these approaches, but the principles still will stand.

When appropriate, we will go into some of the newer technological approaches based on biological processes. Some of these are of as much concern to the regulators and public-interest organizations as the original hazardous materials themselves, and this slows their adoption.

Separation Processes

Separation techniques have many applications in the treatment of hazardous waste. These processes may be both physical and chemical in nature, but they share a common base: the removal of a toxic material from a larger quantity of relatively inert material. This is usually accomplished by creating a much smaller waste stream that is highly concentrated in the toxic or hazardous substance(s) while leaving behind a waste stream that can be handled by routine disposal methods.

The ban on landfill disposal of "free liquids" applies to any solid waste which, when subjected to gravity drainage through a so-called "paint filter" in an EPA test method, produces a liquid fraction that is greater than 5% of the volume of the original crude material. Physical separation of the liquid by screening or pressing could make the dry matter eligible for landfilling or nonhazardous incineration while the liquid might be recycled or disposed of through other channels.

The reverse of this is not an acceptable practice. Adding solid matter to liquid wastes to soak up the liquid became a common practice, but the EPA has noted that this "kitty-litter" system often produces a wet solid that decays or simply has its liquid burden squeezed out under the pressures inside a landfill, which defeats the purpose. Some retailers offer boxes filled with an absorbent material (probably clay) into which they suggest that the consumer pour used engine oils and fluids before disposing of the whole thing in the ordinary garbage. The acceptability of this practice is questionable at best under the rules for handling used oil. Perhaps this system will actually bind the used oils permanently so that they do not reappear in the leachate from a sanitary landfill, but it seems unlikely that it can do so in the presence of the water and weak acids commonly found inside a landfill. Take it to an auto service station for disposal. Anyone who sells engine oil is required by law to provide a disposal station for consumers. This requirement may be met by posting a prominent sign directing the consumer to the nearest approved disposal site.

Under the new congressional ban on the disposal of even nonhazardous liquids in landfills, it is now necessary to prove that any liquid fraction will not reappear under disposal conditions, as we will discuss a little later.

Separation of hazardous substances from nonhazardous wastes should occur at the source and is regarded as a waste-minimization technique. By far the most common applications of separation systems are found in processes which produce a waste stream that is water-based and too concentrated in some hazardous pollutant for permitted disposal as a discharge to public waters or to publicly owned sewage systems. All public sewer systems have limits on the amounts and concentration of toxic pollutants and those which can harm the sewers or the treatment plants. On the other hand, any water pollutant which is discharged in accordance with a discharge permit issued under the Clean Water Act is exempt from simultaneous regulation *in the wastewater stream* as a hazardous waste. The most desirable way to handle aqueous waste streams under these rules is through pretreatment that reduces the concentration or amount of the limited pollutant to a level which will allow the wastewater system to accomplish the major part of the waste treatment task. If a concentrated form of the excess pollutant can be produced for separate destruction or treatment, then the system will be balanced and effective.

If the wastewater stream is only slightly above the maximum permitted concentration levels, one is tempted to dilute the stream with fresh water to make it acceptable. This is not only bad practice that will raise your wastewater charges, but is unlawful and can result in severe penalties since it is a violation of the Clean Water Act and usually of the State and local laws and rules. Dilution is not the solution to pollution!

For water-based waste streams, the usual separation processes are evaporation, filtration, or settling of the stream to produce a clear effluent for treatment or for sewering. The evaporation process is always called distillation when organic solvents are the carrier and the solids must be removed.

Many kinds of equipment are involved, particularly evaporators and stills with their auxiliary units. Filters, centrifuges, and various kinds of presses are also common. In special cases where high costs can be borne because of unique hazards, like radioactive wastes, extraction of a hazardous chemical with another solvent may be justified. Electroplaters use ion exchange systems, with chemical cycles comparable to those in home water softeners, to collect and concentrate precious or hazardous metals for recovery, often in an off-site service. The precious metals industry gained an exception to the rules for classifying hazardous wastes by pointing out the fallacy of labeling a drum of ion exchange resin loaded with gold and worth many thousands of dollars a "waste" at all.

All these methods are expensive to install and to operate. An infinite variety of combinations of process and equipment can be put together by engineers who are knowledgeable in the field. Qualified technical support is essential to the selection of separation techniques and technologies and to interpret the claims of suppliers and vendors.

Destructive Treatment

The ideal solution to all problems of disposal of hazardous wastes, assuming recycling and reuse has already been maximized, would be found in treatment methods which completely "destroy" the waste. No one will seriously argue against this ideal. The problem is with the idea represented by the word "destroy."

It is no mere exercise in semantics or in legal hair-splitting to say that in any technical sense, no system can be devised which will "destroy" hazardous waste, or for that matter, any other physical substance. Reactions in the stars and the related ones in nuclear reactors and weapons on earth can convert small amounts of physical matter into energy, but these are the only exceptions to the basic Law of Conservation of Matter.

Therefore, by "destruction" of wastes, we really mean applying some treatment which can convert the waste into other physical forms which we find safer or simply more desirable in our environment. More specifically, we arrange matters so that the chemical substances which make up the waste (or anything else, for that matter) are changed into other chemical substances which are more acceptable or whose effects on the environment we judge to be more desirable.

For a more detailed discussion of this principle and of the limits on the chemistry of destructive processes, see Chapter 11, "Chemistry of Waste Materials."

While there are many methods of bringing about the chemical changes needed to accomplish the destruction of hazardous wastes, in recent years destructive methods have tended to be identified with processes which apply energy to the wastes from outside sources or by causing combustion-type reactions which supply the energy to bring about molecular rearrangement. Incineration, pyrolysis, and high-pressure oxidation, in their many forms, are the subjects of this part of this discussion. Reactions with other chemical substances (oxidizers other than oxygen gas, chlorine, water, and many others) are encountered in the later discussion of chemical methods of treatment.

Incineration. Chapter 8 on incineration provides an introduction to this fascinating technology and some of the problems one encounters in using it for destruction of wastes, both those formally classed as hazardous and those that are not.

Here, we will simply point out that incineration, while rapidly becoming the most popular technology for destructive treatment of hazardous wastes, is not yet proven for many applications. Its rapid expansion is occurring in the face of widespread public uncertainty. These concerns are exacerbated by continuing public disagreement among "experts" resulting from widely varying interpretations of the very limited performance data which have been obtained on large-scale incinerators.

This uncertainty and reluctance to commit to incineration as a technology of choice for destructive treatment of hazardous wastes is reflected in the long, expensive, and risky business of obtaining all the permits required for operation of an incinerator system. This permit process is difficult for handling municipal wastes and much, much harder when the wastes are regulated under RCRA.

The problem for the public-policy and regulatory agencies is that the precipitate action of Congress in banning or restricting landfill disposals was based on the assumption that alternate means of treatment, mostly by incineration, were readily available and only awaited more attention. After all, European countries have been committed to incineration for a long time.

So, the debate goes on, but gradually more and more incinerators are coming on line. From the standpoint of a generator with wastes to get rid of, having them incinerated is about the only way you can truly be assured that they will not return from the "grave" end of the "cradle-to grave" management system. It seems today that any generator who can arrange for reasonably economic treatment of his wastes by incineration is well advised to do so.

Pyrolysis. Some waste materials are so lacking in energy content that can be used to sustain the combustion process in an incinerator that heat must be added from outside the system to accomplish the disruption of the molecular structure of the waste. In most high-temperature incinerators, like those for chlorinated materials, added fuels like natural gas or even selected wastes like old tires are added to ensure that the temperature is maintained high enough long enough to destroy the feed to a satisfactory degree. Pyrolysis is simply an extension of this approach until all the energy used for the destruction process is supplied from outside the reaction.

The variations on pyrolysis are almost endless. Electrical energy is popular, since it adds no chemicals to the mix being broken down. Resistance heaters are common, sometimes with carbon electrodes, sometimes transferring the energy to the charge by radiation, sometimes by using molten salts or glass. The presence or absence of oxygen or air in the reaction chamber will greatly affect the form of the unburned residues and may define the limits of usefulness of any given system.

Pyrolysis of wastes which leave significant quantities of dangerous residues after all the organic materials are destroyed is often specially useful for highly toxic metals and particularly for radioactive substances when the residue is picked up in a molten glass or ceramic material in the reactor. When cooled, the glass can lock up a hazardous metal in a form that is resistant to leaching by groundwater so that it can be disposed of by long-term burial during which the radioactivity may decay to levels below those of concern.

Sterilization. A special case of thermal processing is found in the treatment of infectious and other medical and hospital wastes before incineration in open incinerators or landfilling with mixed municipal waste.

Autoclaving or other methods of applying moist heat are the methods of choice for this purpose. Untreated wastes should be packed and held for short times in red plastic bags sold for the purpose. Once the material has been treated to eliminate its infectious character, it can enter a more conventional disposal system. There are proposals to exempt infectious waste from classification as hazardous waste and to rely strictly on a specialized regulatory program for these materials. Recent incidents in which medical wastes have appeared in inappropriate places like beaches and ordinary garbage will no doubt result in some changes in the regulatory process for these materials. Incineration is the only really acceptable means of disposing of surgical wastes, and this capability must be maintained by the facilities providing surgical services.

Non-Thermal Destructive Methods. It has been known for a long time that microorganisms of many kinds are capable of changing the chemical character of chemical substances. Indeed, the whole basis of most wastewater treatment was, until comparatively recent times, the management of populations of microorganisms which feed on the waste or related substances in the water and break down the impurities until they are innocuous and acceptable for disposal into streams and other water bodies.

By conditioning or acclimatizing the "bugs" to live in the presence of the waste material, even highly toxic wastes can be handled in biological systems. The sludge in a wastewater treatment system is largely composed of the living and dead bodies of the organisms which consume the waste. Today, much chemical waste is treated in activated sludge systems, and the extension of this idea to trickling filters, where the organisms are growing on beds of crushed stone or on the surfaces of lattice-like boxes of plastic or wood is well established. The regulations affecting effluent discharges under the Clean Water Act generally prohibit the placing of materials into sewer systems that will damage the sewers, injure the microorganisms in the treatment plant, or pass through the treatment plant in dangerous amounts into the receiving water. This broad requirement, in addition to the rules of the particular sewage treatment authority, means that great care must be taken to comply with the regulations and usually that a consultation with the treatment system operator's staff is necessary before making a decision to use a wastewater treatment plant as a destination for chemical wastes.

Generally, the use of various forms of biological treatment systems is limited by the ability of the organisms to survive and reproduce in the presence of the toxic substances which are to be destroyed. In liquid waste situations, they can be remarkably effective, although the costs can be high because of the large-scale equipment and space needed.

Biotreatment of Soils. Use of microflora to digest toxic and hazardous materials in soil and other solid media is rapidly developing into a new technology and holds much promise for cleaning up contaminated soils at spills and other Superfund-type cleanups. The use of genetically engineered organisms is generally believed to have much promise for extending the usefulness of the technique, but the prevailing fear of releasing these "unnatural" organisms into the environment is slowing the pace of the development. The natural extension of these developments into the treatment of solid wastes in composting and other similar solids-processing will no doubt open new opportunities in the future.

A promising new nonthermal destructive treatment has gained impetus from the problems of cleaning up chlorinated dioxins in the environment, such as arose in Times Beach, Missouri. At least some of the dioxins of concern are able to be degraded to nontoxic substances when exposed to ultraviolet light. The rate of "photodegradation" is generally proportional to the intensity of the light. Many organics have this same sensitivity, and work is being pursued to find ways of using this property effectively. Light obviously does not penetrate deeply into solids, and ultraviolet is strongly absorbed by thin films of many liquids, so the problem of arranging a treatment system to expose a solid or bulk liquid to the light uniformly to get complete degradation of the dioxin or other toxic is a difficult one, but not an impossible one if the process is effective.

Chemical Methods of Treatment

Treatment of hazardous wastes with chemicals to alter their chemical behavior and effectively eliminate toxic properties and change their needs for disposal is a highly developed and sophisticated field of work. We will consider some of the more common examples, but it is not possible to do more than introduce some of the possibilities. At the outset, we need to realize that chemical treatment can introduce new possibilities of creating toxic and hazardous properties along with any new chemicals added to the mix. This work should be undertaken only by skilled and highly qualified technical personnel who are trained in the field.

Neutralization. The addition of alkalies to acid wastes and the less common addition of acids to basic substances is a very old method of treatment. The easily observed damage to materials and organisms when they are exposed to corrosive wastes clearly demands that these streams be made less damaging by neutralization. Most sewer systems limit the discharge of acid and bases to a range of about pH 3 or 4 to pH 8 or 9. The definition of hazardous waste is based on corrosivity and covers materials of pH 2 or less and of pH 12.5 or more. The meaning of the term pH is discussed in Chapter 11, "Chemistry of Waste Materials," and in most standard chemistry texts.

When dealing with such wastes, ordinarily simple mixing of the required amount of either acid or base to bring the pH within the acceptable range is all that is needed. Formerly, many acid wastes were dumped into pits filled with limestone for an inexpensive neutralizer, but this practice is subject to regulation under RCRA and is not likely to be allowed in most areas.

Oxidation-Reduction Treatment. An example of the application of oxidation-reduction chemistry to waste treatment is found in the common method of destroying cyanide in dilute wastewater by the addition of controlled amounts of chlorine or hypochlorite solution (similar to household bleach) into the stream and mixing it before discharge. The hypochlorite acts as an oxidizing agent, and the cyanide is changed to harmless nitrogen gas and water.

Similar results can be obtained using, for example, potassium permanganate to oxidize trace quantities of impurities, followed by a filtration step to remove the black manganese oxide formed in the reaction. This is a common way of removing ferrous or "clear water" iron from solutions, but must be closely supervised to avoid the problems that an excess of permanganate solution can cause.

Hydrolysis. Many organics are readily degraded by hydrolysis—reacting with water in solution or even in damp soils and other solids. The usual requirements for promoting hydrolysis are a controlled pH and a proper temperature range. This reaction is the reverse of the salt-forming reaction between acids and bases but is usually possible only for salts of relatively weak acids, particularly the organic acids.

Miscellaneous Chemical Treatments. As we said at the outset, it is impossible even to list very many of the enormous variety of chemical treatments available for the detoxification or other modification of hazardous substances. Chlorinolysis is the reaction of a substance with chlorine gas under conditions which promote such complete reaction that only inert gases and inorganic salts remain. Oxidation of wastes with hydrogen peroxide has become feasible with the improved shipping and handling technologies for that agent now available. In a few cases, hydrogenation of wastes, reacting with hydrogen gas at high pressure, can successfully detoxify materials and drastically affect their reactivity—one of the characteristics of a hazardous waste.

The combination of chemical treatments with physical separations often is necessary to achieve the removal of the hazardous characteristic from a waste stream. Thus, neutralization of a metal-containing stream can result in forming a solid precipitate of the offensive metal, which can be made to separate out of the water with the help of other agents which form coagulates and easily filtered solids which will carry the metal solids with them when they are removed from the liquid.

Reduction of the Potential of Wastes to Migrate

The most troublesome property of hazardous wastes and other related substances is that they simply will not stay where they are put. Even if toxic materials are inherently dangerous, they can't hurt people or things unless they escape confinement to create an exposure of some sort. Thus goes the argument for developing ways for limiting or eliminating the mobility of hazardous wastes after disposal. The basic acceptability of the concept is shown by the EPA approach to the issue of banning substances from landfills: if those who wish to use landfills can show that the material "will not migrate from the site," then the disposal can be allowed. The period during which there will be no migration is not clear, but is by implication infinite. When asked, the question is often answered as 10,000 years or some other equivalent of forever.

Although the regulatory approach to landfills still describes methods of construction and operation that are intended to prevent migration or at least to identify it so that remedial actions can be taken, it is clear that no method of landfill construction alone will suffice to meet the targets set by Congress and its public-interest preceptors. Thus, the same regulations allow for landfills to be used after treatment of the waste—which we have discussed earlier in this chapter. Most of the treatment methods already discussed are focused on the reduction of the amount of a solid waste or the elimination of its hazardous nature. Processing the material in a way that secures it against leaching by groundwater or degradation by natural exposure to living organisms or chemical reactions in the soil can offer an attractive alternative, allowing for inexpensive burial or even reuse under the right technical and economic conditions.

We find two distinct and different applications of the technology of "fixation" of hazardous wastes—the generic term for converting these wastes into permanent nonmigrating substances. The remediation of surface impoundments, ponds, and lagoons has tended to attract the most attention from the standpoint of technical development work, but the application of fixation methods to certain kinds of newly produced wastes, particularly metal-containing and radioactive ones, is becoming more and more attractive as landfilling becomes more expensive. In the impoundments, usually the material is solidified and left in place after it is proved that additional leaching will not occur. In the new wastes the search is usually directed toward a "product" that has some use other than taking up space. A popular objective is the manufacture of something which can be used in construction or road building, activities which we cannot imagine ever stopping.

Sorbent Addition. We have already discussed the "kitty litter" idea for eliminating liquids from waste before disposal. The experience of using prepared clay minerals to soak up spilled oil for safety reasons in many businesses and homes led naturally to this way of managing the messy problems of liquid wastes. It is still popular; at least it is still being promoted for use by consumers.

Today, both the regulations and experience should make us wary of the simpler forms of solidification by admixture. We use the term "sorbent" to name the materials added, since some of them work by *ad*sorption and some by *ab*sorption. Adsorption is a process by which materials are held on the surface of an additive solid by chemical-type binding, which is often quite firm and hard to separate. Absorption is the word used to describe the simple holding of a liquid in the spaces between the particles of a solid which can often be squeezed out by mechanical action. The relative proportions of the two forms of binding is specific to both the solid and the liquid and may be affected by outside factors such as the presence of water or other solvents, acidity, and temperature.

Many solid materials can be used in this form of solidification. Even the EPA suggests that an expanded form of the mineral serpentine (trade-named "vermiculite") be used to ensure that spilled hazardous materials do not escape from a so-called "lab-pack" containing small amounts of regulated material for shipment.

One popular sorbent because of its presumed cheapness is fly ash from coal-burning power plants. Fly ash is specifically excluded from the definition of hazardous waste, but this can be a trap when planning to use it as a sorbent. Once mixed with a nonexempted waste, the entire mixture is subject to management as a hazardous waste unless it can be shown to have none of the defined characteristics. Fly ash can contain enough of some of the heavy metals to make it fail the Toxicity Characteristic Leaching Procedure Test, although this is rare. Still, better to be sure that the sorbent is not itself going to contribute to the toxic nature of the final mixture.

Before using any sorbent to solidify a liquid or partly liquid waste, check carefully with your regulatory agency to determine whether the method you propose is allowed. It may be necessary to go into a testing program on the materials to satisfy both the regulators and the disposal facility operator that the method is acceptable. A test program like this is usually pretty expensive and time consuming and should not be undertaken lightly.

Cement or Lime Solidification. In some cases, particularly where no real recovery facilities exist, water-based wastes, either in dilute form or concentrated by some other process, can simply be mixed with cement or lime to make a hard, resistant solid that will not leach toxic materials in the Toxicity Characteristic Leaching Procedure Test. Some candidate wastes for this kind of solidification might be solutions or slurries of heavy metals like arsenic, barium, lead, etc., which for some reason are not able to be reused and where the disposal sites available have such vulnerable groundwater that it is not good practice to place any leachable metal waste there. Where it is usable, the technology can be pretty simple, involving regular cement or mortar mixers that are commonly found in every community.

An appealing feature of this method is that at least some of the waste materials can be chemically integrated into the cement matrix and become extremely resistant to extraction and leaching.

Encapsulation. As applied to the immobilization of wastes, encapsulation is a term applied to the mixing of waste materials with asphaltic or organic polymers which coat the particles of a solid and hopefully prevent leaching of toxic substances from the mix. Like the related fixation processes we have discussed, this one does nothing about the basic toxicity of the waste itself. The use of organic materials may make the life of such an encapsulated system questionable.

Vitrification or "Glassification." In our discussion of thermal processing of wastes, we mentioned the fairly recent development of incineration and pyrolysis systems which use a molten glass or ceramic material to capture ash and unburned materials. The technique has been applied to many different kinds of waste, from radioactive materials where the organic components can be burned away to the destruction of PCB's in which the entire piece of electrical apparatus is destroyed leaving a number of metals and other inorganic solids to be removed from the kiln or reactor in solution in the glass or ceramic phase.

Interest in this method of fixation of wastes first developed in the nuclear energy program as a means of preparing long-lived isotopes for disposal. When the radioactive materials are dissolved in a glassy material that is of proper chemical composition to resist soil and water environments, it is believed that no significant migration will occur for very long times, perhaps approaching the scale of geologic time. When dealing with nuclear waste containing isotopes with half-lives of the order of thousands of years, such extraordinary steps can be justified.

It is less likely that the "ordinary" industrial wastes that are the subject of this book are candidates for such draconian measures for their disposal. However, if some of the newer methods of incineration and/or pyrolysis at extremely high temperatures prove feasible and desirable for handling our most difficult materials, it may be that liquid glass and ceramics are good heat transfer agents and good sinks for the ash. We would then reap additional benefits by having final residues that are highly resistant to leaching and really satisfactory for land disposal.

Recently, as the problems of long-term cleanup of the Department of Energy's nuclear fuel processing plants have become apparent, a bizarre-sounding application of vitrification has been suggested. The idea is that several electrodes would be buried in the ground where contamination is found. Very large electrical currents would be passed into and through the ground by energizing the electrodes. The result is expected to be a fusing of the inorganic soils into some ceramiclike mass which will bind the radioactive materials forever, like natural ores. When huge amounts of soil and rock and groundwater are contaminated by years of operation where the priority was on production, maybe such incredible efforts are needed. We should watch this development for signs of wasteful attempts to solve an enormous problem simplistically by the application of money without a liberal application of brains first.

SUMMARY AND CONCLUSIONS

The current structure of the hazardous waste regulations and laws make it necessary for every generator of regulated waste to evaluate the possibilities for treatment of any wastes that must be managed.

Many treatment possibilities can be found through simple analysis of operations and the wastes produced. Such analyses should be directed toward the minimization of waste production and the careful identification of any potential for eliminating it. After this step, simple ways of neutralizing, filtering, settling, or other physical processes can be considered.

Application of methods to "fix" or reduce the ability of wastes to migrate or of elaborate chemical reactions to reduce or eliminate toxicity are possible, but need careful research. Those doing the research must be well informed on the regulatory schemes so that the outcome of the work, if successful, can actually be used with minimum risk of unexpected consequences.

Treatment of hazardous waste is an essential kind of technology for the future and should be given appropriate support and attention in both the public and private sectors.

Chapter 10

Waste Management: Minimization and Recycling

The most popular and often the most elusive answer to the problem of hazardous waste is, "Don't make it!"

Earlier, we discussed some reasons that such an easy solution is not likely to be available to us in the real world. Production processes do not always perform at 100% efficiency, and the laws of chemical equilibrium set limits on the purity of reaction mixtures. These limitations force manufacturers to use purification methods that often result in unusable by-products or residuals that have no markets or even known uses.

Hazardous materials are necessary ingredients for many of our essential consumer products, like textiles, paints, fuels, pesticides, and more. Since humans are responsible for the storage and handling of these substances, there will always be spills, leaks, and accidents in the plant or on the road or railroad. If indeed, as we have assumed throughout this book, hazardous waste is any nonproduct material that has the characteristics of a hazardous waste or is on some official list, then we have a large universe, including, for example, cleanup materials from spills, to reckon with.

The basic tenet of the advocates of Waste Minimization and Recycling is an essentially reasonable one: it's better to avoid making waste than to develop methods of disposing of it after it is produced.

An allied and equally valid belief maintains that if wastes can be utilized by recovering their useful components or by finding a way to employ them directly, two good things can happen: new raw material is not needed to the extent that recycled matter is suitable and the reuse naturally reduces the amount of waste which can go only to some disposal facility.

In this chapter we will examine these two major elements of the whole universe of waste generation management and disposal.

WASTE MINIMIZATION

From the beginning of Federal attempts to regulate solid and hazardous wastes, the language of the 1976 Resource Conservation and Recovery Act (RCRA) called for major emphasis on reduction of wastes, but little was done at government levels to forward the objective. In many ways the legislative action was regarded as a pious fraud like the call in the Clean Water Act for the elimination of all discharges of wastewater.

In the 1984 Amendments to RCRA, the Congress at least spoke clearly on this subject:

> The Congress hereby declares it to be the national policy of the United States that whenever feasible, the generation of hazardous waste is to be reduced or eliminated as expeditiously as possible. Waste nevertheless generated should be treated, stored, or disposed of so as to minimize the present and future threat to human health and the environment.

Following this policy statement by Congress, by

1985 the EPA began to require some action related to this subject. In its current Waste Minimization Strategy, the agency foresees positive, voluntary steps by all generators of hazardous wastes to demonstrate a commitment to reduce wastes. The program is given some life by the requirement that each signer of a Uniform Hazardous Waste Manifest has to certify that his organization has an actual plan which is intended to reduce or eliminate the generation of hazardous waste to the maximum extent possible, consistent with sound economic principles. If other mandatory measures can be devised, it seems likely that they will be.

Taken literally, the prevention and reduction of wastes is one of the essential responsibilities of anyone who has responsibility for the use of resources. This includes the ordinary citizen who tries not to throw away good food or who orders the kids to close the door so heat will not be wasted. This stewardship of the earth's resources is an imperative for the minimization of all kinds of wastes and in particular, of wastes that may be hazardous or at least not welcome in the environment.

In industry, accountability for the economic use of resources to produce the plant's products falls on the managers in charge of manufacturing. In any industrial company, the people responsible for producing the product have their job performance, raises in salary, and future promotions riding on their ability to make the product on time, within quality specifications, and within allowed costs. Naturally, this never-ending burden gets their best efforts, and any work that is not readily seen to be helpful in accomplishing the production goal is likely to receive a lower priority. In a typical manufacturing plant, lower-priority work does not get done very often! In short, if programs like waste minimization or energy conservation or other similar ones are to happen, it will be because the manufacturing managers come to understand that these activities are important and will be significant in evaluating their job performance.

In any organization, this kind of involvement will occur only if the top managers make a commitment to the program, whatever it is. Making a commitment like that in the face of all the pressures of a top management job is difficult. The decision has to be based on an understanding that in recent years many things have changed the traditional relationships of costs and public responsibility. Today's managers have to dispose of wastes safely or risk the long-term future of their organization.

The continuing development of the laws affecting solid waste and particularly hazardous waste management puts ever-increasing pressure on the manufacturing system. The sharp restrictions on land-disposal options are forcing the waste generator to use more expensive disposal systems such as incineration or chemical treatment to detoxify wastes. Executives not ordinarily involved with decision making about waste disposal are unlikely to be completely aware that disposal methods have been limited in a continuing series of public policy actions until few options remain, and that the cost of these options is not weighted very heavily in legislative decisions on prohibitions.

Finally, the old habit of turning wastes over to someone who would take them away never to be seen again has become a reckless practice under the "cradle-to-grave" concept of responsibility built into Superfund and its related statutes, as we have discussed earlier in this book. The responsibility for wastes and their effects on the environment cannot be contracted away. Therefore, waste haulers, disposal contractors, and outside recyclers are required to meet high standards themselves and must charge higher prices for their services. A business operator must be aware that the future of the whole company can carelessly be turned over to a "garbage man" whose failure, if any, can result in enormous costs that appear out of nowhere. The nature of these risks can be measured by the extreme reluctance of the insurance industry to assume the risk at any price. The only real protection must come from the company by its own action, monitoring the behavior of contractors and forcing corrective steps when required.

Thus, there is plenty of incentive to find ways to produce less waste and to reduce the hazard associated with those that cannot be reduced further or eliminated. In a 1986 EPA Report to Congress, some of the incentives are listed:

1. Increases in the price of all forms of hazardous waste management

2. Difficulty in finding sites for new waste management capacity

3. Permitting burdens and corrective action requirements

4. Financial liability of hazardous waste generators

5. Sharp increases in the cost of liability insurance with a near-total lack of availability of useful coverage

6. Public pressure to reduce waste production

The determination of Congress (and many State and local officials) to reduce greatly or eliminate landfill disposal, no matter how sophisticated the design of the facility or how well operated, is not to be underestimated. Between November 1986 and May 1990, the EPA was required by law to make decisions on the disposability of all hazardous wastes in and on the land. If the agency could not make a case that land disposal is more protective of "human health and the environment" than any other method of treatment or disposal *regardless of cost*, the particular waste must be prohibited from the land. Many already are, since Congress has legislated arbitrary rules that became effective in 1990, banning many wastes from land disposal.

There seems to be a developing consensus that the various options for management of hazardous waste can be ordered in a hierarchy in which reduction of waste volume at the source is always preferred, followed in decreasing order of preference by recycling, then by treatment to eliminate toxicity, and finally by land disposal. There are some real problems involved in using such a ranking for decision making, particularly the problem of deciding when, for example, all possible waste reduction has been achieved before moving on into treatment.

Before beginning a brief discussion of how a waste minimization effort can be done in real operating facilities, it is useful to review the starting points of the national programs. The best information, at least the most quoted, is from the Congressional Budget Office Report of May, 1985, *Hazardous Waste Management: Recent Changes and Policy Alternatives*.

In Table 6, taken from the CBO report, the twelve largest industrial waste-generating sectors are shown. The quantities of waste on which the rankings were based are supposedly derived from estimates of wastes classified hazardous according to the "characteristic" tests for ignitability, corrosivity, reactivity, toxicity, etc., discussed in Chapter 4, "How to Recognize a Hazardous Waste."

Clearly, from the data in Table 6, which is unlikely to change very quickly because it is based on percentages of the total waste stream, the chemical process industries are the major contributors to the total output of hazardous waste. This is no surprise, but it does put forth the factual base behind our intuitive judgment that hazardous waste and "chemical waste" are pretty well identical concepts.

Table 6

Estimated National Generation of Industrial Hazardous Wastes
Ranked by Major Industry Group[a] (In thousands of metric tons)

Major Industry	Estimated Quantity in 1983	Percent of Total
Chemicals and Allied Products	127,245	47.9
Primary Metals	47,704	18.0
Petroleum and Coal Products	31,358	11.8
Fabricated Metal Products	25,364	9.6
Rubber and Plastic Products	14,600	5.5
Miscellaneous Manufacturing	5,614	2.1
Nonelectrical Machinery	4,859	1.8
Transportation Equipment	2,977	1.1
Motor Freight Transportation	2,160	0.8
Electrical and Electronic Machinery	1,929	0.7
Wood Preserving	1,739	0.7
Drum Reconditioners	45	b
Total	**265,595**	**100.0**

[a] See CBO. *Empirical Analysis*. Table 1 for the master list of specific industry types that are aggregated into the major industry groups presented in this table.

[b] Less than one-tenth of one percent.

Source: Congressional Budget Office

Based on some of these data, the Report from the Congressional Budget Office also made some estimates of the potential for waste reduction on a national scale. These estimates are summarized in Table 7. This table probably had great impact on the congressional decision to require action on waste reduction and minimization in the 1986 and 1987 period. The figures are, however, pretty rough and based on some elaborately constructed methods of extrapolation. Nonetheless, the conclusion is real—waste reduction can make a significant impact on the national

Table 7

Estimated Changes in Waste Generation Patterns, 1983 and 1990, Ranked by Waste Quantity, Under Alternative Cases (In thousands of metric tons)

Waste Type	Quantity in 1983[a]	1990 Quantity With No Waste Reduction[b]	1990 Quantity With Waste Reduction[c]	Percent Change in 1990 Due to Waste Reduction
Nonmetallic Inorganic Liquids	82,261	89,908	71,705	-12.8
Nonmetallic Inorganic Sludge	28,061	28,177	26,768	-4.6
Nonmetallic Inorganic Dusts	21,120	22,214	19,993	-5.3
Metal-Containing Liquids	19,760	19,983	5,995	-69.7
Miscellaneous Wastes	15,415	16,759	15,921	+3.3
Metal-Containing Sludge	14,497	15,147	13,632	-6.0
Waste Oils	14,249	13,932	11,842	-16.9
Nonhalogenated Solvents	12,130	11,869	10,682	-11.9
Halogenated Organic Solids	9,784	11,558	11,558	-18.1
Metallic Dusts and Shavings	7,733	8,112	6,895	-10.8
Cyanide and Metal Liquids	7,383	7,284	1,821	-75.3
Contaminated Clay, Soil, Sand	5,461	5,748	5,748	+5.3
Nonhalogenated Organic Solids	4,578	5,227	5,227	-14.2
Dye and Paint Sludge	4,236	4,112	3,086	-27.1
Resins, Latex, and Monomer	4,018	4,586	4,357	-8.4
Oily Sludge	3,734	3,556	3,200	-14.3
Halogenated Solvents	3,479	3,803	3,423	-1.6
Other Organic Liquids	3,435	3,755	2,817	-18.0
Nonhalogenated Organic Sludge	2,242	2,478	2,478	+10.5
Explosives	720	821	780	+8.3
Halogenated Organic Sludge	715	683	683	-4.5
Cyanide and Metal Sludge	557	593	505	-9.3
Pesticides, Herbicides	26	28	24	-7.7
Polychlorinated Biphenyls	1	1	1	0.0
Total[d]	**265,595**	**280,364**	**229,141**	
Percent change from 1983		+5.6	-13.7	

a Mean estimate of waste generation under previous RCRA policies.
b Assumes no waste reduction efforts by industry in response to 1984 RCRA amendments. Projection of waste generation levels in 1990 based on growth in industrial output levels forecast by CBO's generation model. Decreases in waste quantities therefore result from declining levels of industrial (and waste-producing) activity.
c Assumes waste-reduction efforts by industry. Forecast includes waste growth from increases in industrial output identical to that of the higher 1990 waste-reduction case.
d Columns may not add to totals because of rounding.

Source: Congressional Budget Office

problem of management of hazardous wastes.

Having said this, it is also necessary to say that waste minimization is not an easy or cost-free activity. Even if the people given the task have top management support and all the tools they need to work out a program, there are significant barriers to the widespread achievement of meaningful levels of minimization of waste generation. Simplistic slogans will not help and may actually hinder progress by creating expectations that cannot be fulfilled even given a first-rate effort.

Capital investment funds needed to revise and modernize processes are not always readily available, even though savings may be made through the investments. This is a hard lesson learned from the energy conservation programs so heavily promoted since 1974, which have some common characteristics with the waste reduction/minimization programs under consideration or in place. Even if money and knowledge are available, process changes can affect product identity and quality. If there is an actual change or even the suspicion of a change in product resulting from waste-minimizing process changes, marketing departments often have enough concerns about these changes to slow or stop them.

There are technical problems which may not be overcome even with extensive research. Some products simply cannot be manufactured without producing some residuals which may be hazardous waste. In small companies, every owner, manager, and employee is usually working as hard as he or she can all the time. The time needed to study the waste problem and to acquire the engineering skills needed to make effective changes is very scarce—often so scarce that little can be done without outside assistance, which can introduce formidable costs.

In an effort to overcome at least some of the barriers to an effective program, many of the States are providing technical assistance to small businesses. These assistance programs are usually founded on an informational effort to alert all businesses to the potential for both cost savings and a safer, cleaner operation which can be achieved by better waste management, including reduction or minimization. Some programs provide technical specialists who can identify opportunities for savings in cost or quantity, but it is expected that the business will provide the technical and management talent needed to implement the opportunities or deal with the problems uncovered, even if that means hiring consulting engineers, etc., from outside to do the work. A list of some of the State assistance programs is shown in Table 8, taken from an EPA publication of October 1987.

Practical Programs for Waste Minimization

Every plant, every business for that matter, is unique in its pattern of generation and disposal of wastes. It is not possible in limited space to spell out a program which can be implemented in a cookbook fashion, but we can outline the major activities and goals of a program to reduce or eliminate both regular and hazardous waste in any operation.

State assistance personnel find that many small businesses have no idea that they are subject to the regulations, much less how they can comply. A program for any business must start with the owner or someone in authority learning at least the rudiments of the regulatory requirements. Chambers of Commerce and Manufacturers' Associations often have information available to their members and may provide seminars and workshops that can be very helpful. They can also put the businessperson in touch with the National Association of Manufacturers or the National Chamber for publications in the field. Specialized trade associations, notably that of the metal-finishers, usually can give assistance in this educational process.

A word of caution is in order here. The subject of hazardous waste regulation is incredibly complex as we have pointed out many times in this book. It is unlikely that the established helpers of businesses, the lawyers, consulting engineers, accountants, etc., can be relied on for sound advice in these matters. Still worse is the small business operator's habit of finding informal help from another business owner or even a family member. The penalties of a misjudgment are simply too severe, and usually the high costs of a specialist will be inevitable.

Any program for waste reduction or minimization must start with an audit of the operation to determine the kinds and quantities of wastes being produced and the methods being practiced for their management. Where costs are available, they should be recorded also. An audit of a large plant is complex and usually expensive and best done by qualified outsiders who can evaluate what they find objectively. For a smaller plant, it may be possible for the owner/operator personally to find answers to questions like these:

1. Do you know what kinds of waste you produce?
2. Do you use hazardous substances in your operations that can appear in your wastes?
3. At what point in your operation are wastes produced?
4. What quantities are produced—daily, monthly?
5. Where are wastes stored until treated, reused, or disposed of?

Table 8

State Waste Minimization Programs

Alabama

Hazardous Material Management and Resource Recovery Program
University of Alabama
P.O. Box 6373
Tuscaloosa, AL 35487-6373
(205) 348-8401

Alaska

Alaska Health Project
Waste Reduction Assistance Program
431 West Seventh Avenue
Anchorage, AK 99501
(907) 276-2864

Arkansas

Arkansas Industrial Development Commission
One State Capitol Mall
Little Rock, AR 72201
(501) 371-1370

California

Alternative Technology Section
Toxic Substances Control Division
California Department of Health Services
714/744 P Street
Sacramento, CA 94234-7320
(916) 322-5347

Connecticut

Connecticut Hazardous Waste Management Service
Suite 360, 900 Asylum Avenue
Hartford, CT 06105
(203) 244-2007

Connecticut Department of Economic Development
210 Washington Street
Hartford, CT 06106
(203) 566-7196

Georgia

Hazardous Waste Technical Assistance Program
Georgia Institute of Technology
Georgia Technical Research Institute
Environmental Health and Safety Division
O'Keefe Building, Room 027
Atlanta, GA 30332
(404) 894-3806

Environmental Protection Division
Georgia Department of Natural Resources
Floyd Towers East, Suite 1154
205 Butler Street
Atlanta, GA 30334
(404) 656-2833

Illinois

Hazardous Waste Research and Information Center
Illinois Department of Energy and Natural Resources
1808 Woodfield Drive
Savoy, IL 61874
(217) 333-8940

Industrial Waste Elimination Research Center
Pritzker Department of Environmental Engineering
Alumni Building, Room 102
Illinois Institute of Technology
3200 South Federal Street
Chicago, IL 60616
(312) 567-3535

Indiana

Environmental Management and Education Program
Young Graduate House, Room 120
Purdue University
West Lafayette, IN 47907
(317) 494-5036

Indiana Department of Environmental Management
Office of Technical Assistance
P.O. Box 6015
105 South Meridian Street
Indianapolis, IN 46206-6015
(317) 232-8172

Iowa

Iowa Department of Natural Resources
Air Quality and Solid Waste Protection Bureau
Wallace State Office Building
900 East Grand Avenue
Des Moines, IA 50319-0034
(515) 281-8690

Center for Industrial Research and Service
205 Engineering Annex
Iowa State University
Ames, IA 50011
(515) 294-3420

Kansas

Bureau of Waste Management
Department of Health and Environment
Forbes Field, Building 730
Topeka, KS 66620
(913) 296-1607

Kentucky

Division of Waste Management
Natural Resources and Environmental Protection Cabinet
18 Reilly Road
Frankfort, KY 40601
(502) 564-6716

Maryland

Maryland Hazardous Waste Facilities Siting Board
60 West Street, Suite 200A
Annapolis, MD 21401
(301) 974-3432

Maryland Environmental Service
2020 Industrial Drive
Annapolis, MD 21401
(301) 269-3291
(800) 492-9188 (in Maryland)

Massachusetts

Office of Safe Waste Management
Department of Environmental Management
100 Cambridge Street
Room 1904
Boston, MA 02202
(617) 727-3260

Source Reduction Program
Massachusetts Department of Environmental Quality Engineering
1 Winter Street
Boston, MA 02108
(617) 292-5982

Michigan

Resource Recovery Section
Department of Natural Resources
P.O. Box 30028
Lansing, MI 48909
(517) 373-0540

Minnesota

Minnesota Pollution Control Agency
Solid and Hazardous Waste Division
520 Lafayette Road
St. Paul, MN 55155
(612) 296-6300

Minnesota Technical Assistance Program
W-140 Boynton Health Service
University of Minnesota
Minneapolis, MN 55455
(612) 625-9677
(800) 247-0015 (in Minnesota)

Minnesota Waste Management Board
123 Thorson Center
7323 Fifty-Eighth Avenue North
Crystal, MN 55428
(612) 536-0816

Continued on next page

Table 8
(Continued)

State Waste Minimization Programs

Missouri

State Environmental Improvement and Energy Resources Authority
P.O. Box 744
Jefferson City, MO 65102
(314) 751-4919

New Jersey

New Jersey Hazardous Waste Facilities Siting Commission
Room 614, 28 West State Street
Trenton, NJ 08608
(609) 292-1459 or 292-1026

Hazardous Waste Advisement Program
Bureau of Regulation and Classification
New Jersey Department of Environmental Protection
401 East State Street
Trenton, NJ 08625
(609) 292-8341

Risk Reduction Unit
Office of Science and Research
New Jersey Department of Environmental Protection
40 East State Street
Trenton, NJ 08625
(609) 633-1378

New York

New York State Environmental Facilities Corporation
50 Wolf Road
Albany, NY 12205
(518) 457-4139

Division of Solid and Hazardous Waste
New York Department of Environmental Conservation
50 Wolf Road
Albany, NY 12233
(518) 457-3273

North Carolina

Pollution Prevention Pays Program
Department of Natural Resources and Community Development
P.O. Box 27687
512 North Salisbury Street
Raleigh, NC 27611
(919) 733-7015

Governor's Waste Management Board
325 North Salisbury Street
Raleigh, NC 27611
(919) 733-9020

Technical Assistance Unit
Solid and Hazardous Waste Management Branch
North Carolina Department of Human Resources
P.O. Box 2091
306 North Wilmington Street
Raleigh, NC 27602
(919) 733-2178

Ohio

Division of Solid and Hazardous Waste Management
Ohio Environmental Protection Agency
P.O. Box 1049
1800 WaterMark Drive
Columbus, OH 43266-1049
(614) 481-7200

Ohio Technology Transfer Organization
65 East State Street, Suite 200
Columbus, OH 43266-0330
(614) 466-4286

Oklahoma

Industrial Waste Elimination Program
Oklahoma State Department of Health
P.O. Box 53551
Oklahoma City, OK 73152
(405) 271-7353

Oregon

Oregon Hazardous Waste Reduction Program
Department of Environmental Quality
811 Southwest Sixth Avenue
Portland, OR 97204
(503) 229-5913

Pennsylvania

Pennsylvania Technical Assistance Program
501 F. Orvis Keller Building
University Park, PA 16802
(814) 865-0427

Bureau of Waste Management
Pennsylvania Department of Environmental Resources
P.O. Box 2063
Fulton Building, 3rd and Locust Streets
Harrisburg, PA 17120
(717) 787-6239

Center for Hazardous Materials Research
320 William Pitt Way
Pittsburgh, PA 15238
(412) 826-5320

Rhode Island

Ocean State Cleanup and Recycling Program
Rhode Island Department of Environmental Management
9 Hayes Street
Providence, RI 02908-5003
(401) 277-3434
(800) 253-2674 (in R.I.)

Center for Environmental Studies
Brown University
P.O. Box 1943
135 Angell Street
Providence, RI 02912
(401) 863-3449

Tennessee

Center for Industrial Services
Suite 401
226 Capitol Boulevard Building
University of Tennessee
Nashville, TN 37219-1804
(615) 242-2456

Virginia

Office of Policy and Planning
Virginia Department of Waste Management
11th Floor, Monroe Building
101 North 14th Street
Richmond, VA 23219
(804) 225-2667

Washington

Hazardous Waste Section
Mail Stop PV-11
Washington Department of Ecology
Olympia, WA 98504-8711
(206) 459-6322

Wisconsin

Bureau of Solid Waste Management
Wisconsin Department of Natural Resources
P.O. Box 7921
101 South Webster Street
Madison, WI 53707
(608) 266-2699

Wyoming

Solid Waste Management Program
Wyoming Department of Environmental Quality
Herschler Building, 4th Floor, West Wing
122 West 25th Street
Cheyenne, WY 82002
(307) 777-7752

Source: USEPA, October 1987

6. Do you have a waste hauler/disposer?

7. Do you have records of waste shipped? Manifests?

8. Where do your wastes go for final disposal?

9. How much do you spend on waste disposal?

Other questions will arise and should be answered by the audit. Some more of these are covered in Chapter 2, "A Businessman's Guide to Life under the Hazardous Waste Laws and Regulations."

In a large plant or multiplant operation having trained staff people, this work should have been done in the course of ensuring compliance with the State and Federal laws and regulations. Still, when considering a formal program of waste minimization, it will probably be worthwhile to review all the information afresh or even to make a completely new start. Recently imposed requirements under the Superfund Amendments and Reauthorization Act (SARA Title II) should have forced collection of data for those reports which will provide the basis for a new start in waste minimization.

For the small business, perhaps still in the stage of not knowing or believing that the plant is capable of producing hazardous waste or that the operation is subject to a massive regulatory program of the nature of the one we are discussing here, the first step still is the development of a basic knowledge that is a precursor to action. Once again, State technical assistance program personnel or trade association leaders can help you get over this hump. It is amazing, after all the publicity given to waste hazards over the past few years to hear small business people say, "You are being an alarmist, nobody would make a regulation like—whatever."

The truth is that auto body repair shops, small machine shops, dry cleaners, and countless others are subject to the rules and to the penalties, and that ignorance is no defense, particularly if groundwater or other "public" resources become contaminated. Table 9 lists some businesses typically operated as small independent units and shows some of the materials that they employ in their work that can contribute to waste that is subject to regulation as hazardous.

Table 9

Some Small Business Operations and Associated Hazardous Materials

Waste Type	Typical Businesses Producing
Acids/Bases	Boiler Cleaning, Machine Shop, Painting, Paint Stripping
Heavy Metals	Electroplating, Printing, Paint Making, Petroleum Bulk Storage
Paints, Inks, Organic Residuals	Metal Fabricators, Auto Body Repair, Construction, General Manufacturing, Printing
Pesticides, Herbicides	Agriculture, Lawn Service, Golf Courses, Exterminators
Solvents	Auto Repair, Dry Cleaner, Floor Stripping, Printing, Furniture Finishing and Refinishing
Used Oil	Service Stations, Truck Service, Engine and Compressor Service
PCB (polychlorinated biphenyl)	Any operation using transformers made before 1979

Once the answers to the audit questions (and others that will develop) are in hand, then specialized technical assistance is essential. Sometimes the entrepreneur will be so steeped in the technology that he or she will know alternative ways of doing things, but often someone is needed to determine whether the Emperor is in fact wearing clothes. Large companies will or should have in-house experts or can retain consultants and attorneys to advise the manufacturing staff on the judgments needed about the classification of wastes and to help ensure that existing and proposed handling and disposal complies with the regulations.

Small companies will also need to seek professional help, either from retained consultants or through governmental assistance programs. It makes little sense to literally bet the future of your business on any advice but the best that can be found. Federal laws, implemented by the States, impose financial penalties of many thousands of dollars for *each day* of a violation, and if a jury decides the violation is "willful," jail sentences are becoming routine.

Having identified, quantified, and classified all wastes from the operation, it is now possible to begin making judgments about the possibility and feasibility of reducing the amounts of waste produced. If the amounts cannot be reduced, perhaps ways can be found to change the character of the waste in-plant so that it is not subject to the most stringent levels of regulation.

Issues of plant supervision, housekeeping, and maintenance can be evaluated early in the process:

1. Are hazardous wastes being segregated at the point of production? The *mixture rule* requires that when a hazardous waste is mixed into a nonhazardous waste, the whole mixture must be subject to regulation as hazardous. Segregation also allows separate measurement of quantity for determination as to whether the plant is a small-quantity generator and therefore not subject to all the paperwork requirements of a larger generator.

2. Are leaks, small spills, etc., contaminating soil or cleanup materials like absorbent clays in a way that creates hazardous waste in the form of contaminated soil or even polluted water in the storm drains? Fixing small leaks can avoid big trouble if soil or water is contaminated.

3. How are empty containers of raw materials and supplies treated, stored, and disposed of? There are specific requirements for handling empty cans, drums, etc., after hazardous products like paints, solvents, pesticides, adhesives, and the like are removed.

4. Does boiler blowdown and steam condensate exhaust into storm drains? These streams often contain chemicals needed for boiler operation but are not allowed in surface water discharges without treatment.

These are merely examples of administrative and operational things that can be managed to reduce waste production and exposure of the plant to potential violations.

When it is established by the audit and ongoing supervision that the plant is running as cleanly as possible, then it may be time to look into the opportunities for reducing waste by changing basic processes. Altering basic manufacturing methods and processes to minimize waste is, as we have already discussed, a difficult, often traumatic, sometimes expensive venture. Examples range from the metal-products fabricator who can convert from painting small parts to a powder-coating operation that uses no solvents, to the massive conversion of the auto industry to water-based paints. The wastes that can be reduced by such a change may include cleanup materials, used filters, and an enormous reduction in air pollutants discharged, even after treatment with expensive end-of-system adsorbers and fume incinerators.

The most difficult barrier to process change is rarely technical, although there are plenty of technical problems. Not all technical barriers are solvable, at least within reasonable economic limits. People and attitudes always form the most intractable opposition, first to work being undertaken to improve manufacturing and even more to the installation of changed systems and equipment. Powerful incentives are needed to get the people involved and to maintain at least an open mind toward change for a good purpose.

Examples of this kind of program are to be found in the "3P" program of the 3M Company, where the name is derived from the slogan "Pollution Prevention Pays." The main elements of this program show the direct support of the top management and a major commitment of technical staff personnel at both plant and headquarters, all strongly promoted throughout the company by an excellent employee and public relations program. The 3M program has been in place for twenty years or more, and can cite substantial accomplishments. Savings in pollutants discharged, reduced use of raw materials, and direct monetary contribution to profits by avoiding costs that would have been incurred by older manufacturing methods have helped keep enthusiasm for the program high.

A number of other companies have similar programs, but the 3M approach has been widely recognized and admired. It is often cited as an example of the possibility of this kind of achievement in international environmental organizations promoting the concept of "Low and Non-Waste Technologies." The State of North Carolina has adopted the "3P" symbol in its programs of assistance to state industries.

A key feature of all successful 3P-type activities is their requirement that proposals for change be economically sound. Under the stimulus of good leadership to improve operations and to reduce impact on the public, company people can be taught to include in the economic analysis all the costs of a particular course of action.

In a company which understands the hidden (or delayed) costs of cleaning up leaking landfills, or worse, of cleaning up after careless and unlawful disposals, the managers can justify higher process costs or the costs of new investment to replace equipment that is still operating. Present and contingent future costs for waste disposal can tip the scales to the new systems. One of the most difficult attitudes to overcome is the working engineer's or plant supervisor's failure even to request funds for new equipment because he has already judged, "They'll never approve the project anyway." Strong leadership is needed to give these programs credibility among the working management.

In the smaller company, the problems may be more difficult. The changes may well affect the only significant operation in the business. The sense of ownership of the established methods and the fear of displacement of people by new methods reaches a peak in such operations. The probable need for downtime for the installation of new equipment raises a specter of loss of income, and sometimes union leaders see change as an opportunity for accomplishing some long-desired contract changes, also.

Fortunately, the original entrepreneurial spirit and leadership that started the business in the first place is often still involved. The owner/operator has to take the lead, using whatever help can be found to seek better methods. In smaller organizations, there is no bureaucracy to persuade, no corporate bean-counters to mollify, and the job can be as interesting as the original startup. Better yet, it can improve competitive strength and make money!

Waste minimization, then, can often produce major benefits to both company and society through reduction of quantity or toxicity or other unwanted characteristics of residuals. An effective program requires sincere leadership and demonstrated willingness to give personal recognition, investment funds, and adequate staff resources to understand the opportunities, design improved systems, install them, and stick with them until they become routine.

The main barriers to be overcome are the mind-sets of individuals:

1. "We've always done it *this* way."

2. "This is only an excuse to get around the labor contract."

3. "It's a waste of capital to replace good [that is, operable] equipment."

4. "We'll do it next year when things are better."

5. "We shouldn't spend stockholders' money without a return on investment until we are under court order."

6. Add your own, out of thousands available!

RECYCLING

Like waste minimization, recycling is in many ways the simplest concept in the often-bizarre world of waste management. In fact, while recycling has other applications, it can be a primary tool in waste minimization. We consider recycling separately here because the word itself is in danger of becoming almost like an all-purpose answer or even a mantra which, repeated often enough, will make it true. Public meetings are often deluged with demands that difficult waste issues be settled by mandating recycling so that the waste presumably will no longer exist.

There are many reasons why reuse of once-used materials wherever possible is a preferred way of doing things. The first and most-cited reason for recycling materials is that by reusing something like the metal in aluminum cans, we can often substantially reduce the need for new materials. Since new materials can be made only by the use of natural resources, it is true that the pressure on those resources will be reduced. The same logic applies to the recycling of auto bodies in steel mills. A few years ago, it was reported that the conversion of the telephone network to fiber optics cables with electronic switching would release enough copper from obsolete installations so that little or no new copper production would be needed in the United States until well into the twenty-first century. Whether strictly true or not, the idea makes sense, since recycled copper wire is clearly economically and technologically proven.

Recycling of all these metals (aluminum, copper, and iron) makes good sense because of a common characteristic. That characteristic is that the winning of pure metals from natural ores consumes huge amounts of energy and capital investment. Remelting and refining of used metal consumes much less energy than the original manufacturing process did. The main problem in these recycling industries is found in gathering, storing, and delivering the used cans, autos, and wire to the reprocessor. This barrier is overcome by the essentially free work of citizens preprocessing waste to separate the useful metals and put them back into the recovery system. It's still a good deal for the country, but the low-paid efforts of the public are needed to make aluminum can recycling feasible.

In similar fashion, we recycle newspaper for the Boy Scouts or the local church to benefit when they provide the service of collection and accumulation of a large enough quantity for sale to a paper broker. Some people regularly separate glass and metal from the household garbage, even when not required to do so by local ordinance.

Large businesses are built on this recycled material. Most consumer food cartons in which the food does not contact the carton itself (because of a liner or inner package) are made from recycled newspaper. Turn the carton of your favorite breakfast cereal inside out and notice the gray cardboard with occasional dark flecks which shows that not all the ink was removed in the recovery process. This particular recycling process and system saves more than a ton of wood from the forests for each ton of newspaper processed and avoids the use of many tons of water in the paper mill that must be treated and discharged, while also avoiding the consumption of a lot of energy along the way. Less energy means less air pollution, less coal to be mined, less acid mine drainage in streams, and so on and so on.

All the recycling systems so far discussed have a common way of doing things. Each material is made fresh, used by consumers of some kind, recollected, and converted again into useful products. This is called *external recycling* and has an infinity of examples. The major examples have another common feature: *none of the used materials is hazardous waste under most regulatory schemes.* External recycling of materials that are now classified as hazardous waste introduces great complications, not the least of which is the waste generator's continuing liability for the wastes even though delivered or even sold to a recycler.

Some of the most spectacular incidents in recent years, in which hazardous materials were released by explosions and fires and often spilled into surface waters or groundwater, have occurred on the premises of companies or even individuals who promised to dispose of wastes, presumably safely. Instead, they accepted both wastes and fees, kept both, and spent the money without doing safe disposal. Such incidents have created much of the public and congressional outrage about hazardous waste management.

Today, as discussed elsewhere in this book, waste generators are responsible for the effects of their material wherever it goes and must assure themselves that the recycler is doing the job within acceptable guidelines, including final disposal of residues that cannot be further used. Most companies use these outside services for solvent recovery, sending out used or mixed degreasing, extraction, or otherwise-contaminated solvents and receiving purified materials in exchange. Properly handled, this, too, is a good deal for the country, since waste solvents are a major problem in the environment.

Delivering wastes to outside reprocessors can often be the best solution to a waste management problem. In making this choice, great care must be taken to ensure that the recycler is of good reputation, is licensed for the work at hand, and has a history of good operations and good relations with regulatory agencies. Continual monitoring of the reprocessing operation is essential for the life of the arrangement.

An important variant of outside recycling has developed in recent years in the form of Waste Exchanges. These organizations are often put together by interested generators of waste products that may have some use to another company. An example might be a company that uses acid for cleaning or other use that leaves it contaminated but still of reasonable strength. Acid wastes like this are difficult to dispose of; so the generator places a notice in the regular listing published by a Waste Exchange, giving the characteristics of his waste, the quantities available, and any other information that might be helpful in deciding whether the material is useful to someone else. Some exchanges require that the generator be identified, but some permit "blind" listing so that the staff acts as initial intermediaries. No Waste Exchange takes custody or ownership of wastes being offered through its auspices, but it simply puts people in contact to work out their own deals. Costs of exchange operations may be borne by membership fees or through charges based on completed transactions.

The crucial function of a Waste Exchange is accomplished when another member or outside subscriber to the listings has an interest in the used acid cited as an example. Perhaps a strongly alkaline wastewater must be neutralized before discharge to a sewer system, or some similar use in which the acid purity is not a major consideration to a user. In any case, if the two parties can agree on price and other terms, the waste acid can be sold for useful purposes and is no longer a component of the waste stream.

Most areas of the country are served by one or more Waste Exchanges. Information on them is usually obtainable through solid waste regulatory offices, almost all of which encourage this form of recycling. Whatever form outside recyling may take, it is a major activity spawning new industries and becoming attractive to cities which are running out of landfill space. However, it is far outmatched in volume and economic value by in-plant recycling.

In-plant recycling, sometimes called "on-site," is practiced on a huge scale in many industries. Steel mills based on ore reduction often recycle large amounts of "mill scrap" that is made in the ordinary operation of the finishing sections of the mill. Open hearth furnaces, now almost extinct because of their excessive air pollution, worked best when about one-third of the charge for a new batch was scrap steel, usually mill scrap, while the remainder was raw iron, usually in liquid form.

This internal scrap comes from trimming ingots, blooms, or other rough or finished products along with off-specification products that cannot be sold without retreatment. Modern steel-making furnaces may not absolutely require scrap in the charge but often make it possible to recycle poor-quality scrap like chopped-up auto bodies that contain unwanted metals like aluminum and copper.

Many chemical processes incorporate recycling as a means of controlling and stabilizing a reaction or for maximizing the yield of a particular product as compared with others that might form in the reactor. When possible, such process systems allow rework of off-specification products without removing the material from the stream.

There are few if any statistics on the amount of potential hazardous waste streams that is converted into useful or salable product in the chemical industry. The definitions of hazardous wastes do not apply to these in-process materials until they are abandoned or released into the environment. This possibility is greatly reduced by in-plant recycling. The materials are regulated, however, under provisions of the Toxic Substances Control Act and other related statutes.

A proper waste stream and process audit as was discussed earlier in this chapter can turn up opportunities for on-site recycling that can be of substantial value in waste reduction/minimization programs.

The Limits to Recycling

Enthusiasts sometimes get carried away with the idea that everything unwanted can be recycled for some use. If we accept burning with heat recovery as a valid form of recycling, the idea gets nearer to reality. Industries often overlook real opportunities and stonewall the activists. In neither view is the confrontational approach realistic or productive.

There are many recycling opportunities like the aluminum can, auto body, and copper wire examples that, along with newspaper and glass, are obvious to a sensitized person. Ignoring their potential is downright irresponsible for the public and the government officials having jurisdiction. Such examples are exciting and a source of satisfaction and profit to many. They should be studied to see why they work.

In the case of the metals, markets for the recovered materials are already in existence, if the metals can be produced within objective standards of quality. Supplied by many producers of both new and recovered materials, the market is reasonably efficient in the economic sense; so prices can be determined in the market, not only by contract with a user.

Equally important, the public (through aluminum recycling appeals in national emergencies and a long-term follow-up on environmental grounds) accepts the real possibility of recycling an aluminum can, for example. When small financial incentives are added by deposit laws, a positive response is pretty well assured. It is interesting that despite several major attempts to develop public recycling of tin-coated steel cans for the recovery of tin, the good economic justification is not enough to gain public participation. These programs must achieve credibility before the public will get involved.

In the case of the other materials—steel, copper, newsprint, and even glass—similar patterns are found. The major need, a competitive market in which commodities can be sold, has been developed for each. Glass furnaces often work better with some scrap glass in the feed, and similar conditions prevail in the paper converter's operations.

Second, for most materials (glass still excluded), systems are in place to find the used material and to see that it gets to the recycler. Old auto bodies were involved in the commercial scrap-metal industry to only a limited degree for a long time. Under the stimulus of Lady Bird Johnson's drive for highway beautification, many states became acutely conscious of the miserable aesthetic impact of the growing auto graveyards all over the country and instituted systems designed to encourage the reclaiming of old cars. Many States now pay bounties on wrecked and abandoned cars, using funds provided by a fee on new car registrations.

Almost overnight as such things go, the electric minimill designed to process auto scrap emerged as a major force in the domestic steel industry. Many believe the older steel producers simply overlooked a fantastic opportunity which others found ways to exploit. In the beginning, the minimills were limited to the production of steel products for uses in which the chemical analysis of the steel is relatively unimportant, like concrete-reinforcing bars. This limitation is being overcome as the business becomes established and better control of extraneous metals in the finished products is being achieved.

Such exciting stories certainly make us want to increase recycling to the maximum. Regrettably, the convergence of institutional structures (bounties on old cars), technology (the minimill concept), and markets (scrap-based steel at a price competitive with foreign producers) rarely occurs.

Sometimes, markets can be created by forces having little to do with the waste itself. Incineration of municipal waste with heat recovery in the form of steam for heating buildings or for process uses, is an example. The technology of burning mixed municipal waste was improved and developed over many years in Europe, although the Europeans have notably been less concerned about air pollution than have the people, at least the vocal ones, in the U.S. The pressures on land-use created by dense population and the need for conserving farmland have altered the balance between available technology and other needs in many countries.

The market for municipal wastes has been artificially created by restrictions on landfill, aided by increasingly distant locations of the ones allowed. Politicians having the power to coerce waste collectors to dump at the incinerators set both policy and price for the waste—something less than zero. The hauler pays the incinerator to take his load, presumably a fee based on the cost of alternatives, but actually determined by straightforward political decisions on acceptable ways of raising revenues for all purposes. At the same time, the political process sets up the institutional structures (licensing of haulers, rates charged to households and other generators, foreclosing of alternative disposal opportunities) needed to ensure a steady supply of waste at a predictable cash flow.

From observation of a number of projects for the incineration of municipal waste, it appears to make little difference to the public bodies investing whether the waste is burned pretty much as received and produces steam at the incinerator site or is converted into a refuse-derived fuel (RDF) for burning in a power plant some distance away. Recovery of metals, glass, plastic, etc., while often used as additional justification for the project, has so far not been a spectacular success.

All this about incinerators for municipal waste brings up two points about this method of disposal. First, little protection is built into the collection system to exclude hazardous materials often thrown out by households but prohibited from incineration in industrial waste except under stringent precautions; and second, the problem of disposal of the final residual, the ash, remains an ambiguous one, but most such ash has been classified as hazardous.

Recycling of hazardous waste is subject to all the limitations we have been discussing. We are finding as time goes on that nearly all organic wastes can be destroyed by thermal processes of some kind—some pretty high-tech. Where the heat content is high enough, recycled energy can be recovered for use, but many of the most difficult wastes are hard-to-handle sludges and solids that may need a lot of help in the form of supplemental fuels to be burned at all. For a more detailed look at these factors, see Chapter 8 on incineration.

Aside from their use as a heat source, chemical wastes must be considered for recycling on an individual basis. A great many chemical wastes are themselves residuals from the separation and purification of marketable chemicals from complex reaction mixtures. In general, in the process of separation, just about every marketable product is removed, leaving a gunk that is unlikely to have many uses or much value. Indeed, a look at the listed hazardous wastes shows many such stillbottoms, sludges, tars, and various other terms.

Stillbottoms are the residue left in the pot after all the good stuff is boiled out. It is not unlike what happens when the teakettle is boiled dry or somebody forgets the roasting pan in the oven when starting to preheat for another meal.

Sludges are usually thick semisolid materials that may contain all kinds of chemicals. One example is the sludge produced when making white mineral oil. White mineral oil has been found to be a necessary product, not only for the familiar use as a laxative, but as a lubricant for food and pharmaceutical machinery and as an ingredient in baby oils and many cosmetics. To prepare it, a special fraction of a selected crude oil is repeatedly brought into intimate contact with strong sulfuric acid.

The acid reacts with the unwanted parts of the raw oil, mostly unsaturated hydrocarbons and aromatic compounds, that contribute color and odor (the raw oil is straw-colored to light brown or even greenish). When the mixture is allowed to separate, the now-decolored oil floats and can be skimmed off the acid sludge (concentrated sulfuric acid is much more dense than light oil) for further neutralization, washing, and final treatment before packaging. The sludge that is left is a product that no one could love, dark in color, odoriferous from the sulfonated hydrocarbons dissolved in it, and yet it is commonly recyclable where the proper facilities exist.

Acid sludges (often made in other refining processes, too) can be burned in special burners and the resulting sulfur dioxide mixed with that prepared by burning fresh sulfur. The combustion process consumes all the organic part of the sludge, and the sulfur oxides are converted into fresh sulfuric acid for new uses. A sulfuric acid unit requires a massive investment and must have large throughput to be economical; so these kinds of sludge-burning plants are usually found only in areas in which large-scale petroleum refining is conducted.

Today, acid sludge disposal is not possible except through recycling, to any significant degree. This forces small-scale acid treaters to close down or move or to accept the risk of shipping the acid sludge to a recovery plant. Most would not consider any of these as a good option.

A principle derived from the study of thermodynamics states that when substances are mixed with each other, even if no chemical reaction occurs, the components of such mixtures can be separated only by the application of energy in the form of work. The amount of energy needed is characteristic of the substances themselves, but it cannot be zero. In theory, too, complete separation would require an infinite amount of energy! Thermodynamics simply describes in mathematical terms the dilemma that the would-be recycler encounters because essentially all wastes are mixtures of a few to hundreds of compounds.

By definition, solvents are used to dissolve something, which is a form of making a mixture. In most processes, the solvent eventually will dissolve all of the solute it can hold in solution. Usually, it must be kept at a small fraction of this "saturation" concentration to be useful. Recovering the solvent requires separating it from the solutes or even from other solvents which may have become part of the mix. This is the point at which the technology of recycling comes into the picture.

A local dry cleaner uses perchlorethylene or trichlorethylene to wash clothing, having been denied use of the traditional hydrocarbon solvents by regulations designed to reduce ozone pollution of the air. The solvent does the job—it dissolves and carries away the oils and fats that hold soil to cloth. After repeated use, eventually the solvent becomes so concentrated in the undesired fats, oils, and soil materials that, in accord with the laws of equilibrium discussed earlier, it leaves too much undesired soil on the clothes and has to be replaced. The cure for the problem is distillation of the solvent. In the simplest case, the dirty solvent is placed in a boiler and heated until vapor comes off like steam from the teakettle. The vapors are condensed on cool surfaces, and the solvent is ready to reuse. Sounds easy.

The basic picture is correct, but this is what really happens. In a small dry-cleaning operation, the "spent" solvent (spent is a word for exhausted or dirty) is exchanged with someone, usually a supplier, for fresh material. At a use level of about 250 gallons per day, some say less, it becomes financially practical for the solvent to be distilled in the shop. In either case, the cleaner is up against another of the basic principles of thermodynamics and of some facts of equilibrium chemistry and molecular stability that transform him from a dry cleaner into a producer of hazardous waste!

At any given pressure, mixed substances will boil at some temperature between the boiling points of the pure constituents. There are exceptions to this, but that need not concern us here. The boiling point temperature is determined by the relative amounts of the components of the mixture. In general, then, as our mixture of solvent, fats, oils, and dirt boils in the still and the pure solvent is removed for reuse, the amount of the high-boiling point contaminants will increase compared to the amount of the solvent that is left, and the boiling point in the still will slowly rise.

As the temperature in the boiler rises, well before all the solvent has left, thermal decomposition of the solvent and other matter will begin. In the case of the chlorinated solvents under consideration, the decomposition becomes serious at about 175° C (350° F) with hydrochloric acid released, and the remains in the pot begin to convert to a tarry substance. The distillation must be stopped short of recovering all of the solvent to protect the equipment. What's left is a stillbottom, in this case a listed hazardous waste (EPA No. F001).

The only reason for all this tiresome detail is to make the point that even with the best intent and the right equipment and procedure, nature sets limits to many everyday activities. That does not mean we should not try to do those things which would be of value if they work, but we should be prepared to believe that no legislature can require what nature prevents. Slogans are good motivators, and a positive mental attitude can drive the research that overcomes serious technical difficulty, but some things are not possible.

Recycling, then, is a special case of waste reduction/minimization. When the facts are right and the will, technology, and markets are in being or can be developed, recycling is a clearly preferred means of managing any waste, and particularly hazardous waste.

In the most attractive combination, useful product can be recovered and the residue left in a condition that allows it to be disposed of in a nonhazardous waste disposal facility. The process, often called *delisting* in the jargon of the environmental engineer, is so named because it moves the waste off the lists of "listed hazardous waste," and if it possesses no characteristic of a hazardous waste, it can be handled like any routine junk. Delisting cannot always be achieved, and sometimes one feels that the regulatory process is moving inevitably to eliminate it as a goal, but it should always be evaluated as an option for serious hazards.

SUMMARY AND CONCLUSIONS

Waste minimization and recycling are closely related activities which can be implemented by just about any manufacturing or other industrial operation.

Both of these activities require thorough knowledge of the processes and systems in use and of the local availability of markets for waste and recovered materials.

With the strong commitment of the senior management of the enterprise, sound technical and operational skills, and a willingness to keep an open mind, much can be done to reduce the amounts of unusable wastes requiring disposal. In general, this objective can be accomplished best when sound and realistic requirements for economic return on the investments needed are set in advance.

Neither waste minimization nor recycling is a panacea for all the ills of the hazardous waste management system. Properly applied, both concepts form very useful parts of that system.

Part III

Basic Reference Materials

Chapter 11

Chemistry of Waste Materials

Chemistry is defined by Webster as "the science dealing with the composition and properties of substances, and with the reactions by which substances are produced or converted into other substances."

Every material thing, object, or organism in the universe is, or is made up of, a substance or combination of substances. This simple, self-evident fact should establish chemistry as one of the basic ways to organize knowledge about ourselves and our universe and as a subject with which each person should have some familiarity. Regrettably, even in the absence of deep research, common experience shows that that familiarity is not achieved by most of the population.

In reaction to the Russian launch of Sputnik in 1957, there was a brief surge of interest in the sciences, including chemistry and related physics and materials science, but that soon faded to something like its earlier level or lower. Recently, the Secretary of Education and other prominent educators have publicly warned that a student can often complete a formal education in the United States without ever having studied chemistry or physics or other scientific subjects in depth.

Chemistry, physics, and mathematics have long suffered from an image problem among students, probably beginning with their emergence as distinct scholarly disciplines in the eighteenth century. Chemistry as a serious science emerged slowly and fitfully from the mists of alchemy and magic. Merlin the magician is prototypical of those who possessed this awful knowledge—the "secrets" of chemistry. Merlin was also widely feared and avoided by most right-minded folk.

The arcane symbolism for the elements, at least the metals which were known before "chemistry" began, is a residue of the mystery-origin of chemistry:

Pb	Lead
Hg	Mercury
Na	Sodium
Fe	Iron
K	Potassium

All have perfectly plain roots in non-English languages but are too often used to intimidate the seeker after chemical knowledge by people who ought to know better. Indeed, the lack of systematic terminologies which requires the student to memorize enormous loads of facts has been a barrier to wide, effective dissemination of basic knowledge of chemistry in the population.

Since, as we have discussed, the world is made up of "substances," then it is plain that hazardous waste, being part of the world, is also comprised of chemical substances. In fact, this is the precise language of the congressional acts establishing the regulatory system for the control and abatement of abusive waste disposal practices.

An informed citizenry is a primary requirement for an enduring democratic society. If citizens want to play their proper part in the development of policy and programs to control the risk to the population which may be caused by hazardous waste generation, handling, and disposal, then they must gain some knowledge of the chemistry of these materials. The only other option is to place one's faith in those who claim expert knowledge.

As already mentioned, experts may reach widely differing conclusions, starting with the same facts or opinions. It is far better if the affected citizens at least know what questions to ask the experts. This ability can often clarify the answers being so positively stated by active participants in the adversarial process that is defined as justice in the United States.

Finally, despite these brave words, the discussion that follows cannot possibly cover much of even the preliminary course in general chemistry. Readers will need to excavate the old high school textbook or borrow one from the kids, if any are so aggressive as to be taking a course in chemistry. At the back of the book in "Suggested Reading," a few references are listed that can expand this brief summary and place it firmly in context.

CHEMICAL NATURE OF SUBSTANCES OF CONCERN IN WASTE MANAGEMENT

A song writer once said "fish gotta swim, birds gotta fly ... and it's a wonderful world!" Like all good songs, even if the ideas might be old, the truth of them becomes clearer in good poetry.

All living organisms must be someplace, and that someplace both defines and confines the organism. This is so because that place—we call it an "environment"—provides a *source* of the input the organism needs to live and at the same time, a *sink* to which wastes can be discharged. This *exchange* of materials is, as we have seen in earlier chapters, essential to life. Plants both consume and release oxygen from their surroundings, either air or water. Whether consuming or releasing, the process is controlled by light intensity, temperature, and other factors affecting the photosynthetic basis of plant growth. Animals consume plants, directly if they are vegetarians, indirectly if they are meat-eaters. All animals (with a few possible exceptions in the deep sea) consume oxygen from their surroundings to sustain life.

No one knows all the intricacies of the life of even the simplest living organism—what it needs in addition to the obvious water, air, light, and space to thrive. Some clues are available, but they are only clues. Mammals, as well as some other animals, require substantial amounts of iron in their blood to permit that blood to transport the oxygen that's needed in each cell. Any major deprivation of this essential input will result in sickness and eventual death. Other metals—potassium, sodium, and even zinc and cobalt—can be shown to be needed for complex animals to grow and develop properly in the environments in which they are found or kept.

All of these input and surrounding substances—the components of air, water itself, its minor impurities or major components—are chemical substances; in common language, they are chemicals.

The making, combining, and separation of chemical substances and their behavior under changing circumstances is the field of science (read "knowledge") that we call *chemistry*. There are no sinister "secrets" of chemistry, but there are still large areas of ignorance even among the most advanced chemists. Much of the knowledge of chemicals and their behavior in the terrestrial environment is incomplete and desperately needs the application of research. For this reason, chemists will sometimes differ on the interpretation of information, which leads to confusion in the less technical public.

While there are important exceptions and ambiguities to be handled, for nearly all practical needs, chemical substances can be divided into two broad classes, each of which has its unique behavior patterns. The first class and the oldest in systematic knowledge is the *inorganic chemicals,* often (erroneously) thought to be lifeless and not directly related to living organisms (therefore *in*organic or *not* organic). The Earth itself—the rocks, soil, water—is made up of inorganic substances, and we shall see how this is important to the study of wastes in the environment and the hazards therefrom.

The second grand class of chemical substances is the organic chemicals—so named because they were always associated with life and life processes and therefore with organisms. While it is true that living things are largely composed of organic chemicals, there are a lot of organic chemicals around today which are made only in artificial reactors and not in life processes. Notably, the origin of these substances does not alter their conformity to the principles of chemistry, but it is believed by many to make them somehow sinister and even harder to understand. "Synthetic organic chemicals" are often cited in the popular press as a lurking danger or in similar language. The definition of organic chemicals for our purposes is historically based but broader than the old one. We define organic compounds (chemicals) as those in which carbon is the fundamental structural element—occasionally incorporating sulfur, silicon, and a few other elements, notably nitrogen, in their molecular skeletons.

OF ATOMS AND MOLECULES

Humankind, always seeking to cast light into the darkness of ignorance and often driven to find new materials and new ways for meeting its needs and wants, has understood something of the nature of chemicals for a very long time. Anthropological work has shown that some groups of people had a substantial base of chemical technology in metallurgy, glass, and ceramics before recorded history began.

The isolation and purification of metals marked the anthropological rungs of the ladder of humanity's change from dependence on things found to a new form of dependence on things made by people. The progression from stone tools to bronze to iron to steel was accompanied by a gradual shift from dependence on physical skill involved in flaking stone or shaping bone toward dependence on "experts" who had knowledge that was not readily available to ordinary people. Surely, early technology that brought charcoal into contact with red earth to produce iron must have seemed magical, and it was not in the interest of the iron makers to make the knowledge accessible to the ordinary run of people who needed knives, axes, and swords to replace those made of cruder stuff. At the technological level, we have come a long way, but the advances are largely in scale and efficiency—millions of tons instead of a few pounds—and in purity and consistency of the finished product.

Through all the millennia since the early Bronze Age, founded on refining of actual metal nuggets in the earth, until now, copper remains copper, iron is iron or steel, and easily recognized. We even still choose bronze for casting long-lasting statues or other memorials to our modern heroes. The ancients left few cookbooks describing their technologies, but the slag heaps and remnants of human activities in Palestine and Egypt and even the remote land of Nubia prove that they practiced "industrial chemistry" many thousands of years before our modern world. The insatiable need of the classical Greeks to understand and to organize knowledge drove those philosophers to begin to organize the early accomplishments in chemistry and technology into a body of knowledge which would grow into our modern science.

The ancient Greeks visualized all matter as a collection of tiny nonreducible particles called *atoms*. As chemical knowledge grew, the atomic theory became the basis for much of our understanding of the chemical behavior of substances. No one has ever seen an atom, but the evidence that the assumption of atoms can explain nearly all we know has piled up so that there is no longer any serious disagreement about the idea.

These same Greeks thought that there were four "elements" in the natural world: Earth, Air, Fire, and Water, and that the observed differences between actual things could be explained by differences in the relative amounts of each of these elements in the atoms of which the substance was made. Later thinkers changed the list of elements, and our modern understanding of the elements began to emerge in the work of the alchemists in much later times and other places in history. The atomic theory has survived, though with a much more sophisticated structure beneath it.

Many minds were devoted to the inductive analysis of the combinations of these elements. At the same time, anonymous craftsmen (symbolized in myth by Haephestus [or Vulcan], the fabled "smith" who could make anything needed by the Olympians using his godly power and art) continued to improve their practical knowledge and to pass it on to apprentices. Many primitive people learned to isolate and work metals into objects of use to them.

Forming the leading edge of modern science, deriving their identity from sorcery, from inductive thought, and, most crucial for the future, from experimental knowledge, scholars who came to be called *alchemists* arose in Europe during the Dark Ages (approximately A.D. 500–1000), although they may have been around much earlier. The classical alchemist sought, among other things, something called a "philosopher's stone" which, when used properly, was believed able to convert rough and valueless materials (usually metals) into gold. Since earliest times, gold has enjoyed a mystical value because of its rarity, its beauty, and, perhaps more important, its observable resistance to all the forces that dull the luster of all other substances known in those times. Even today, when we can enjoy the brilliance of the platinum metals in many forms, gold remains the object of most everyone's affections.

Today, we know that the philosopher's stone was a hopeless goal, but the secret search of the alchemists gave us much new knowledge of the components, the elements, and the compounds that make up the earthly environment. They named most of the metals and many of the other elements, and using a lot of private language which we might think of as magical, they described the behavior of many.

In the intellectual enlightenment that overtook Europe in the seventeenth and eighteenth centuries, much new understanding was added. Some of the accomplishments of the scholars of Islam and of the Jewish scholars of the Diaspora became accessible in translations. Robert Boyle studied the behavior of gases, Henry Cavendish found that oxygen was a component of air essential to combustion and to respiration. In the midst of the French Revolution, Lavoisier worked hard in chemistry until the guillotine called, developing one of chemistry's primary tools, the extremely sensitive weighing instrument usually called a chemical or an analytical balance. The ability to make high-precision comparisons of weights gave access to the combining rules of chemicals in ways not possible before this instrument was available. Work continued in the mills, smelters, and the new "laboratories" of professors and wealthy "philosophers," discovering and studying the behavior of more and more elemental substances. During this time, chemistry probably reached its peak as a subject of mystery or at least of baffling complexity.

Figure 10

PERIODIC TABLE OF THE ELEMENTS

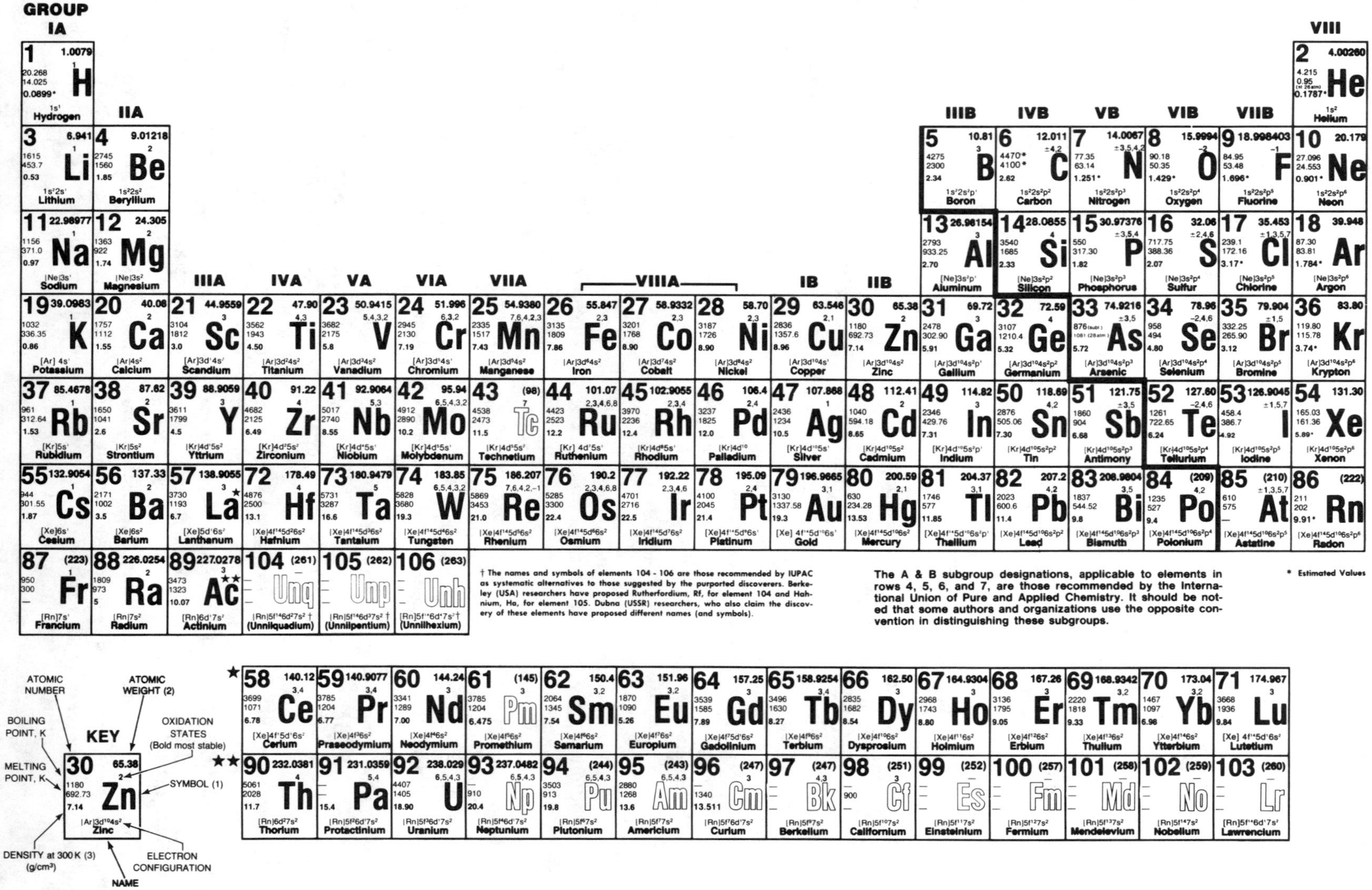

† The names and symbols of elements 104 - 106 are those recommended by IUPAC as systematic alternatives to those suggested by the purported discoverers. Berkeley (USA) researchers have proposed Rutherfordium, Rf, for element 104 and Hahnium, Ha, for element 105. Dubna (USSR) researchers, who also claim the discovery of these elements have proposed different names (and symbols).

The A & B subgroup designations, applicable to elements in rows 4, 5, 6, and 7, are those recommended by the International Union of Pure and Applied Chemistry. It should be noted that some authors and organizations use the opposite convention in distinguishing these subgroups.

* Estimated Values

Finally, as the twentieth century loomed, a Russian chemist named Dmitri Ivanovich Mendeleev proposed a major way of ordering all the disordered information about the elements—something called the Periodic Law of the Elements, expressed in graphic form in the Periodic Table of the Elements, shown in Figure 10. Mendeleev did not, of course, know of all the elements in the Periodic Table shown here. Indeed, some of these were found to exist only after the Atomic Age began to realize some of the hopes of the alchemists and to transmute elements into ones new to science. This phrase "new to science" is chosen to emphasize that these elements are not new in the universe, but that man has only just added them to his knowledge. The Periodic Table has proven to be a useful tool for predicting the properties and indeed even the existence of elements not yet available for study on Earth.

The power of the Periodic Law is based on grouping the elements into patterns which are based on their abilities to combine with each other and on some of their physical properties. It was the observation that when all the elements are listed in this particular way, chemical and physical properties reoccur in a periodic sequence. Each atom in the universe has fundamental properties that identify it as one of the atoms of one of the elements. Elements are the pure substances which cannot be further purified or refined without altering their nature by transforming them into other elements or to energy. We know over 100 elements, some of which have been found in nature only since nuclear reactors produced enough to allow them to be identified; others are still obtained only as products of nuclear reactions.

Current theory says that all the elements heavier than helium are or have been produced in the nuclear furnaces that we call stars and then distributed through the universe by the explosion of stars, ending their lives—novas and supernovas. This means that the rocks, the tools, and even the atoms of our human bodies came into being in some dying star an unimaginable distance away in space and time. New resources are constantly being created, if only we knew how to get them. A lesson of the environmental philosophers is that we had better make good use of those resources which we can reach and use.

The complete basis for this method of classification and listing is beyond the scope of this discussion, but it does illustrate some groupings of elements which are useful in understanding our subject. In each vertical column, the elements have similar chemical natures. Often their physical properties progress in an orderly fashion up or down the column. For example, Fluorine (F) and Chlorine (Cl) are gases at ordinary conditions, while Bromine (Br) is a volatile liquid and Iodine (I) is a crystalline solid. At the extreme right we find the so-called noble gases: Helium, Neon, Argon, and Krypton, which share the property of being nearly completely unreactive with any other elements, at least in any normal environment.

Next to that column, we find one that starts with Hydrogen and continues with Potassium, Sodium, Cesium, etc. These are the most reactive of metals. For years, it was difficult to understand why hydrogen, the most common substance in the universe, should be grouped with metals in this way. In recent years, with new methods of study, it has been found that hydrogen does indeed show metallic properties when cooled to a temperature low enough to condense the gas into a liquid and even to a solid. The power of tradition shows even here. The oxides of these reactive metals, when in water, have old names: caustic soda, lye, and a name derived from Arabic—"al qaliy" meaning "of potash." As a result, the group is still called the "alkali metals."

At the right side of the table, another group of very reactive elements is of great importance in practical chemistry and in the control of environmental hazards. Starting with Fluorine, then Chlorine, Bromine, and Iodine, etc., make up the "Halogens," so-called because of an old Greek word *halos*, meaning salt. The compound of sodium and chlorine, common table salt, is essential to life and is one of the most common substances on our Earth.

In the past few paragraphs, we have used the words "reactive," "compound," and some related ones in connection with the significance of the Periodic Table of the Elements. This reemphasizes the definition of "chemistry" which opened this chapter: it is the science of the properties of the elements and of their combinations. The combinations occur in many forms which make up both the visible universe and many parts of it that are invisible to the eye and may be hard to find even with sophisticated instruments.

With over 100 elements, the possibilities of combinations of two or more are numerous. When the various ways of arranging the substances are considered, the possibilities are endless. Fortunately for the student, there are rules which must be obeyed in the formation of compounds; unfortunately, the rules are themselves complex and often subject to exceptions.

The atoms of the elements in the Periodic Table are rarely able to exist as pure materials in an earthly environment. They interact with each other in millions of ways to produce *more stable* substances which we call *compounds* that are made up of *molecules*. A homely and familiar example is found in our experience with objects of pure or nearly pure iron. Iron and its common alloys, or mixtures with other metals, can be cleaned and polished until it appears to us that only the pure substance remains exposed. We have all experienced the rapid growth on such a surface of a dulling film of iron oxide, which we call rust. Left unprotected, the whole object will eventually be reduced to a pile of red or black flakes which bear little resemblance to the original object. Similar *reactions* occur between elements and between elements and molecules. Sodium, a bright, shiny metal when freshly cut and protected from air, reacts furiously with water (and numerous other substances) with the release of much heat and gas.

Most of the elements shown in the Periodic Table are capable of forming molecules by coupling in some way with themselves or with many of the others. In the process, a fundamental effect is illustrated that will be referred to many times in thinking about chemicals in the environment. That effect is the reduction of total energy that occurs when elements form molecules and even more when the molecules contain two or more elements—and become *compounds*. When a compound contains less chemical energy than equivalent amounts of its constituent elements would if they were free and isolated, it follows that energy would have to be applied to break the compound apart. Such a compound is said to be *stable*, and in general, the more energy released in its formation, the more stable the compound.

As we shall see, the stability of compounds is an important idea in environmental contexts. Knowledge of these energy relationships can often be used to predict the behavior of compounds when exposed to treatments or when adrift in the water, air, or other media of environmental exposure. It is apparent that once the compounds are formed, some kind of energy or bonding must operate to keep them together. This bonding has been the subject of entire book-length studies, both theoretical and experimental, and defies simple definition.

Various lists of known chemicals contain around five million pure substances which are considered "known," meaning someone has made or found a reasonably pure sample and measured some of its physical or chemical properties. (In this work, the term "pure" has very specific meaning, which is beyond our scope.) The International Register of Potentially Toxic Chemicals, maintained by the World Health Organization, expects to list the data for about 60,000 chemical substances thought to have some potential for human or environmental toxicity. This register will be able to provide information to government officials who have to know more about the potential harm such materials can cause in their population.

Making sense of the vast numbers of chemicals, even though most of them will never be encountered by even the most dedicated student of chemistry, much less the ordinary citizen, helps dispel the mystery and even fear that develops because many people feel that chemistry is inaccessible to them. To balance this perspective, we should remember that anyone who prepares food recipes from scratch, who does laundry and removes stains, or even mixes up the old family recipe for window-cleaning solution, is practicing chemistry at its most significant level to himself.

In casual conversation we use the words "compound" or "chemical" freely to refer to those substances we use in everyday life. Such things are a part of the culture in most developed countries, although they would have little meaning in the more primitive nations. In this level of conversation, the knowledge of the rules by which compounds can be formed is of little value, but for an understanding of the impact of chemical substances on us through environmental pathways, it is necessary to go a little deeper.

It is a major principle that some elements cannot be combined with each other. It is a fact of observation that many of the elements are so rare, so inert, or so bound up with other elements with which they form strongly bound stable compounds that they will almost never be of any significance to the ordinary citizen in his or her ordinary life. The Earth itself is largely constructed of iron, aluminum, calcium, and silicon, combined in many ways with each other and with enormous quantities of the gas oxygen. To us the compounds are merely rocks and rarely gain our attention.

The real study of compounds begins with the simplest ones—because they are also very common and can illustrate most of the principles needed in coping with environmental issues. If it were possible to say that any one substance is the most important in and on the earth, then the choice must fall on one that is so familiar that its unique and often amazing properties are often overlooked. That compound is *water*, a combination of two atoms of hydrogen with one of oxygen. Say "H Two O" in almost any conversation, and the person or group involved will know what is meant without thinking.

Water is found everywhere on the Earth. It has been fashionable since the space program gave us new perspectives to call the Earth the "Water Planet." Every elementary-school geography class learns that most of the planet's surface is covered with water in liquid or solid form. We accept one of water's most useful and, indeed, saving properties for life—the solid form, ordinary ice, floats on the surface and does not sink to the bottom of the oceans and lakes where it might be locked up and remain forever. Very few substances show this behavior when changing from liquid to solid (freezing), but it is perfectly tailored to maintain the Earth's suitability for life.

Water is a fascinating substance, one whose properties seem uniquely designed for the survival needs of the biological population as we know it on our planet. As already described, ice floats—if it did not, it would soon cover the ocean floors with ice that would be insulated from solar radiation and might never melt, eventually depriving life of one of its essentials. Water vaporizes readily in the range of temperatures within which most life as we know it thrives, but conveniently forms clouds, and returns as rain to maintain the distribution of the essential resource while acting as an enormous engine of energy storage and transfer needed to maintain equable climate conditions for life all over the globe.

The wonders of water can be the subject of whole books of information. For the purposes of this book and of this discussion on chemistry, however, we will focus on the function of water as a *solvent*, a material into which many things can be mixed to form *solutions* from which the added material, the *solute*, will not separate without the application of energy in some form. These italicized words are very important in understanding the nature of our chemical world. Water is the major transporting agent in both living and nonliving systems on the surface and to a large extent under the surface of the Earth.

Common experience tells us that water can act in two ways to accomplish this transportation: as a *solution*, clear and often indistinguishable from "pure" water, or a *suspension*, like the muddy waters in a gully or stream flowing through erodable land. This difference between a solution and a suspension arises from differences in the solubility of the added substances in the water. If, for example, we add table salt (pure crystals) to pure water and slowly stir the mixture, the salt crystals will gradually disappear. Finally, the solution will accept no more salt, and any extra crystals will slowly settle to the bottom of the container. The liquid becomes clear, as clear as the original water.

It is a mark of *soluble* materials that they make clear solutions through which a light beam can be shone without any definable boundaries of the beam in the liquid. The complete definition is more complex, but for our purposes, this test will almost always be enough.

More commonly encountered in environmental exposures is the class of liquid suspensions in water. A sample dipped from a clear mountain stream will appear perfectly clear, but close observation of the bottom of the sample container will often show fine particles of sand or ground-up rocks which rapidly settle to the bottom. Any other substances carried by the water are in solution.

Following the stream until it reaches a slower, meandering course through silt and clay, a sample now shows different behavior. Immediately after the liquid is dipped out of the stream, settling begins, and particles of solids show up in the bottom as they did in the clear-stream sample. However, it will be quickly noticed that something remains in the water. It may, after standing a long time, appear to be clear, but a narrow beam of light shone through the liquid will leave a trail like a sunbeam in a dusty room—the same phenomenon. Such suspensions, if they must be periodically stirred or agitated to keep them mixed are just that—suspensions, and will usually settle or can be filtered to restore them to clarity if needed. Those which will not settle and which often pass right through a filter are called colloidal suspensions, part of a large class of colloids which are commonly found in all real systems.

The transport role of water is primarily a physical process, moving solutes and suspended matter from place to place. The processes of formation of these suspensions and solutions are often in themselves complex chemical phenomena requiring intensive study for their unraveling. In addition, however, water is possessed of unique properties as a chemical, and it interacts with just about everything on our world by reason of this aspect of its nature.

As we have said, water is simply a compound of two elements: two atoms of hydrogen and one of oxygen. Checking the data in the Periodic Table, we find that hydrogen has an atomic weight of about one, which is no accident, since originally it was hoped that life would be so simple that all the elements would have weights that were a multiple of that of hydrogen. Regrettably, nature just doesn't cooperate in these human attempts to simplify . In the same way, we find that oxygen has an atomic weight of about 16, making the *molecular weight* of water about 18 (well, it really adds up to 18.0152 or thereabouts). For the curious, this means that 6.022×10^{23} molecules of water will altogether weigh just about 18 grams. This number is 6 followed by 23 zeroes—a respectably large number in any context. It is called Avogadro's Number as a memorial to a famous Italian chemist.

Our current understanding of this concept is that many, but not all, of the molecules exist at any moment as just that—molecules of water. They look on the average, like this:

```
  O
 / \
H   H
```

but this is an average, while the actual parts are vibrating around their pictured positions, and it's impossible to identify any two hydrogens with any specific oxygen atom. In fact, one of the great principles of physics says that we can never know both the position and the speed of motion of any particle (read atom or molecule) at the same time. This Uncertainty Principle has suffered from much nonsense from mystics and others, but this is all it really says.

Accepting this basic principle of Dr. Heisenberg, molecular scientists look at water (indeed every substance) as a mixture of all possible states in which it might exist, with the amount of each form controlled by the immediate environment of the molecules or atoms. Such a system can only be described in statistical terms, using the likelihood or probability that a substance might be found as a whole molecule, for instance, or as a collection of pieces of molecules or even partial atoms under any particular conditions and surroundings.

One of the things that can happen to a water molecule (and to others, as well) is that it can separate or *dissociate* into two parts. The water molecule, as all good molecules must be, is electrically neutral. That is, all the electrical charges which hold it together are in balance. Closer examination, however, shows that in every liter of pure water, one in 10^{-7} (one in ten million) molecules is, on average, separated into one hydrogen ion (H^+) and one hydroxyl ion (OH^-). If we add these together, we get water that is neutral electrically and still has a ratio of two hydrogens to one oxygen. But we have introduced two new and important terms for ideas that are of vital use in the understanding of chemical phenomena.

The word *ion* is an old Greek word for a *traveler*. Its use comes about this way: if we subject pure water to an electrical voltage by, say, hooking opposite sides or poles of a battery to platinum or other nonreacting metal pieces on opposite sides of the container of water, then the ions (remember there are a small but real number of them) will travel or move toward one or the other of the two metal pieces, which we can now name *electrodes*.

Since, in chemistry as in life, opposites attract, the positive ions like H^+ move toward the negative electrode, called the cathode, the negative ones toward the positive electrode, called the anode. Logically, then, positive ions are called *cations* and negative ions are called *anions*. These words are of fundamental importance in understanding communications about chemistry and chemical effects in water solutions. It is meaningless to attempt to read an analysis of groundwater, for example, unless one comprehends the significance of the terms used to write the report.

Continued application of the electrical current will make hydrogen and oxygen form at the electrodes, but that is not the point here. The language of chemistry is laced with words and terms that have origins like those for anions and cations. This terminology survives because it provides a shorthand way of talking and writing about molecules in solution and their ways of combining and recombining to create the enormous variety of compounds in the chemists' handbook.

The absolute law in solutions is that for every positive ion charge, there must be an equal negative ion charge. In simple compounds, it is easy to see:

$NaCl$ (Salt) exists as $Na^+ + Cl^-$

HF (Hydrofluoric Acid) exists as $H^+ + F^-$

The idea of ions in solutions becomes more useful and helpful in dealing with more complicated solutions. For example, calcium, principal mineral ingredient of limestone, teeth, and bone, carries two positive charges in solution. Similarly, the sulfate ion, main constituent of sulfuric acid and of gypsum, carries two negative charges in solution. When we combine them with other common ions, we might get:

$CaCl_2$ exists as $Ca^{++} + 2Cl^-$

and

Na_2SO_4 exists as $2Na^+ + 1\ SO_4^{=}$

since the anions are also to be found with varied charges—called *valences* in chemical parlance. Valence is defined as the capacity for an element, or a combination of elements called a radical, to combine with others. Valence of an atom is determined by its fundamental electron structure and its state of ionization, controlled by its environment. For the purposes of customary problems in environmental contexts, the ordinary valence states of the ions in water solution will nearly always be sufficient. Most common substances tend to be found in only one or two possible valence states, and these are readily deduced from the formulas of common inorganic compounds.

Another way of measuring combining abilities is the use of oxidation numbers, which in the general case can be used to establish valence by simple rules while using fundamental knowledge about atomic structures as a basis. This way of thinking will be discussed later under the heading of oxidation-reduction reactions. In the realm of organic substances, the way in which these rules apply is quite different, and we shall consider them at that time.

The few ideas we have been discussing, combined with the necessity that solutions will, under all ordinary circumstances, be electrically neutral, makes it easy to talk about solutions and to begin to understand other chemicals in water—a principal way in which the parts of our world interact with each other. It's important, however, to make a distinction between the electrical neutrality of solutions and the idea of acidity and alkalinity, in which the term neutral means a solution that is neither acid nor alkaline.

We need not worry about electrical neutrality for most purposes, but it is easy to see that we need some way of determining whether a solution is acid (probably corrosive to metals) or basic/alkaline (possibly damaging to skin and other tissues) or neutral (probably not corrosive). As we shall see, however, neutral solutions can be toxic or even acutely poisonous; so this criterion alone does not tell us much about that form of hazard.

To find the measurement that will identify acidity/alkalinity and determine how much acidity or alkalinity there is in a solution, we return first to the character of water—the nearly universal solvent. As was discussed earlier, research has shown that in each liter of pure water at 20° Centigrade there will be found 10^{-7} grams or one ten-millionth of an atomic weight of hydrogen ion. The water is electrically neutral because of the balancing presence of hydroxyl ions and is neither acid nor alkaline—this is the definition of a neutral solution.

To build our measuring scale, we need a benchmark, a starting point. In this case, as unusual as it may seem, we begin the scale at the middle, a fixed point that can be found repeatedly: take the logarithm (base 10 for mathematicians and elementary-grade students of New Math) of the hydrogen-ion concentration of the neutral water as that midpoint and reverse the sign so we don't waste time and ink on negative signs; the result is 7. Needing a name for such a number, it was originally called the *potential of hydrogen*; now, in an age of shortened names, the term is *pH*. The idea of pH is one of the most commonly used ones in chemistry. It is also one of the most often misused and misinterpreted ones.

It is clear that any measuring scale needs not only a middle but ends, or at least ranges around the middle. How do we establish the ends and what do they mean? After all, neutral solutions are of interest, but it is always the extremes that generate the most concern.

The zero point of the scale is at least conceptually straightforward, because of the mathematical nature of logarithms, the basis of the scale. The number whose logarithm is zero is 1, and this is the lowest meaningful pH measurement. Lower ones are mathematically possible, but there are good physical chemical reasons for not extending the scale.

At a concentration of 1, of which the logarithm is zero, the solution contains *one atomic weight of hydrogen ion* in each liter of solution at the stated temperature. There are means of compensating for temperature changes in the measurement, within reason.

The other end of the scale is found by establishing the hydrogen ion concentration in a solution that contains a chemically equivalent amount of a strong base or alkali to the one atomic weight of hydrogen which defines zero pH. That concentration amounts to 10^{-14} atomic weights of hydrogen ion in each liter of solution. The amount of alkali needed to produce this concentration is the weight of the compound that will provide the amount of hydroxyl ions needed to react with one atomic weight of hydrogen ion.

Thus we have a measuring scale for acid/base properties of solutions, readily measured today in lab or field with hand-held instruments and easily interpreted—maybe.

These pH meters are not the only means available for the measurement of acidity/alkalinity. In earlier times, solutions were neutralized precisely in the laboratory by adding a standard solution of acid or base to the sample until a colored chemical also present (an indicator) changed color to indicate that neutrality had been reached. This procedure, called *titration*, is still in occasional use and is sometimes needed when pH meters give poor results. An extension of this method is the use of paper tapes saturated with indicator chemicals that when wet with a solution, change color to indicate acidity or even, by a range of color, a pH value for the solution.

It is not generally appreciated that indicator papers are of limited value in the measurement of the phenomena in environmental media. All indicators have a range of pH values over which they complete their color changes, and a highly trained eye may have trouble reading the results with precision. Many well-meaning amateurs have found to their dismay that environmental information recorded at great personal sacrifice of time and energy is valueless because the indicators used were not reliable in the poor lighting and sometimes difficult field conditions.

The problem, on the other hand, with pH results is that the center-neutral scale is not natural to many users. We are used to measuring sticks, tapes, etc., that start somewhere and go to an end, while pH starts in the middle and goes both ways.

Acid	Neutral	Base
pH 1	pH 7	pH 14

Also, the scale is logarithmic, and most of us prefer a nice evenly graduated scale on which one unit or division means the same amount of change wherever it may be found on the scale itself.

We soon find that the logarithmic scale is essential to a sensible system of measurement because the range is enormous. A solution at pH 1 is ten million (10^7) times as concentrated in hydrogen ions as a neutral solution at pH 7. The hydrogen ion concentration of a solution at pH 14 is not half what it would be at pH 7, but one ten-millionth of it. A linear scale to cover such a range would be thousands of miles long!

Even so, we need to maintain some perspective on the meaning of common ranges of pH measurement in our real world. In chemist's jargon, a solution of pH 1 or pH 14 is a *normal* solution, containing either one atomic weight of hydrogen ion or an amount of base that would react with that amount, called an *equivalent*.

For hydrochloric acid, this solution contains about 36.5 grams of hydrochloric acid per liter; for sulfuric acid, having two hydrogens per molecule, about 48 grams. Each solution will rather rapidly attack steel and dissolve a 1/4-inch plate in under a year, if there's enough solution for the amount of steel available. Each is commonly used for special cleaning work on metals, brick, and tile, and similar things, with ordinary rubber gloves and eye protection preventing injury.

On the other hand, if such powerful solutions are found where they are not wanted or needed, their presence is very serious. However, many high-quality drinking water supplies show pH ranging from perhaps 6 to 12, largely because surface waters pick up carbon dioxide from the air to form a solution of pH about 8.5 or 9, and water flowing over or through limestone will react with it to become alkaline.

One final caution about pH—it is a valid measurement in *water solutions only*. Any measurement in a sample that is not a water solution is meaningless as a pH value. There are similar methods available for the measurement of certain ions other than the hydrogen ion, but these cannot be called pH either.

An interesting and vital question arises from the idea of pH, representing acidity, neutrality, and alkalinity. That is, if one has a solution of some pH value, how can that value be changed, and what are the consequences of such a change? The process of shifting pH is often called neutralization, acidification, or some other term that is derived from the chemicals used, which can get confusing.

The simplest case, illustrating the general approach, is one in which base reacts with acid to form a simple salt, which remains in the solution while the pH changes as desired. We begin with two solutions, one containing one molecular weight (36.47 grams) of hydrochloric acid in each liter of solution, and one containing one molecular weight (40.01 grams) of sodium hydroxide in each liter of solution. The hydrochloric acid has a pH of 0 and the sodium hydroxide solution a pH of 14. Assuming the solutions are properly made, these pH values result from the definition of pH and are exact.

If we slowly add the acid solution to the sodium hydroxide solution, the pH of the mixture begins to fall from the original value of 14 in direct proportion to the amount of acid solution added. When the equal volumes are reached, the solution is said to have been neutralized, and the resulting pH is 7. In the solution, a process of mixing is taking place, and a chemical reaction is occurring. For one thing, if the mixing container is insulated, the solution gets warm and needs to be cooled to return to the original temperature. The chemist's shorthand for this whole process is:

$$HCl + NaOH \text{ become } NaCl + H_2O + \text{heat.}$$

What's happening is that as the HCl solution is added to the sodium hydroxide, more hydrogen ions are added, while the Cl^- and Na^+ ions offset each other in electrical charge. Since pH measures the concentration of H^+, the pH will fall toward the neutral point, pH 7, and will stop there when a mole of HCl is added for each mole of NaOH in the solution. The term "mole" is a useful idea and represents a quantity of a chemical, the weight of which in some unit is equal to the molecular weight of the substance. We ordinarily think of a "mole" as a gram-mole, but ton-moles and pound-moles are equally valid measures.

All the water in the solution could now be evaporated leaving behind a deposit of crystalline salt. Evaporation of the water removes all the H^+ and OH^- ions. The remaining salt is an ionic crystal made up of sodium ions (Na^+) and chloride ions (Cl^-), not of molecules of NaCl. Each sodium ion is surrounded in the crystal by six chloride ions and each chloride by six sodium ions, none of them being specifically identified with any other single ion, but the whole crystal being in balance electrically and locked into a fixed arrangement of the atoms that is characteristic of the "halite" crystal identified with salt.

Most of the metallic elements and the common nonmetallic ones form ions, either alone or in combination with other elements. The study of the combining rules and ways in which the elements can combine in these inorganic compounds is very interesting, but for our purposes, we will just list in Table 10 some of the more common ions likely to be encountered in environmental media.

OXIDATION AND REDUCTION IN IONIC SYSTEMS

A short inspection of Table 10 will turn up some unusual-looking behavior of some of the elements. Chromium, for example, one of the common metals in our flashy society, appears in several guises:

1. As Chromium Ion Cr^{3+}
2. As part of Chromate Ion CrO_4^{2-}
3. As part of Dichromate Ion $Cr_2O_7^{2-}$

This versatility of chromium, while interesting and illustrating a property shared with a number of elements, might be of merely passing interest except for one key observation. The "ordinary" chromium in Cr^{3+} is generally considered a mildly toxic and relatively

Table 10

Some Ions of Interest in Environmental Studies

Positive Ions		Negative Ions	
Ammonium	NH_4^+	Acetate	$C_2H_3O_2^-$
Copper (I) (Cuprous)	Cu^+	Bromide	Br^-
Hydrogen	H^+	Chloride	Cl^-
Potassium	K^+	Cyanide	CN^-
Sodium	Na^+	Fluoride	F^-
Cadmium	Cd^{2+}	Bicarbonate (Hydrogen Carbonate)	HCO_3^-
Calcium	Ca^{2+}	Bisulfate (Hydrogen Sulfate)	HSO_4^-
Copper (II) (Cupric)	Cu^{2+}	Bisulfite (Hydrogen Sulfite)	HSO_3^-
Iron (II) (Ferrous)	Fe^{2+}	Hydroxide	OH^-
Lead (II)	Pb^{2+}	Hypochlorite	ClO^-
Magnesium	Mg^{2+}	Iodide	I^-
Manganese (II)	Mn^{2+}	Nitrate	NO_3^-
Mercury (II) (Mercuric)	Hg^{2+}	Nitrite	NO_2^-
Nickel (II)	Ni^{2+}	Thiocyanate	SCN^-
Zinc	Zn^{2+}	Carbonate	CO_3^{2-}
Aluminum	Al^{3+}	Chromate	CrO_4^{2-}
Arsenic (III)	As^{3+}	Dichromate	$Cr_2O_7^{2-}$
Chromium (III)	Cr^{3+}	Silicate	SiO_3^{2-}
Iron (III) (Ferric)	Fe^{3+}	Sulfate	SO_4^{2-}
Manganese (IV)	Mn^{4+}	Sulfite	SO_3^{2-}
Arsenic (V)	As^{5+}	Phosphate	PO_4^{3-}

"safe" material, while the Chromate and Dichromate forms contain chromium in the Cr^{6+} form that is considered a proven carcinogen in mammals, including humans exposed in the working environment.

We know that the elements in the first two columns, Groups IA and IIA of the Periodic Table, are true metals and have valences or oxidation numbers that are the same as their Group numbers, always +1 or +2. Generally, the oxidation numbers of the elements can be derived from their electron structures, being the number of electrons given up or added to an atom during compound formation to "complete" an outer shell of electrons on the atom—usually adding up to eight as in the noble gases.

Some of the elements in the Transition section of the Table (Groups III, IV, V, and VI) can exchange electrons in several ways, depending on the surroundings of the particular atom. This is called *speciation*, a change of ion species in response to certain properties of the solution in which the ion is found. The controlling feature of the solution is called the oxidation-reduction potential, usually shortened to *redox potential*.

The response of metal and complex ions to oxidation-reduction factors in the environment has been found to be of vital importance in understanding the pathways by which metals may travel to expose living things to any effects we may wish to understand. Metals may exist in or be released to environmental media by natural processes and from man-made sources like wastewater or leakage from storage systems or dumps.

Most people are familiar with the term "oxidation" as it is commonly used to describe the interaction of materials with the oxygen of the air. This meaning, while usually accurate, has become too narrow for the precise language of chemical science. In chemistry, the term oxidation means the removal of one or more electrons from an atom in the process of combining that atom with others to form a compound or the free element. Reduction is the opposite process, adding electrons to an atom of an element. The production of iron is based on the reduction of iron oxide ore to the free element. The terms themselves are derived from experience in the laboratory and are used also to describe reactions in which oxygen is not involved at all. In fact, in environmental contexts microorganisms are common agents for bringing about oxidation and reduction of metals in solution.

A notorious example of this phenomenon was found in the biological oxidation of mercury metal in the beds of streams, forming organic compounds of mercury that can readily be absorbed by organisms living in the water, more readily than when the mercury is in metallic form on the bottom of the water. This discovery of a chemical-biological pathway for exposure of people who ate fish from waters contaminated by small amounts of mercury metal which escaped from some kinds of industrial plants came only after several tragic impacts on large groups of people had occurred. Once the reaction and pathways of exposure were discovered, mercury in wastewater proved to be much more dangerous than had been believed on the basis of the earlier incomplete information and simple chemistry.

In another sense, reduction of metal ions in groundwater may help bind the metals firmly to soil so that it may become nonhazardous, a process exactly the opposite of the process of mercury oxidation just described.

All inorganic chemical reactions are redox reactions. The balance between oxidation and reduction is assumed in the balancing of the equations describing chemical reactions.

In the simple reaction between hydrogen and oxygen to form water,

$$2H_2 + O_2 \rightarrow 2H_2O$$

the original free elements hydrogen and oxygen have, by definition, oxidation numbers of zero. In water, we know that the hydrogen exists at least partially as H^+ with an oxidation number of one; therefore, the oxygen must have been reduced to an oxidation number of -2.

Another example, not involving oxygen, is that between hydrochloric acid and zinc metal:

$$\text{Zn (metal)} + \text{2HCl (in water)} \rightarrow$$

$$\text{ZnCl}_2 \text{ (in water)} + \text{H}_2 \text{ (gas)}$$

Here the zinc is a reducing agent, giving up two electrons (leaving the atom with two positive charges as an ion in solution), while two H^+ ions are oxidized to form gaseous hydrogen. The zinc ion associates with the chloride ion already supplied by the HCl.

In use, this concept measures the tendency of a solution to oxidize or reduce and can be measured in most solutions with common instruments. In dilute solutions, changes in redox potential can usually be made by the addition of specific reagents. Among the most common oxidizing agents is chlorine in slightly alkaline solution in water—sodium hypochlorite or household chlorine bleach. Perhaps the most common reducing agent in similar applications is sulfur dioxide gas in solution—sulfurous acid. Both of these substances are dangerous and can do severe damage to persons not protected.

ORGANIC COMPOUNDS

By far the largest number of chemical substances known are organic chemicals. The distinction between the inorganic compounds just discussed and the organic ones is often arbitrary and sometimes ambiguous. However, for our purposes in dealing with the chemistry of waste materials, a simple definition will almost always be enough:

> Organic chemicals are characterized by a structure in which direct carbon-to-carbon bonds form the primary framework of the molecule. Many other substances may be attached to or even incorporated into the structure. Among these, the most common are hydrogen, oxygen, nitrogen, and sulfur.

This definition, slightly expanded from the one earlier in this chapter, covers an immense number and variety of chemicals. Some observers estimate that the known organic chemicals are approaching 5 million distinct compounds!

As chemistry emerged from the eighteenth century, many chemists were aware that there was a class of substances that were not apparently subject to the same laws of combination and compound formation that had been observed in the metals and the ionic compounds, and that compounds in this class were held together by then-mysterious forces that had no measurable ionic components. For some time, it was held by many that such "organic" materials must require some unknown "vital force" derived from living organisms to be formed, and that they could not be made from simple nonliving substances.

This mystery doctrine proved false when Frederick Wöhler accidentally made urea from simple known substances. Urea was already clearly understood as a natural substance theretofore derived only from metabolic processes. This single experiment in the year 1828 showed that the distinction between natural and synthetic or artificial organic chemicals was meaningless. Since then, practitioners of chemistry have realized that the properties of a molecule have no relation to its origin, and depend solely on the makeup of the molecule itself. This observation still remains unconvincing to some people not engaged in chemistry.

Today's vast production of organic chemicals is almost entirely based on petroleum and natural gas, with some role for coal and its by-products. Although the identification of some products as "coal tar chemicals" still is used in some contexts, all those products can be and usually are made from petroleum fractions today. Carbohydrates, fats, and oils, while potentially sources of just about any organic chemical, tend to be limited by economic factors to uses in foods where few chemical conversions are needed before they can be used.

The modern world is not imaginable without the organic chemical substances that are so pervasive in our daily lives: foods, fuels, fibers for fabrics, paper, paints and coatings, plastics, rubber products, soaps and detergents, medicines and cosmetics, and thousands more. However, in the use of this high technology to produce the products we need and want, we also find wastes being produced which may be harmful if not carefully managed. Finished products, like solvents, paints, and plastics, can be and often are thrown away in unsatisfactory and dangerous ways.

Understanding the risks and evaluating the benefits of the immense and growing family of chemicals is one essential for effective participation in the modern, democratic, decision making process.

THE NATURE AND CLASSIFICATION OF ORGANIC CHEMICALS

It is not likely that anyone, in a lifetime, will study any significant fraction of the millions of known organic chemicals. For the purposes of this book, the actual number of compounds of concern is much smaller, probably a few hundred, certainly no more than a few thousand that could be encountered in any significant way in managing hazardous waste. The problem is, which ones should we study? How can the knowledge needed be gained without a lifetime of scientific training devoted to chemistry?

The answer is that the study of the classes of organic compounds will show orderly and consistent relations between members of the various classes or types of chemicals and provide the basis for becoming conversant with the general properties of a vast number of compounds, or perhaps more important, how to find more information when it's needed. Too often, citizens are intimidated by the language of technology, and this seems particularly true of the language of chemistry.

It is essential, as was said earlier in the introduction to this chapter, that the reading of this discussion be supplemented by reference to a current textbook in chemistry and by access to one of the standard handbooks; for example, a recent edition of *Lange's Handbook of Chemistry* or the *Handbook of Chemistry and Physics.* These hefty volumes contain an enormous amount of accurate information but need to be studied in themselves in order to learn how to use them.

This brief discussion will begin with carbon-carbon bonds and the nature of hydrocarbon chains and rings. We will then take up the role of *functional groups* and *substituents* which, when added to a hydrocarbon framework create the variety of chemicals we have already discussed. Along the way the concept of *homologous groups* and their usefulness in predicting chemical behavior will be brought in. We will finish with some of the reactions organics frequently undergo in the environment when they are released and with a tabulation of the common names and handbook names for some of the common chemicals of concern that are too often referred to by acronyms or other short names that are less than informative to persons not devoting their full-time attention to these kinds of matters.

A final word of caution: this is not a book dealing with the making of organic compounds. We will describe the substances and some of their behavioral properties, but not how they are made in a lab or factory, except perhaps to illustrate some point about the chemical itself.

CARBON, THE FUNDAMENTAL BUILDING BLOCK

The element carbon, found in nature in many forms, among them diamond, graphite, and soot, is the basis, indeed the identifier, of all organic compounds. It is made in the helium-burning nuclear furnaces of old stars and spread by cataclysmic events throughout the observable universe.

Because of its electronic structure, carbon can exist in oxidation states of from +4 to -4 and forms bonds with a wide variety of other elements. Of most interest here is the special ability of carbon to form covalent bonds with itself and to form long chains thus:

–C–C–C–C–C–

A few other elements can be arranged this way, but with more limited chain length and other factors than the carbon chains that are found. In addition, each carbon atom has the ability to form four covalent bonds. This results in methane, the primary constituent of natural gas, with one carbon atom bonded to four hydrogen atoms:

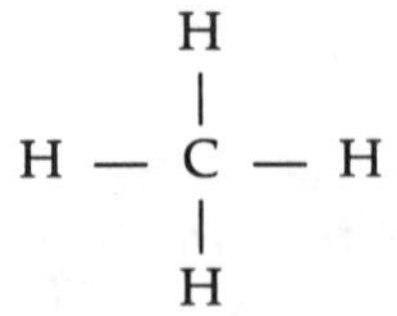

Methane is the simplest of the *hydrocarbons*, compounds of hydrogen and carbon that form chains and other frameworks for constructing molecules. As we shall see, any truly representative description of a hydrocarbon molecule must show in some way the arrangement of the atoms with respect to each other in space. These *structural formulas* are sometimes difficult to print in a document; so we often use condensed formulas which show the groups making up the molecule in a special way that can be interpreted pretty well with some training, although the shortened formulas are not always sufficiently precise to identify unambiguously a complex molecule. The first three members of the series of hydrocarbons starting with methane are:

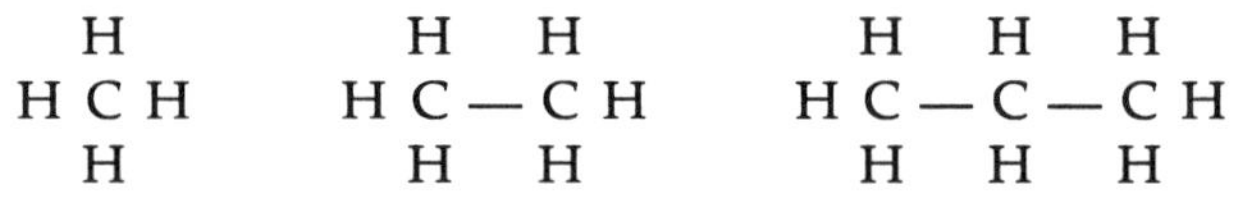

CH_4 Methane | C_2H_4 Ethane | C_3H_8 Propane

Once the number of carbon atoms reaches four or more, the probabilities become more varied. Butane, for example, has four carbon atoms and can exist as two *isomers*, two compounds with the same chemical analysis, C_4H_{10}, but different structures:

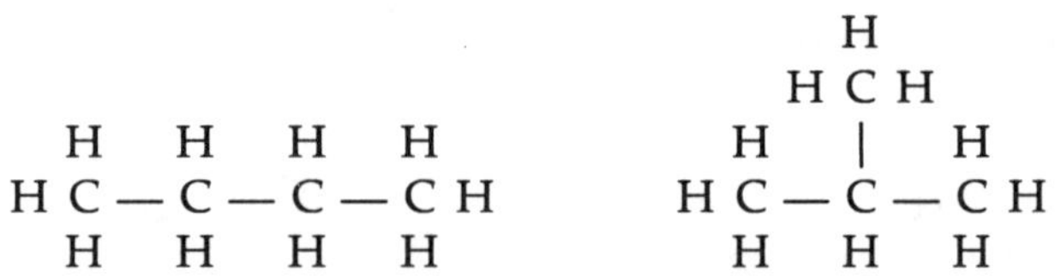

Butane
Boiling Point -0.5° C

Isobutane
Boiling Point -11.7° C

The boiling points are shown to indicate that one difference between the straight-chain butane and the branched-chain isobutane is that the isobutane boils at a temperature more than 11° C lower than butane at atmospheric pressure. This kind of subtle difference in properties or behavior is one of the observational tools chemists use to identify different compounds and to separate them when necessary.

A little thought will show how many variations can exist in a larger chain molecule. For instance, with 8 carbon atoms, there are 18 different isomers of octane possible. This emphasizes the need for precision in language when naming compounds without showing structure explicitly. The term isooctane, formerly in frequent use, tells one little about which octane is being considered. In the example of butane-isobutane, while there are only two isomers, it is clearer to name isobutane *2-methyl propane* using the longest chain (3) for the molecular name and adding the name for the group consisting of one carbon atom and three hydrogen atoms—the *methyl group*. We will see this pattern repeating itself as we go further into the building of organics.

The hydrocarbons we have been discussing, in which each carbon atom is bonded to four other atoms, either carbon or hydrogen, are called the *aliphatic hydrocarbons*, and sometimes the *alkanes*. This name is convenient in dealing with compounds in which a hydrocarbon chain is attached to another, when the cluster of carbon and hydrogen atoms of this type is referred to as an *alkyl* group, a functional group like others that will appear later. Still another way of referring to these simplest hydrocarbons is that they are the *saturated hydrocarbons*, since all possible bonding of carbon atoms is filled, and none is left to react further.

Two other groups of hydrocarbons are *not* saturated; that is, the carbon atoms may be bonded to each other more than once, a consequence of the tetrahedral shape of the carbon atom, which allows either one, two, or three points of the four-pointed atom to be near enough to another point on another atom to share electrons and form a bond. A molecule in which two carbon atoms are bonded to each other by two bonds is called an *alkene*, one in which two carbons are bonded three times, an *alkyne*.

An example is the progression in the two-carbon series:

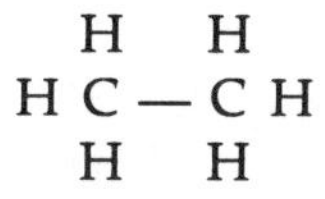
Ethane Alkane

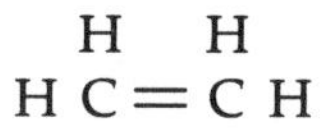
Ethylene Alkene

Acetylene Alkyne

The one (–), two (=), and three (≡) line symbols stand for multiple bonding between the carbon atoms. We have seen how the alkane series can continue to be extended in chain length with physical and chemical properties changing in some regular ways. In principle, the same applies to the alkenes and alkynes.

With the double- and triple-bonded molecules, we enter a new kind of isomer definition. Propylene, as an example, has one doubly bonded pair of carbons and a singly bonded pair. Since moving the double bond to the other position just results in a mirror image of the original positioning, placement of the double bond makes no difference in this simple molecule. However, for butane, having one more carbon, it does make a difference:

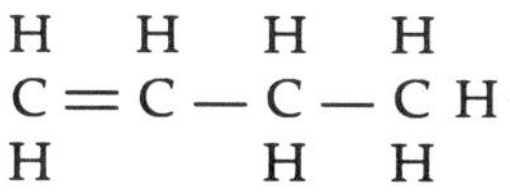
or
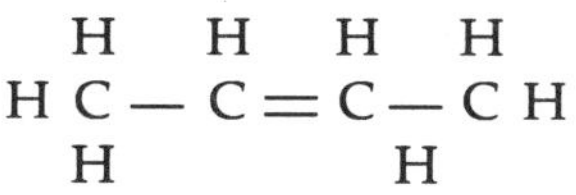

1,2 Butene 2,3 Butene

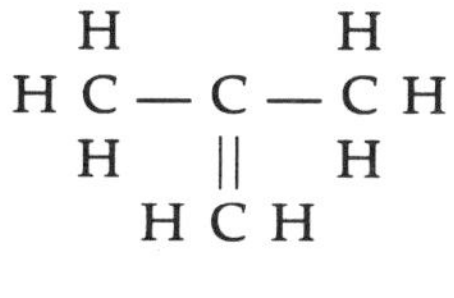

Isobutene

The most important thing about the unsaturated compounds is not the complexities of naming them but their tendency to react with other chemicals at the unsaturated bonds to form other chemicals, notably plastics and other substances like rubber. Ethylene and propylene are among the highest-tonnage industrial chemicals produced in the world for use as raw materials for the familiar plastics polyethylene and polypropylene and their relatives. Butadiene, a four-carbon (remember butane) hydrocarbon with two double bonds, is one of the essential ingredients in synthetic rubber and many plastic materials. Less recognizable is the role of the unsaturated hydrocarbon chains that make up the fats and oils in producing paints and varnishes that harden or "dry" by complex reactions with each other and with oxygen and other trace materials found by chemists to accelerate the process. These "unsaturated fats" are identical to those that doctors have found are less harmful in the diet than the saturated ones commonly found in certain kinds of protein foods. All this is by way of example, showing how the general language of chemistry applies to the same materials wherever they are found, and can be helpful in clarifying the meaning of these words that have become commonplace in today's regular communication between literate people.

Ethylene, while usually made in huge quantities by subjecting other hydrocarbons to extremely high temperatures in "cracking" furnaces, is also a product of plant metabolism that is responsible for the ripening of fruit at the right time. Methane, usually found in quantity in natural gas, is also a product of metabolism, bubbling from the bottom of swamps and ponds in which microorganisms break down plant matter, and also released in large amounts by colonies of termites and from the digestive systems of the millions of ruminant animals, both wild and domesticated, on the globe.

Another group of hydrocarbons which, although seldom encountered in environmental contexts, should be mentioned is the cycloalkanes. In these compounds, the long chain of the carbon backbone bends back on itself and links up into rings of various sizes. Rings from the smallest possible, three carbons, to very large ones have been found, but the most important is the six-membered one, cyclohexane, C_6H_{12}, which is a valuable raw material for some plastics, notably nylon, but is rarely seen outside of the manufacturing plant. Its properties are not greatly different from the other saturated hydrocarbons.

AROMATIC ORGANIC CHEMICALS

Another ring-structured family of organic chemicals is the foundation of another major world of chemistry—the aromatic compounds, which today are often referred to as the benzenoid compounds. The substance we call *benzene* was known to early chemists who worked with the chemical by-products of the coking of coal for metallurgical uses. Some of its properties were unusual, and the early attempts to find out its structure were frustrating. These structural studies were the major preoccupation of chemists in the nineteenth century and still continue today because the process of "elucidating" structure provides all kind of useful information about the compound being studied.

In 1858 the chemist Kekule proposed one of the great scientific achievements of the nineteenth century: a structure that was a 6-carbon ring, but unusual at that time because the molecule of benzene contained only one hydrogen atom for each carbon atom, meaning that the molecule should be highly unsaturated, although on the basis of lab work, it was by no means sure that its properties were in accord with that idea.

The Kekule picture of the benzene molecule was basically this:

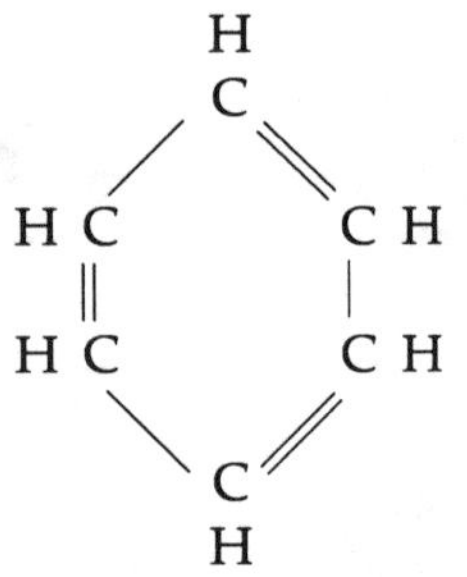

but it is clear just looking at the drawing or a model that the assignment of positions for the double bonds is arbitrary, and they could just as properly be assigned to positions that would create a mirror image. Today, we believe that the structure is in constant motion; it *resonates* between several states of varying probability without being accurately represented by any one fixed structure. This is a great departure from the conventional thinking of Kekule's time, and it was many years before a consensus was developed that accepted the resonant ring structure with improvements. Today, the ring system is often represented by a hexagon with a circle in the interior, suggesting that all parts of the ring system are subject to some uncertainty about their position and therefore their reaction behavior.

While the structure of benzene is of interest and the understanding of that structure essential, for the practical world, benzene is the simplest member of a series of chemicals usually called the *aromatic* compounds—a name chosen years ago because so many of the good-smelling oils and perfuming substances contained benzenelike rings.

Having established the new framework for the aromatic chemicals, we can then proceed with compound construction as we did in the alkanes and their relatives.

If we add to benzene one new group, say the methyl group (CH_3), we get a substituted benzene:

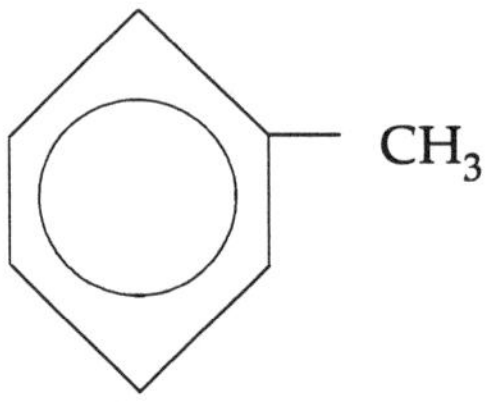

Toluene

Many groups can be attached one at a time to form familiar chemicals:

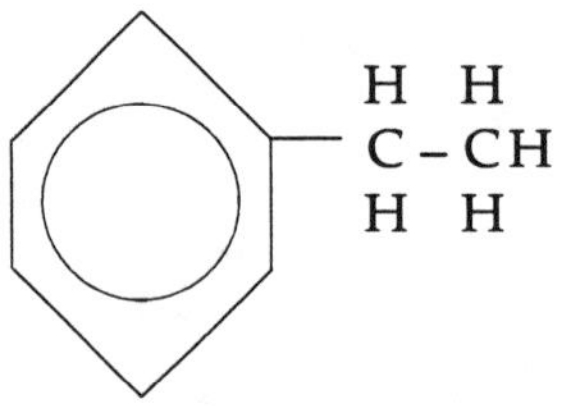

Ethylbenzene

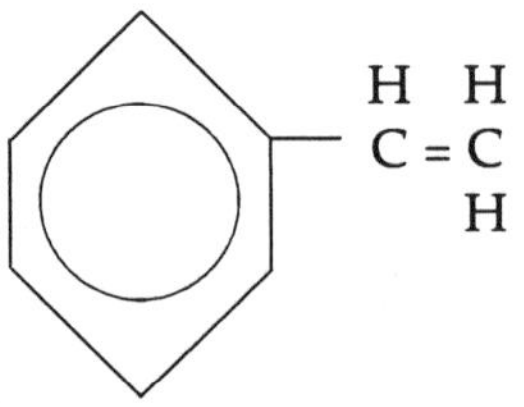

Styrene

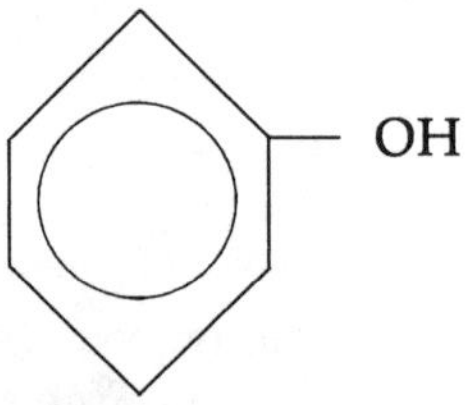

Phenol
(Carbolic Acid)

Complications arise when two groups are added to the benzene ring. The groups may be attached at different points on the ring. In drawing these compounds, it is necessary to number the carbon atoms as shown:

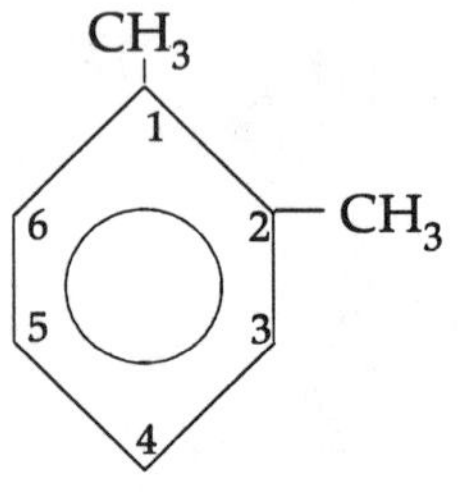

1,2 Xylene
orthoXylene

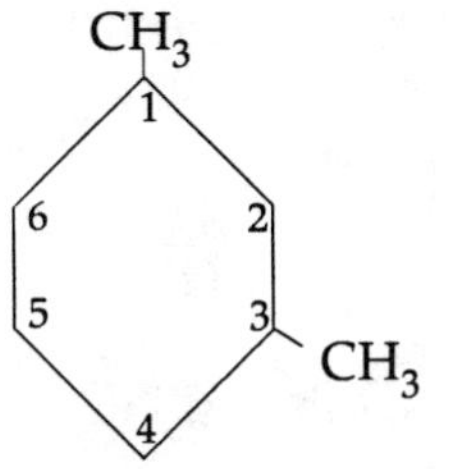

1,3 Xylene
metaXylene

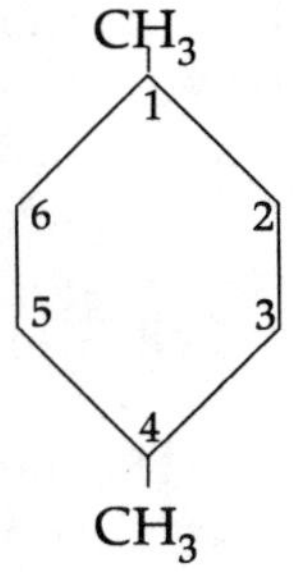

1,4 Xylene
paraXylene

These are all xylene—benzene with two methyl groups, but their properties are distinctly different—so different as to allow them to be separated and used as raw materials to make completely different finished chemicals.

The positions of the carbons in the ring are numbered beginning with the top and proceeding clockwise. The use of these numbers is a much-preferred way to describe structure than the older names shown below the more rigorous ones above. Many of us have studied the ortho- meta- para- system, but the newer nomenclature is much more definite and has the virtue of not taxing the memory.

There are a number of aromatic ring systems in addition to benzene. These are commonly called polycyclic aromatic hydrocarbons or PAH in the jargon of the regulators and are of major importance in some toxic discharges to both air and water. In environmental issues, these chemicals are almost always found in complex tarlike mixtures, often with other unidentifiable substances.

PAH's are made up of fused aromatic ring structures, in which much sharing of carbon atoms takes place. Because of the very large number of possible ways these compounds can be structured, their number is myriad, and an exhaustive list, if it exists, would be impractical to print. Some of the simpler members of the class have ring structures like these:

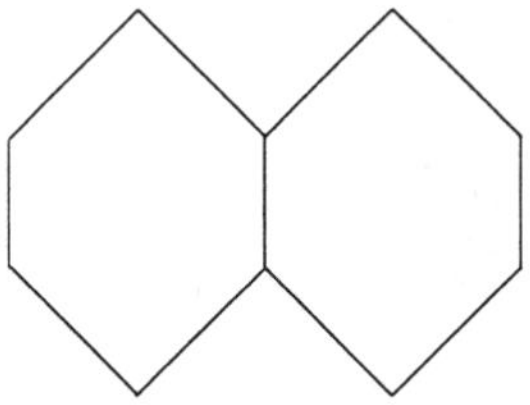

Napthalene $C_{10}H_8$

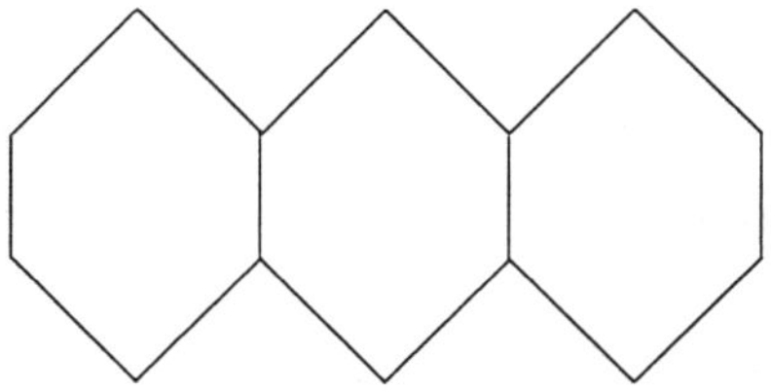

Anthracene $C_{14}H_{10}$

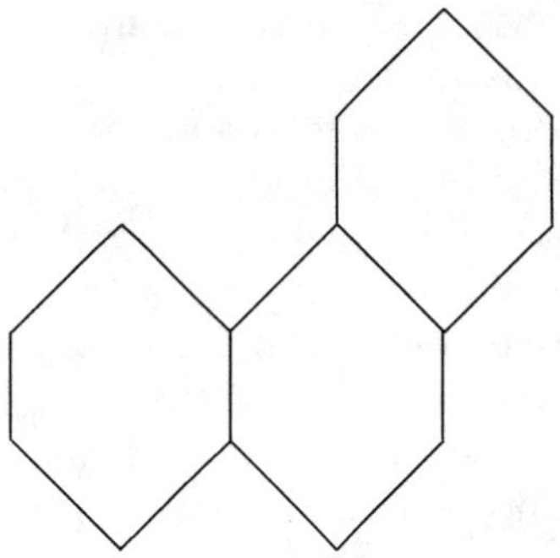

Phenanthrene $C_{14}H_{10}$

These complex molecules can be the framework for substitution and the construction of an infinite number of families of chemicals.

One final group of framework molecules contains some atom or atoms other than carbon in the ring itself. Some of the classes likely to be encountered in waste management are:

Pyridine

```
   C
 /   \
C     C
|     |
C     C
 \   /
   N
```

Furan

```
C  —  C
|     |
C     C
 \   /
   O
```

There are, of course, many others, particularly those used as the basis of complex pesticide and herbicide chemicals. Anyone interested in seeing the structural formulas of these relatively rare materials is advised to contact the Office of Pesticides and Toxic Substances at the United States Environmental Protection Agency in Washington.

FUNCTIONAL GROUPS AND HOMOLOGOUS SERIES

The term *functional group* is applied to atoms or groups of atoms that are recognizable for their contribution to the chemical and physical nature of compounds formed when they are attached to the various framework systems just described. It has been found that certain groups of elements, when added to the structure of a series of hydrocarbons, for example, make a new series of compounds whose properties also change in an orderly way as the framework structure changes.

An example of the construction of a homologous series of such compounds sharing a single functional group is found in the *chlorinated hydrocarbons* in which chlorine is a functional group replacing hydrogen in a variety of hydrocarbon frameworks. These compounds, not found in nature in any significant quantity, have become nearly indispensable in modern society, but because of some of their properties, have come under suspicion as bad actors when released into the environment.

Starting with the simplest hydrocarbon, methane, we can substitute 1,2, 3, or 4 chlorine atoms for the four hydrogen atoms in the methane molecule; each chlorine will attach directly to the central carbon atom. The homologous series that results, the *chloromethanes*, is as follows:

Methane	CH_4	
Methyl chloride	CH_3Cl	(monochloromethane)
Methylene chloride	CH_2Cl_2	(dichloromethane)
Chloroform	$CHCl_3$	(trichloromethane)
Carbon tetrachloride	CCl_4	(tetrachloromethane)

The other *halogens* (remember Group VIIA in the Periodic Table), fluorine, bromine, and iodine, can in principle form comparable series. Thus, in addition to the progression in number of halogen atoms substituted in the molecule, an additional series can be found that goes chloroform, bromoform, iodoform, with similar properties in that series also. Fluorine is a special case and is left out of this list for that reason.

The homologous series of halogenated alkanes, referred to as the alkyl halides, can be expanded almost without limit. Starting with ethylene, the series produces the familiar solvents: dichloroethylene, trichloroethylene, and perchloroethylene, where the prefix *per* means four. Using bromine in the series gives among others ethylene dibromide (EDB), an agricultural chemical that has been under much suspicion in recent years.

A similar approach reveals the homologous series of *alcohols*:

CH_3OH	Methanol	(methyl alcohol)
C_2H_5OH	Ethanol	(ethyl alcohol)
C_3H_7OH	Propanol	(propyl alcohol)
	Isopropanol	(isopropyl alcohol)
C_4H_4OH	Butanol	(2 isomers)
	Isobutanol	(3 isomers)

and so on.

It is not possible to describe the construction of every possible or even every common series here. Table 11 provides a list of the most common functional groups that contribute their names and their properties to whole families of organic compounds, many hundreds of them in daily use.

This list should be used with caution. Not all alcohols behave alike; in fact ethyl alcohol is the only member of the series that is low enough in toxicity to be tolerated in the human system without immediate and disastrous consequences. In a comparable way, not all ethers are exactly alike, nor all esters, etc. It is true, though, that with knowledge of the chemical and often the biological effects of a number of members of a homologous series well understood, a well-informed person can make some useful predictions about the properties of other, less familiar members.

CHEMISTRY IN THE ENVIRONMENT

The behavior of chemicals once released into the environment is often described as the study of the transport and fate of those substances. It is apparent that, to affect anything, a chemical must find its way to the organism to be affected. Anything that may intercept, change, or divert the material from any possibility of affecting the "target," affects the fate of the chemical.

We have discussed chemicals to this point in a descriptive way. No attempt has been made to get into chemical reactions as a way of producing chemicals whose properties and names have been discussed. Substances in the environment outside the laboratory and plant are, however, the real focus of this book. In order to understand the processes that affect chemicals once released from control by events natural or man-made, some study of the nature of chemical reactions is needed.

To this point, where reactions have been discussed, we have implied that the initial material or materials came together and instantly formed the product or derived material, without complications. In the real world of chemical reactions, the complications are numerous, and we cannot take up many of them here. One of the first realities which has to be brought up is that most chemical reactions take time to happen. It is natural to speak of the *rate* of reaction—some measure of the speed with which we can expect the products to appear or the raw materials to change.

One other factor is that many reactions are *reversible*, that is, they can be started from either side:

Reaction 1: $A + B \rightarrow C + D$
Reaction 2: $C + D \rightarrow A + B$

or, in chemist's shorthand

$$A + B \rightleftharpoons C + D$$

Table 11

Some Classes of Organic Compounds with Their Functional Group Structures

Class	General Formula	Structure of Functional Group	Structural Formula of Typical Member of Series	Common Name of Typical Compound
Alkyl halide	R*X	–X X can be F, I, Cl, Br	CH_3Cl ClHC=CHCl	Methyl chloride Dichloroethylene
Alcohol	ROH	–OH	CH_3OH C_3H_7OH	Methyl alcohol Propyl alcohol or Isopropyl alcohol
Ether	R–O–R	R–O–R	CH_3–O–CH_3 CH_3CH_2–O–CH_2CH_3	Dimethyl ether Diethyl ether
Aldehyde	H R–C=O	H –C=O	H H–C=O H CH_3–C=O	Formaldehyde Acetaldehyde
Ketone	R–C–R ‖ O	R–C–R ‖ O	CH_3–O–CH_3 ‖ O CH_3–C–CH_2CH_3 ‖ O	Acetone Methyl ethyl ketone
Carboxylic Acid	O // R–C \\ OH	O // –C \\ OH	HCOOH CH_3COOH	Formic Acid Acetic Acid
Ester	O // R–C \\ OR	O // –C \\ OR	$HCOOCH_3$ CH_3COOCH_3	Methyl formate Methyl acetate
Amine	R–NH_2	–NH_2	CH_3NH_2 $CH_3CH_2NH_2$	Methylamine Ethylamine
Amide	O // R–C \\ NH_2	O // –C \\ NH_2	O ‖ CH_3C–NH_2	Acetamide
Nitrile	R–C=N	–C=N	CH_3C=N H | H_2C–C–C=N	Acetonitrile Acrylontrile

*R represents a hydrocarbon chain (a *radical*) which is attached by one bond to the functional group.

A common example of a reversible *process* is the freezing of water. By simply adjusting the temperature of a cooling fluid, water exposed to contact with that fluid can be made to freeze into ice as the temperature is reduced below 0° C. Raising the temperature of the ice will result in its resuming the liquid form. Careful adjustment can produce slushy forms containing a wide range of ratios of solid to liquid.

This process of freezing and thawing is completely *reversible* in the ordinary sense, and the same water can be frozen and thawed repeatedly without change. To avoid any misunderstanding, there is another use for the term *reversible* in thermodynamics that is much more strictly defined than the word we use here. Hopefully, those who need to know the difference already can handle it; the rest of us interested in environmental protection will have little use for the thermodynamicist's version!

Reversible *reactions* may be thought of as competing to combine and regenerate raw materials. The reaction in each direction proceeds at a rate which is controlled by the temperature at which it occurs and by the concentration of the ingredients available to react. Our intuitive sense that adding more of a raw material will increase the speed of a reaction is often true, but it can cause other complications not relevant to this discussion.

Raising reaction temperature by 10° C usually will double the rate of a chemical reaction, although not all are affected equally. This effect is described by a fundamental mathematical relationship called the Arrhenius Equation, although the rule of thumb about doubling of reaction rates is a handy one for guessing how long to heat something when the available heat source is too hot or too cold.

Obviously, when the first reaction (A+B) starts, there are no products (C+D) in the reactor. As these products of the first reaction appear, however, they can begin to react with each other to reform the original raw materials. At first, there are too few new molecules to "find" each other easily and react. As the concentration builds up, though, the "collisions" between molecules wanting to react become more frequent, and the rate of the reverse reaction begins to increase, reforming A and B.

Eventually, the two reactions will reach a balance. The formation and disappearance of (A+B) and of (C+D) are taking place at rates that are equal, since the buildup of products will increase the rate of returning to the original reagents, while the decreased amounts of (A+B) will slow that reaction down.

The resulting *equilibrium condition* is like the experience we all enjoyed as children when we played with the faucet that filled the sink or lavatory. Running water into a sink with the drain open, we can adjust the flow until a constant level is reached and maintained above the drain. Increasing the flow will raise the level, but since the higher water level forces more water out of the drain, a new steady level is established. In the opposite sense, a slight decrease in water input will lower the equilibirium level.

Of course, what would childhood have been if we had not put in so much water that the drain would not handle it all and an overflow occurred? The same general effect can happen in chemistry with sometimes disastrous results!

If the drain is of higher capacity than the inflowing water, then the water will not accumulate in the sink, since the water is removed as fast as it enters and the reverse pressure needed to overcome the restriction of the drain does not develop. This, too, is analogous to a chemical reaction in which one product is removed from the reaction mixture as a gas or a nonreactive solid, a precipitate. Such reactions are not reversible in the practical sense because the loss of one or more products continues to drive the main reaction until all the ingredients are used up.

A common experience with this kind of reaction is the use of baking soda to neutralize spilled battery acid in the car. When the soda solution or paste contacts the sulfuric acid spilled from the battery, it foams actively because carbon dioxide is released in the reaction:

$$H_2SO_4 + 2NaHCO_3 \rightarrow Na_2SO_4 + 2CO_2 + 2H_2O$$

This kind of dynamic equilibrium in which the balance point can be altered by adjusting inputs and outputs is a characteristic of our physical and chemical world. A French chemist, LeChatelier, summarized the way equilibrium processes behave:

> If a stress is applied to a system in equilibrium, the equilibrium will shift so as to relieve the stress.

Stress can be said to be applied to a reacting system when temperature, pressure, or reactant concentration is changed, among other ways. In this sense, formation of a precipitate or removal of a product from the system as a gas or separated liquid amount to altering the concentration of one of the reactants.

REACTIONS OF CHEMICALS IN THE ENVIRONMENT

Once released to or placed in the general environment, chemicals are affected by all sorts of factors which can sometimes change their composition and form, and often their effects on "target" organisms. Some of the most significant kinds of processes can be discussed in general, but the complications are such that it is not possible to describe all the impacts of environmental factors on chemical substances.

Many materials are biodegradable; that is, left unprotected under normal exposure to air, water, and soil, they gradually decay, break down, or otherwise disappear. It is often said that "natural" things are biodegradable, while "synthetic" ones are not. The real world is not so simple, and natural objects made of organic materials such as bone, horn, or even wood, are occasionally found even though they may be thousands of years old. The Environmental Protection Agency defines as "biodegradable" any substance that is broken down by the action of microorganisms.

Some chemicals are particularly resistant to biodegradation. Among the best known of these are the halogenated hydrocarbons, represented by trichloethylene, chloroform, and a variety of others, most notably DDT and the chlorinated dioxins.

At one extreme of this group, these large molecule compounds contain much chlorine and are so complex that many isomers of each compound are possible. Polychlorinated biphenyl (PCB) has more than thirty isomers and the halogenated dioxins exist in at least as many isomers, some having unique levels of toxicity and others apparently not.

The structures of these halogenated materials make them resistant to attack by microorganisms, and it may be that the removal of a halogen atom from a molecule creates toxicity for the microorganism responsible. This resistance of many chlorinated substances to breakdown after use as a pesticide or herbicide or even as a dry-cleaning solvent created one of the classic confrontations between competing values that have marked the environmental movement's impact on the public consciousness.

When DDT (Dichloro Dibenzo Trichloroethane, a poorly constructed name) was introduced in the 1930s, it was the first persistent pesticide—one that would remain effective for a long time because it did not easily degrade after application. For this it was hailed as a great boon to mankind and widely used for insect pest control in the war zones and all over the world. Authorities predicted the elimination of several insect-borne diseases and pushed for the increasing use of this miracle chemical. It worked, for a while.

Since DDT does not readily degrade, it was later found to *accumulate* in the tissues of many animal species, while the insects which were its targets gradually evolved new metabolic systems that resisted the toxicity of the pesticide. Similar things occurred with the other persistent chlorinated and brominated pesticides—aldrin, dieldrin, dibromochloropropane (DBCP), and their relatives.

More important, the accumulated pesticides, while still not proven to damage complex organisms like humans, were found to interfere in the life cycle of birds. In her seminal book of the modern environmental movement, *Silent Spring*, Rachel Carson clearly identified the issue to the public. Although years were needed to reach a consensus for public action, by 1972 DDT was prohibited from production and use in the United States. Several other actions to ban related compounds were taken in the same time period.

The example of DDT illustrates two significant aspects of the effects of the environmental fate of a chemical intentionally or accidentally introduced into the general environment: bioaccumulation and biological adaptation. The first, *bioaccumulation*, results from the ability of a chemical to persist without major degradation and can lead to vastly increased concentrations of a chemical in a particular animal or plant or in particular organs or tissues. In Japan, for example, a special grass is used to selectively remove copper from drainage waters from old mines. It is so selective that the copper can be recovered.

DDT was shown to accumulate in fatty tissue and to be transmitted in the milk of mammals to their young—a frightening discovery that may or may not cause harm, but which no one wishes to test. The main effect on birds arises out of an effect on the strength of egg shells, some being so weakened that the normal brooding process destroyed the egg before the young could hatch.

The development of *resistance* to pesticides by flies, mosquitoes, and other insects illustrates another important aspect of the fate of some chemicals in the general environment. This aspect may be useful or may make a relatively innocuous product more dangerous. It turns out that many organisms have the ability to adapt to exposure to toxic materials and some can even "learn" to utilize unlikely chemicals as nutrients or as sources of energy for growth.

In the case of microorganisms, the ability to evolve and adapt to the presence of usually toxic chemicals can be a very useful tool in waste treatment. Gradual conditioning of a population of microbes to exposure to one or more chemicals or wastes can often develop the ability in the microbes to detoxify the waste. New experiments with genetically engineered organisms have shown real promise for this kind of work, but this work is still primarily in the laboratory scale.

In addition to the biodegradation of chemicals, a number of ordinary chemical reactions are likely to occur in groundwater, soil, or air. Groundwater moves very slowly and is in contact with many kinds of minerals and natural organic compounds derived from soil and rocks. The slow migration of the water underground provides the time needed for reactions to proceed toward equilibrium under the conditions affecting the groundwater. Even though temperatures are low and concentrations are often infinitesimal, over a long period of time reactions do occur. Conditions like pH variation, changes in redox potential, and even direct reactions with minerals like limestone are found underground.

The EPA has used a typical migration rate of flow of groundwater of 10 meters, a little more than 30 feet per year. Even at barely discernable reaction rates, many chemicals can undergo changes, becoming more or less toxic than in their original form at the time of release from improper disposal, spills, or leaking landfills. Probably the most common reaction of chemicals in groundwater is *hydrolysis*, the reaction of a chemical with water to produce one or more new chemicals.

A simple hydrolysis reaction is represented by the chemical equation:

$$C_2H_5OONa + H_2O \rightleftharpoons C_2H_5OOH + Na^+OH^-$$

Sodium Acetate	Water	Acetic Acid	Sodium Hydroxide

It will be seen that, if the environment is alkaline, the equilibrium will be driven to the left and little hydrolysis will happen. This reaction is characteristic of many organic compounds, each in its own fashion. It is often necessary to evaluate whether the fragments of the original molecule are more or less toxic or otherwise of concern in deciding how hazardous a spill or leak can be.

In a contrasting kind of reaction, some compounds, notably the chlorinated dioxins, which have been of great concern in recent years, are extremely stable and resistant to biodegradation or most chemical reactions in groundwater or similar locations, but are degraded fairly rapidly by sunlight or other sources of light of selected frequency. This process, called *photolysis*, can occur when the energy of photons of light is just right to be absorbed by the key bonds of the molecules, which can be excited to a level at which the chlorine atoms may be kicked out of the structure. The result, a change in the composition of the pollutant, can dramatically change its properties, including toxicity. There is an obvious difficulty in arranging for light to be applied to anything on or in soil, but some trials have shown useful results. Compounds less hard to handle than dioxin can be treated this way in infrared or ultraviolet tunnels.

Finally, much of the original justification for in-ground disposal of wastes was based vaguely on an idea that soil was able to hold chemicals and other wastes firmly in place and so prevent them from becoming sources of exposure of people and animals. This too-simple model has failed, as shown by experience when landfills and disposal sites have been found to be sources of contamination of soil and water far from the original point of disposal. It turned out that this process of migration has both physical and chemical elements that are of importance in understanding exposures from land disposal sites.

Nearly everyone has had experience with applying some liquid, usually water, but maybe gasoline or oil to soil. At first, the soil merely changes color and may have the characteristic odor of the spilled liquid. Adding more liquid will make a pool of free liquid, that spreads the spot further as the liquid moves away either horizontally or vertically. This basic physical process, in which the spaces between the grains of soil are filled with liquid, is one process by which contaminants can travel through soil to other locations at which they may create hazards.

When underground liquid storage tanks leak, the physical migration of gasoline, for example, is a major pathway by which hazardous materials can enter basements, manholes, and other underground spaces. This physical process of *absorption* occurs to some extent with all contaminants, and is a significant pathway of exposure, as we have just discussed. However, it is water in the zone of *saturation* where the interstices between soil particles are filled with water, that is the most significant carrier of contaminants over long distances. The absorption process through the dry soil is most important for substances that are chemically relatively inactive, like the hydrocarbon components of gasoline and fuel oil.

Many chemicals, however, have functional groups that can form bonds to chemically active spots or sites on the surface or in the irregularities of the surface of soil or sand, or even with organic substances often found in soil: peat, coal, wood, charcoal, or humic acids. This kind of bonding is called *adsorption* or sometimes *chemisorption* to emphasize the role of chemical bonds in the phenomenon.

The bonds which form between the functional group or groups in the molecule and the soil vary in strength from very weak to so strong as to be practically unbreakable, in which case what amounts to a new chemical compound has been formed and is as resistant as any other molecule to disruption. Many of the weak ones are mere associations easily broken by dilution or other changes that can occur in the groundwater environment.

The effect of chemisorption is to first adsorb the chemical as it arrives at a point in the soil, reducing the concentration of the contaminant in the groundwater and therefore preventing further migration of the pollutant. Actually, an equilibrium develops in which the more contaminant adsorbed, the higher the concentration of the pollutant left in the liquid in equilibrium.

If there is sufficient supply of the pollutant, the soil at the point of contact will gradually adsorb all it can in equilibrium with the concentration in the flowing groundwater, until the passing water is up to the original concentration and begins to equilibrate with another layer of soil. Even though the water provides a continuing supply of the chemical, these new supplies are held back and delayed, giving a rate of migration slower than if there were no interaction with the soil. This process, an important one in designing and monitoring groundwater protection systems, is called *retardation*.

When the flow of a chemical is retarded, even more time is provided for hydrolysis and other breakdown reactions to occur. Often, too, the retardation will be so great as to substantially eliminate any migration beyond some arbitrary point for a time long enough to reduce risk to persons and property to an acceptable level. In these exceptional cases, land disposal does work as if the old ways of thinking were correct.

The retarding of migration through soil is measured by a *diffusion coefficient* which can be determined in a laboratory. Values of the coefficient are dependent on both the chemical and the soil type through which it migrates. The values are used in various mathematical equations and / or numerical calculation models to make predictions about migration. This is preeminently an expert field, in which untrained or half-trained persons can get greatly confused. It is possible, however, for the layman to ensure that the values of the diffusion coefficients are appropriate for the kinds of soils under consideration.

Finally, the interaction between released chemicals and the natural environment is usually slow and deliberate. Frightening pictures on the evening news give the impression that apocalyptic things have happened or are about to happen. While this is true of the spectacular events surrounding a train or truck wreck or a massive spill, such events are rare and out in the open for all to see and take precautions. The contamination of groundwater or soil and often even of the air by improper disposal of wastes is very rarely going to create a dramatic emergency. This is no excuse for complacency, but the discussion in this chapter should show that there is a vast array of knowledge about these chemicals and a wide variety of ways of studying them. The myth of some mysterious threat from chemical disposal is just that—a myth—and can be dispelled by deliberate study and carefully developed actions based on the orderly laws of chemistry.

Chapter 12

The Resource Conservation and Recovery Act and Extensions

BACKGROUND OF THE ACT

During the decade 1968–1978, the "environmental decade," an uncountable number of laws were passed at Federal, State, and local levels to address the perceived problems of air and water pollution. Finally, it seemed, the circle was closed and all waste brought under public control by the passage of the Resource Conservation and Recovery Act (always referred to as RCRA) in 1976. Still to come was the Superfund law and numerous amendments and expansions of the basic environmental statutes, but with RCRA the essentials were in place.

It's difficult, after the passage of years, to remember the public interest, at least in some of the public, that found expression in opposition to the Vietnam War and in a nearly equal but less articulate popular movement opposing profligate use of the earth's resources and all forms of pollution of the environment. A new environmental movement found its voice in militant, even radicalized organizations. Some were of venerable age, like the Sierra Club and the Wilderness Society. Some were formed out of the awakened concerns—Friends of the Earth, Natural Resources Defense Council, and many others planned to deal with local issues, often lasting only so long as the issue itself lasted.

To anyone proposing a new development, whether of public infrastructure or private investment for profit, it seemed that every neighborhood sprouted vocal groups opposed to just about everything. In this highly polarized confrontation, it is not surprising, at least in retrospect, that technical limits, economic factors, and even human nature got lost in the reforming zeal.

Faced with this near-revolutionary public movement, Congress moved hesitantly at first, then with more confidence and less concern for reality, into the political mine field of environmental legislation. It is ironic that the course chosen was basically dependent on technology as a fix for the perceived errors and abuses of the technological age. Few institutions in modern society have less competence in technology and its applications, but Congress is immune to self-examination in such matters. Congress-watchers sometimes marvel at the cavalier way in which the legislative branch imposes heavy demands on all sectors of society in such areas as environmental policy, discrimination, even employee safety and health, then blithely exempts itself from the same inconvenient laws.

In a similar way, Congress is almost never deterred by advice that its requirements are divorced from reality. In the case of the environmental laws, a new term—"technology-forcing legislation"—was invented to stifle all such carping. The real basis of technology-forcing is an infinite faith in technology itself, surely the most involuted way of controlling the impact of technology yet devised! At any rate, in the progression which began with the National Environmental Policy Act, and proceeded with the Clean Air Act and the Clean Water Act, the eventual arrival of a "Clean Earth Act" was inevitable and demonstrably needed to provide some control over the contamination that was bound to come when the pollutants were removed from air and water and needed some final resting place.

When it arrived, finally, it bore the less-than-clear title of the Resource Conservation and Recovery Act to reflect a political consensus to shift away from the narrow goal of preventing pollution and to claim the high moral ground of conservation and restoration of resources. These higher purposes still were to be accomplished by punitive laws restricting disposal of residuals. The idea of technology-forcing legislation reaches a new height of expression in RCRA: if disposal is prohibited, the generator will (must) find a way to avoid the production of the interdicted waste. There is generally no recognition of economic or even of technical limits on the actions ordained. Congress legislates as if it had the power to change physical laws and lets the chips fall where they may.

The first time the law was amended, its name was changed to the Solid Waste Act (SWA), later to the Hazardous and Solid Waste Amendments of 1984 (HSWA), but it remains RCRA in the hearts and minds of those involved in its implementation.

The result of this approach has been years of conflict and litigation over the interpretation of a law that is in many ways impossible to implement in literal compliance with its language. No one really knows how many printed pages of regulations have been issued in the wake of RCRA, but an estimate of ten thousand pages is current and may be low. Interpretations and internal documents of the EPA for the guidance of its own people probably are many times as voluminous. The Permit Writer's Guide for the location of waste disposal facilities, for example, is nearly four inches thick in typewritten form and positively bristles with acronyms and shorthand jargon intended to keep down its bulk, making it totally incomprehensible to any but the most initiated readers.

To illustrate the enormity of the RCRA statute, Table 12 shows the main headings of the law as amended through 1984 by the Hazardous and Solid Waste Amendments (HSWA). This information was prepared by Dr. Ernest Rosenberg of the E. Bruce Harrison Company.

A full-length book is required even to begin a detailed analysis of the RCRA statute and the regulations as they stand at any given moment in the attempt to make it work. Readers who want such detail will have to seek it in publications of the Environmental Law Institute, the major trade associations like the National Association of Manufacturers and the Chamber of Commerce of the United States, or those of large consulting organizations working in the field. Curiously, the major environmental public interest groups do not appear to have publications intended to assist the public in understanding the law.

Here we will attempt only to discuss the structure of the act and programs related to it. Anyone responsible for compliance with the regulations is urged to obtain and study the available publications, to attend quality seminars and training courses, and to establish a relationship with an attorney with special competence in the field. The consequences of even well-intentioned mistakes are simply too severe to allow for any carelessness or casual response.

STRUCTURE OF THE ACT

The law which spawned "the most complex regulatory program ever devised" is itself long and complex, as we have repeatedly said. There are nine Subtitles of varying significance. Subtitle E, for example, provides for the Department of Commerce to cooperate with the EPA in the only real bow to resource recovery and recycling to be found in the plain language of the law (other activities are implied and "encouraged" in other sections). The same Congress that wrote this mandate has never provided more than token funding for Commerce to participate in a meaningful way. Subtitle F just pronounces some pious words about the use of recycled materials by the Federal government and not unreasonably makes the Federal agencies subject to all the provisions of the act as well as to requirements set by State, interstate, and local officials. Since the President is given power to exempt an agency from this law and all others with similar objectives, it is not surprising that the behavior of some Federal agencies in response to the regulatory program has rivaled the response of the most recalcitrant private companies and individuals.

Policy and Goals

In the first two Subtitles, A and B, Congress expresses its rationale (findings) for the legislation, then announces a pair of goals as national policy:

1. The generation of hazardous waste is to be reduced or eliminated as soon as possible, and

2. Any hazardous waste that is still generated is to be managed in ways that minimize the present and future potential harm to "human health and the environment."

After this statement comes a series of objectives (really strategies in conventional planning terms) for the achievement of the policy goals. Among these are proper management of hazardous waste (which this book is about!), waste minimization, elimination of land disposal of hazardous waste, prohibition of open dumps, etc.

Among the key features of the opening Subtitles is the definition of solid and of hazardous waste in

Table 12

Index to Sections and Subsections
The Resource Conservation and Recovery Act of 1976

As Amended by:
The Hazardous and Solid Waste Amendments of 1984 (Pub. L. 98-616, November 8, 1984)

continued on next page

Table 12
(Continued)

Sec. 7003.	Imminent hazard.
Sec. 7004.	Petition for regulations; public participation.
Sec. 7005.	Separability.
Sec. 7006.	Judicial review.
Sec. 7007.	Grants or contracts for training projects.
Sec. 7008.	Payments.
Sec. 7009.	Labor standards.
Sec. 7010.	Interim control of hazardous waste injection.
[Sec. 7011.	No section.]
Sec. 7012.	Law enforcement authority.
Subtitle H—Research, Development, Demonstration, and Information	
Sec. 8001.	Research, demonstrations, training and other activities.
Sec. 8002.	Special studies; plans for research, development and demonstrations.
Sec. 8003.	Coordination, collection, and dissemination of information.
Sec. 8004.	Full-scale demonstration facilities.
Sec. 8005.	Special study and demonstration projects for recovery of useful energy and materials.
Sec. 8006.	Grants for resource recovery systems and improved solid waste disposal facilities.
Sec. 8007.	Authorization of appropriations.
Subtitle I —Regulation of Underground Storage Tanks (All New)	
Sec. 9001.	Definitions and exemptions.
Sec. 9002.	Notification.
Sec. 9003.	Release detection, prevention, and correction regulations.
Sec. 9004.	Approval of State programs.
Sec. 9005.	Inspections, monitoring, and testing.
Sec. 9006.	Federal enforcement.
Sec. 9007.	Federal facilities.
Sec. 9008.	State authority.
Sec. 9009.	Study of underground storage tanks.
Sec. 9010.	Authorization of appropriations. [Refers back to sec. 2007(g).]
Uncodified Provision	
§701, Pub. L. 98-616:	National Ground Water Commission.

Public Affairs/Issues and Planning January 14, 1985

Information prepared by Dr. Ernest Rosenberg of the E. Bruce Harrison Company. Used with permission.

Section 1004 (27). Congress obviously decided that everything that is waste would be in the purview of this law unless it is specifically exempted by statute. Clearly, Congress at this point trusted no administrator or judge to make the decision as to which waste materials are controlled and simply went for it all:

> The term "solid waste" means any garbage, refuse, sludge, from a waste treatment plant, water supply treatment plant or air pollution control facility and other discarded material, including solid, liquid, semi-solid, or contained gaseous materials resulting from industrial, commercial, mining and agriculture activities and from community activities but does not include solid or dissolved material in domestic sewage, or solid or dissolved materials in irrigation return flows or industrial discharges which are point sources subject to permits under Section 402 of the Federal Water Pollution Control Act, as amended (86 Stat. 880), or source, special nuclear, or byproduct material as defined by the Atomic Energy Act of 1954, as amended (68 Stat. 923).

The punctuation of the paragraph results in some ambiguity, a not-unusual situation. However, it leaves no doubt that all forms of waste except a few specified ones are solid waste even if they would leak out of a hole in your bucket or spew into the air if their containers were punctured.

Another key definition which, when taken together with the broad definition of solid waste, covers a whole universe in a few paragraphs, is the one for hazardous waste found in Section 1004(5):

> The term "hazardous waste" means a solid waste, or combination of solid wastes, which because of its quantity, concentration, or physical, chemical, or infectious characteristics may—
>
> (A) cause, or significantly contribute to an increase in mortality or an increase in serious irreversible, or incapacitating reversible illness; or
>
> (B) pose a substantial present or potential hazard to human health or the environment when improperly treated, stored, transported, or disposed of, or otherwise managed.

The obligatory legal authorizations to the administrator of the EPA together with organizational instructions in wearying detail follow. An innovation in environmental law enforcement was introduced—

armed criminal investigators were authorized to seek out responsible parties and initiate prosecution for violation of proper disposal practices.

HAZARDOUS WASTE MANAGEMENT

Subtitle C of RCRA takes up forty pages of fine print in *Environmental Statutes*, a publication of the Environmental Law Institute. It also has required around 100,000 pages in the *Federal Register* to propose, announce, and promulgate its regulatory framework. The best efforts of thousands of professional and technical people in Federal, State, and local governments, industry and commerce, the courts, and the halls of Congress itself have been applied to try to understand and carry out its objectives. Notably, there has never been a vocal opposition to the objectives of this act, but the disagreement over the means of accomplishing those objectives is loud, serious, and continuing.

At this point, we can examine a charge often made regarding the devisers and authors of this immense program. "They" are often accused of operating from an anti-industrial perspective, of being latter-day Luddites, spiritual descendents of those English workers who smashed the steam-driven textile machines to protect their cottage-industry life-style from the soulless factory system as it developed in the eighteenth century. Certainly, there have been advocates of "stringent" controls (a favorite term) who sound like reborn romantics wishing to return to a simpler life in a simpler world.

Experience under these programs, however, seems to indicate that the complexities arise out of simpler, perhaps less noble, thinking, usually called "practical politics" in the American system. First, the advocates of stronger action simply do not care how much the program costs, and when pursuing a dream, costs, at least to others, are not a factor. Second, politicians learned in the 1960s and 1970s that a hard-nosed stance on protecting the environment is worth many votes; it has catapulted relative unknowns into positions of great power as spokesmen for their followers in political life, particularly for those in elective office. Few elected officials will, at least publicly, abandon the theme that got them their office. Responding to the pressure from their opposition, they often decide that the only course open to them in achieving reelection is more regulation—regardless of costs or even feasibility.

Having said all this, we are still, in Subtitle C, faced with the reality of a law that in 19 Sections identifies, instructs, requires, permits, prohibits, investigates, and punishes, all in the name of protection of that grand complex always named and never defined: "human health and the environment."

The usual description of the guiding principle of RCRA is that it provides "cradle to grave" management of dangerous wastes. This slogan oversimplifies the scope of the act, which also extends responsibility for wastes into all eternity for any generator who fails to carry out every detail of the constantly changing regulatory program. If every step is accomplished and fully documented, if all the records are maintained intact, and *if there is no new information justifying a change in the requirements for disposal of the particular waste*, it is possible to be free of further responsibility for the waste after thirty years from the time the disposal site is finally closed. Even then the provisions of the Superfund law (CERCLA) will apply, as discussed in Chapter 13 on Superfund.

As we shall see, the law is designed to establish and to maintain forever the link between the generator of waste and the waste itself wherever the waste may be placed for disposal. To accomplish this permanent link between the generator and the waste and its consequences in the environment, Congress devised a system that forces the generator to track the material through all its later history. To further ensure the permanent accountability for the waste, the generator is made responsible for the performance of every actor in the history of the waste. As we shall see, if the waste ever reappears with negative effects on humans or the environment, the government will always look for the original generator to accept responsibility for cleaning up or "mitigating" the effects.

A generator of hazardous waste is, under the law, any "person" who by act or process produces hazardous waste as identified in the regulations, and each site at which wastes are made is a separate generator. A large corporation with many locations can have many that are separately defined as generators.

A homely example of a generator is this: in the garage or basement, most of us have several cans or buckets of paint which we believe we might someday use. So long as the cans remain quietly on the shelf, they are simply consumer products. They may be of concern to the local fire marshal or to our insurance underwriters, but the hazardous waste laws have no relevance to them. One day, however, in a moment of enthusiasm, we decide to "clean out the stuff" that we just decided "we will never use." At that moment—"we will never use it"—the paint and the cans become hazardous waste and we have become generators. In fact, since nearly all paints become hazardous waste when discarded, we have become a storage facility!

The example begins to fall apart at this point because household wastes are treated very differently in the regulations and are generally exempt from the management requirements. It is important, though, to realize that the decision to classify material which meets the definition of hazardous waste as a waste, even if it remains on the site where it originated, triggers the regulations under RCRA.

Generators who are not exempt under one of the numerous but diminishing lists of facilities and materials must examine all their wastes and classify them to determine whether any of their "solid" wastes are hazardous, and if so, how much is produced. In this process, it is essential to recall that just about any abandoned, recycled, or "inherently waste-like" material must be considered and evaluated. Many things not previously or commonly thought of as wastes are covered by these regulations as solid wastes and must be tested and evaluated against the formal criteria before any disposal is even thought about.

A solid waste, assuming it does not qualify for one of the exemptions (this may be a serious study in itself), will be deemed a hazardous waste if it can be found in any of the lists in current use by the EPA, published in the Code of Federal Regulations, Volume 40, Part 261, Subpart D. This heady volume, found in most larger public libraries, is indispensable for learning what wastes are considered hazardous.

If the waste from a given site is not found on one of the lists—is not a "listed waste"—then it is necessary to test the waste to find out whether it exhibits any of the four "characteristics" which are described in the same Code of Federal Regulations, Part 261, Subpart C. These "characteristics" are ignitability, corrosivity, reactivity, and toxicity, as fully defined in many pages of regulatory rhetoric. (Our Chapter 4, "How to Recognize a Hazardous Waste," also goes into more detail on these characteristics.) Running these tests is not for well-meaning amateurs or even trained chemists who have not been especially checked out for these methods—they are not, in the judgment of many, routine procedures.

Finally, we need to determine if any listed wastes have been mixed with a solid waste—if they have, the whole mixture is classified hazardous. If "characteristic waste" (that is, waste as generated that has one or more characteristics that make it hazardous) is mixed with other waste, the mixture itself must have a characteristic that is in the hazardous range for the whole to be classified hazardous. Not all authorities agree on this last statement, and it is always better to avoid mixing wastes unless the mixing is done as a part of the system of management.

If after all the searches of lists and all the testing of candidate wastes are complete, any one of the wastes is found to fit the regulatory criteria for hazardous wastes, the facility is a generator. If the quantity of wastes produced or accumulated on the site in one month is 1,000 kilograms (2,240 pounds) or more, the full standards applicable to a generator must be met. If the amount is less than 1,000 kilograms per month, then the standards applicable to small quantity generators (SQG) apply. In the case of wastes containing some extremely hazardous materials, the regulations may apply even when waste production or accumulation is less than 100 kilograms per month. We will discuss the requirements later in this chapter.

The Heavy Responsibilities of a Generator

Once it is decided that a site is a generator of hazardous waste, the first required step is the filing of a notification to the USEPA Regional Office for the area including the facility. This notice is filed on a form (EPA 8700-12) which applies for the required Identification Number, which is essential to all further operations under RCRA. State environmental control agencies may and often do have similar or quite different requirements which only a contact with the one having jurisdiction will define. The notices are required immediately on determining the status of the site as a generator, but no more than 90 days after the decision is made.

Figure 11 shows the form in effect in 1990, although it is not practical to show the two-page set of line-by-line instructions that must be followed in completing the notification. Most copies of this form will be State-specific, in that the definition of hazardous wastes that must be listed will vary from state to state. Anyone planning to fill this out who has no background in the field had better seek some advice, starting with the State regulatory agency having jurisdiction.

Actually, any person who manages a hazardous waste in any way (unless exempt) must notify the EPA and the affected State on the same forms as are provided for generators. All these notices are public documents but contain a minimum of information about the activity.

After the notice is filed, the generator becomes orchestrator of the entire "cradle-to-grave" system and must perform all the steps of a complex "dance" without fail all the time to avoid future liability of unknown but likely large magnitude.

The first step in operating under the system is the decision whether a permit is required for treatment, storage, or disposal of hazardous wastes on the site. Even if all wastes are shipped away for eventual disposal, a permit will be needed if the wastes are stored for a longer period than is allowed without a permit. The limit is usually 90 days but in special cases can be more or less; so again, classification of the wastes must be taken into account. In particular, small quantity generators (SQG's) have special rules for storage which take into account the distance to a disposal site and other factors.

Any application for a treatment, storage, or disposal permit is a major venture. Part A is relatively simple and is made on a printed form that can often be completed with

Figure 11

Please print or type with ELITE type (12 characters per inch) in the unshaded areas only

Form Approved. OMB No. 2050-0028. Expires 10-31-91
GSA No. 0246-EPA-OT

Please refer to the *Instructions for Filing Notification* before completing this form. The information requested here is required by law *(Section 3010 of the Resource Conservation and Recovery Act).*

EPA

Notification of Regulated Waste Activity

United States Environmental Protection Agency

Date Received (For Official Use Only)

I. Installation's EPA ID Number *(Mark 'X' in the appropriate box)*

A. First Notification ☐ B. Subsequent Notification ☐ *(complete item C)*

C. Installation's EPA ID Number

II. Name of Installation *(Include company and specific site name)*

III. Location of Installation *(Physical address not P.O. Box or Route Number)*

Street

Street (continued)

City or Town | State | ZIP Code

County Code | County Name

IV. Installation Mailing Address *(See Instructions)*

Street or P.O. Box

City or Town | State | ZIP Code

V. Installation Contact *(Person to be contacted regarding waste activities at site)*

Name *(last)* | *(first)*

Job Title | Phone Number *(area code and number)*

VI. Installation Contact Address *(See Instructions)*

A. Contact Address: Location ☐ Mailing ☐ | B. Street or P.O. Box

City or Town | State | ZIP Code

VII. Ownership *(See Instructions)*

A. Name of Installation's Legal Owner

Street, P.O. Box, or Route Number

City or Town | State | ZIP Code

Phone Number *(area code and number)* | B. Land Type | C. Owner Type | D. Change of Owner Indicator: Yes ☐ No ☐ | (Date Changed) Month Day Year

EPA Form 8700-12 (07-90) Previous edition is obsolete.

only modest technical background and research into the nature of the operation itself. The form is basically the first part of the application and has little function in today's scheme. Part B, however, is a technical report for which the applicant is provided little guidance by the EPA on the format or the data to be furnished, on the theory that no government form can possibly allow for the immense range of conditions encountered in the thousands of applications filed. Most companies not engaged in professional services in the waste disposal field will find it essential to retain experts to get through this process.

If wastes are to be shipped off-site for treatment, storage, or disposal, then the generator has to deal with another major feature of the regulatory system—the need to assure someone else's compliance with the rules. Transporters of hazardous waste must meet their own special requirements under the regulations, which we will discuss later. For now, we will simply say that the generator is responsible for ensuring: (1) that the transporter has its own EPA Identification Number, and (2) that the waste is accompanied by the Uniform Hazardous Waste Manifest at all times from the point of generation to the point of final disposal. Some states have their own Manifest forms that may also be needed, although a major attempt has been made to standardize on the national form exclusively.

The version of the Manifest shown in Figure 12 is representative. It is the Minnesota version, as shown by the State Manifest Document Number and some spaces which are clearly shown as "not required by Federal regulations, but required by State law and rules." This particular version is an eight-copy form, with each copy identified as to its required routing—Copy One, for example, is to be mailed by the TSDF to the generator after the shipment is received at that facility.

The Manifest, in addition to the required description of the waste itself, must be filled out by the generator with a list of every authorized transporter for the particular shipment with each transporter's EPA Number, and must specify the permitted facility which will accept the waste for final disposal with its EPA Number.

Not content with the information on the face of the Manifest, the rules require that the generator write some specific certifications on the Manifest:

1. That the waste is packaged and labeled in compliance with EPA and Department of Transportation rules;

2. That the generator has a positive program designed to reduce the volume or quantity and toxicity of the kinds of wastes being shipped, insofar as is economically practicable; and

3. The proposed method of treatment, storage, or disposal (selected by the generator, but presumably applied at the destination) is that practicable method currently available to minimize the present and future threat to human health and the environment resulting from the disposal of the waste described in the Manifest.

The Manifest must be signed by hand and given to the transporter with enough copies to allow one to be signed and left at each change of custody of the material. The deceptively simple requirement that the Manifest be signed by hand is not so simple. The Manifest is a legal document, and the person who signs on behalf of the generator is presumed to be an agent of the generator, individual or company, no matter what rank that person holds in the company. False certification as to the three items required may result in criminal liability for making false statements to the government. If someone certifies that the company has a program in operation which is designed to minimize or eliminate the waste being shipped, that program had better be there and if necessary, available to investigators who may want to validate the certifications. In short, these papers are not routine ones that can be relegated to the plant garbage crew—they may in all good faith subject the company to large penalties, as we shall see later in this chapter.

Permits

Before undertaking any more discussion of transportation, we need to look further into the subject of permits which are required to be obtained by "all persons" who treat, store, or dispose of hazardous waste. Permits may be obtainable through State agencies in those States having authority delegated to them under RCRA regulations. In the few States not fully authorized, separate permits may be needed—from the Federal EPA and from a State agency. The Federal legislation is designed to have the day-to-day administration and enforcement of the program done by the States, requiring that their program be at least as stringent as the Federal rules.

The only generators who can be and remain free from the need of a permit under Federal rules are those generators who generate less than 100 kilograms per month of listed or criteria waste and who never exceed that rate. Furthermore, those conditionally exempt small generators cannot have more than 1,000 kilograms of waste in storage for more than 6 months (or 9 months if the waste has to be shipped more than 200 miles for disposal) without having obtained a permit. Even these small generators, if their waste is classified as "acutely hazardous," are required to have a permit as a treatment, storage, or disposal facility regardless of the magnitude of their

Figure 12

MINNESOTA POLLUTION CONTROL AGENCY
HAZARDOUS WASTE DIVISION
520 LAFAYETTE ROAD
ST. PAUL, MINNESOTA 55155
ATTN: HWIMS

For MPCA use only

OMB No. 2050-0039
Expires 9/30/91

Please **TYPE** (Form designed for use on elite (12-pitch) typewriter) or print **LEGIBLY. Instructions on cover page.**

UNIFORM HAZARDOUS WASTE MANIFEST	1. Generator's US EPA ID No.	Manifest Document No.	2. Page 1 of	Information in shaded area not required by Federal law. Minnesota rules require Items H. and I.

3. Generator's Name and Mailing Address (Also location of waste generation if different from mailing address.)

A. State Manifest Document Number
MN 117153

B. State Generator's ID

4. Generator's Phone ( ) **County:**

5. Transporter 1 Company Name	**6. US EPA ID Number**	C. State Transporter's ID D. Transporter's Phone
7. Transporter 2 Company Name	**8. US EPA ID Number**	E. State Transporter's ID F. Transporter's Phone
9. Designated Facility Name and Site Address	**10. U.S EPA ID Number**	G. State Facility's ID H. Facility's Phone

	11. US DOT Description *(Including Proper Shipping Name, Hazard Class, and ID Number)*	12. Containers		13. Total Quantity	14. Unit Wt/Vol	I. Waste No.
HM		No	Type			
a.						
b.						
c.						
d.						

J. Additional Descriptions for Materials Listed Above	K. Handling Codes for Wastes Listed Above

15. Special Handling Instructions and Additional Information

16. GENERATOR'S CERTIFICATION: I hereby declare that the contents of this consignment are fully and accurately described above by proper shipping name and are classified, packed, marked, and labeled, and are in all respects in proper condition for transport by highway according to applicable international and national government regulations and all applicable state laws and regulations.

If I am a large quantity generator I certify that I have a program in place to reduce the volume and toxicity of waste generated to the degree I have determined to be economically practicable and that I have selected the practicable method of treatment, storage or disposal currently available to me which minimizes the present and future threat to human health and the environment, or, if I am a small quantity generator, I have made a good faith effort to minimize my waste generation and select the best waste management method that is available to me and that I can afford.

Printed/Typed Name	**Signature**	Date: *Month Day Year*

17. Transporter 1 Acknowledgement of Receipt of Materials

Printed/Typed Name	**Signature**	Date: *Month Day Year*

18. Transporter 2 Acknowledgement of Receipt of Materials

Printed/Typed Name	**Signature**	Date: *Month Day Year*

19. Discrepancy Indication Space

20. Facility Owner or Operator: Certification of receipt of hazardous materials covered by this manifest except as noted in Item 19.

Printed/Typed Name	**Signature**	Date: *Month Day Year*

GENERATOR — TRANSPORTER — FACILITY

In case of spill or emergency, immediately call the Minnesota Pollution Control Agency at 612/296-8100; the Minnesota Division of Emergency Services at 612/778-0800; the National Response Center at 800/424-8802. Outside Minnesota, call the National Response Center and appropriate state agency.

Minnesota Form PQ-00371-04 (9/90) **(Previous editions obsolete)** production. Research facilities have special rules because of the frequent occurrence of small quantities of many different kinds of hazardous waste in their operations. Many States provide no special treatment for small-quantity generators and are not required to do so by the Federal rules.

The basic approach for obtaining a permit for treatment, storage, or disposal (TSD) is the same whether one is operating a plant seeking to deal with its own waste or operating a for-hire service to others. Commercial operators of hazardous waste TSDF are nowadays aware of all the complexities; so for our purposes in this book, we will focus on the issues and procedures which a business must face in storing, treating, and disposing of its own waste on its own property.

Earlier, we referred to the need for a generator to develop a waste management plan, based on a study of waste quantity and classification. This plan is the primary input to the permitting process.

Part A of the application is shown in its current form (1988) in Figure 13. This form amounts to little more than a statement that the plant or facility is going to apply for a full permit. Originally, this Part A was needed to allow existing facilities to continue in operation under "interim status," but this status disappeared in November, 1985, unless a Part B application has been filed. Some 1,200 facilities closed as of that date, since they had no permits allowing continued operation. New operations must now file both parts simultaneously.

Part B is a completely different kettle of fish from Part A. The EPA published no form, only a detailed set of standards for TSDF which are almost all performance-based; that is, the applicant is expected to define what waste will be the subject of the application, then to develop, on a basis specific to the wastes and to the site, detailed designs for the facilities required that will, with proper operation and maintenance, manage the wastes in accordance with the regulations. The regulations themselves usually require extensive interpretation in the light of the specific site conditions to know (or guess) their meaning.

The design basis of an application must take into account location factors peculiar to the site. As an example, even if local land-use regulations allow the industrial or commercial facility to operate and if that operation necessarily produces regulated wastes, it may well be found that the hydrogeology or flood plain limitations in Federal regulations will not allow any TSDF within the plant boundary.

A few large industries with large and well-informed technical, regulatory compliance, and legal staffs might undertake the preparation of a TSD permit in-house. Today, nearly everyone must employ specialist firms familiar with every detail of the permitting process.

It must be remembered that until the EPA formally accepts the application as "complete," nothing will occur toward the preparation of the permit itself. Since the EPA has no easily described way of deciding when an application is "complete," the applicant can be really flying blind without the advice of professional advisers who have prepared and submitted enough applications to have a feel as to when the data and the presentation are likely to be acceptable. Even then, one or two cycles of submission and rejection are not unusual.

For all these reasons, it is virtually impossible to predict the time needed to obtain a permit, build the facilities, and begin operations. Periods of years are to be expected, and any opposition that develops during the required public notice and hearing process can make it virtually impossible ever to obtain a permit. Permits can be issued for ten years, but they are reviewable whenever a major change in the regulations occurs.

As we have mentioned, virtually all the States have been authorized to administer the Federal requirements as well as their own; so while these comments apply in all cases, many States apply additional requirements and limitations beyond those in the Federal regulations. Abuse of this power is common, and some States have managed essentially to prohibit any hazardous waste disposal within their boundaries.

In addition to the waste management plan and the technical design and operational factors, the applicant must develop a plan which describes the criteria for closing the facility—that is, for stopping receipt of waste and completing any operation. This closure plan must describe in detail the steps to be taken to place the site into a safe condition that will ensure no significant impact on human health or the environment from the wastes left there for an indefinite but long-time period.

When the original application is being put together, one of the major technical design factors is a thorough study of the geology and hydrogeology of the proposed site. This work, often involving drilling of many wells for the interpretation of the presence and flow patterns of groundwater in the various geological formations underlying the site and often extending for great distances, is the foundation of the design of any disposal facility that involves burial or placement of wastes in or on the land.

This geological work is essential to monitoring the performance of the disposal facility throughout its active and its post-closure existence.

To ensure the expected degree of safety, the post-closure plan must, based on the geology and hydrology and the kinds of waste involved, describe monitoring procedures, suggest contingency plans in case of a failure of the containment system, and must provide financial guarantees that all this will be done over a period of thirty

Figure 13, page 1

Minnesota Pollution Control Agency
Hazardous Waste Division
520 Lafayette Road
St. Paul, Minnesota 55155
(612) 296-6300

Hazardous Waste Permit
Part A Application

To be filed by owners/operators of hazardous waste facilities operating in the State of Minnesota.
See instructions on attached sheets before completing this application.

I. A. **General Information** (See Minnesota Pollution Control Agency (MPCA) Hazardous Waste Rules, Chapter 7001, parts 7001.0500 through 7001.0730)

Identification Number: ☐☐☐☐☐☐☐☐☐☐☐☐

Applicant (Owner)	Name:	Street: City: State: Zip:	Telephone No.: ( )
Prepared by			
Operator			

Status:	Federal	State	Public	Private	Other
Owner	☐	☐	☐	☐	☐
Operator	☐	☐	☐	☐	☐

Hazardous Waste Facility Name:

SIC Codes ☐☐☐☐ ☐☐☐☐ ☐☐☐☐ ☐☐☐☐

Facility Description:

See Table I attached to complete the following:

Type Facility:	Storage (units)	Treatment (units)	Disposal (units)
Onsite ❑	SO1 ❑ ____	TO1 ❑ ____	D79 ❑ ____
Offsite ❑	SO2 ❑ ____	TO2 ❑ ____	D80 ❑ ____
	SO3 ❑ ____	TO3 ❑ ____	D81 ❑ ____
Storage Areas	SO4 ❑ ____	TO4 ❑ ____	D82 ❑ ____
Indoors ❑			D83 ❑ ____
Outdoors ❑			

B. Topographic Map; a topographic map within a one mile radius of facility is attached. Yes ❑ No ❑

C. Facility Location: Latitude ____ Longitude ____

Street ____ City ____ County ____

Township ____ Range ____ Section ____

PQ-00399-01 (02/85)

Figure 13, page 2

2. Identification of Waste and Processes Used

Waste Number	**Annual Quantity**	**Unit of Measure**	**Process Code(s)**
____________	____________	____________	____________
____________	____________	____________	____________
____________	____________	____________	____________
____________	____________	____________	____________

3. Application Information:

A. **Reason for Application:**

❑ Proposed New Facility

❑ Existing Unpermitted

❑ Permit Modification

❑ Permit Expiration

B. **Environmental Review Completed:**

❑ EIS

❑ Environmental Assessment

4. A. Attach Drawing of Existing Facility and Any Proposed or Past Management Areas.

B. Attach Photographys of Existing Facility and Proposed Areas for Expansion.

5. Other Permits Applied for or Received. Enter "A" if applied for, "R" if received.

☐ Federal Hazardous Waste

☐ Federal Air Emissions

☐ NPDES Disposal System Permit

☐ Dredge Material

☐ State Solid Waste

☐ State Air Emissions

☐ State Disposal System

☐ Other ____________

II. Certification

I certify under penalty of law that this document and all attachements were prepared under my direction or supervision in accordance with a system designed to assure that qualified personnel properly gather and evaluate the information submitted. Based on my inquiry of the person or persons who manage the system, or those persons directly responsible for gathering the information, the information submitted is, to the best of my knowledge and belief, true, accurate, and complete.

I am aware that there are significant penalties for submitting false information, including the possibility fo fine and imprisonment.

A. Owner	Name	Title	Signature	Date:
B. Operator				

PQ-00399-01 (02/85)

years after closure. The financial guarantees are rarely for less than one million dollars and may well involve many millions. The type of guarantee is designed to provide funds even if the original operators of the disposal facility are bankrupt, liquidated, or otherwise not to be found in the event of a problem arising from the wastes many years after the site is closed.

In recent times, the obvious method—the purchase of insurance to cover the possible future costs—has become difficult or often impossible because underwriters have no real way of estimating future liabilities in a constantly changing regulatory and public opinion environment.

Operational Factors

Once a permit has been obtained, the generator/operator must deal with a long list of operational requirements which only a review of current regulations can define completely. These range from required analyses of each shipment of waste before treatment, storage, or disposal; to security provisions designed to prevent unauthorized persons or livestock from entering the facility; to detailed training and testing of waste management personnel to ensure that they have competence in the areas of their work. These are only examples, with literally dozens of detailed requirements spelled out or implied in the regulations.

Based on the elaborate pre-permit site geology and groundwater studies, impact of the facility on all environmental media must be monitored and reported to the regulatory agency in great detail and at regular intervals. In particular, land-based disposal units (and often storage units) must monitor quality of groundwater and must collect and analyze leachate—the liquid which can accumulate in a landfill. Usually, a facility is required to provide satisfactory treatment of the leachate so that it does not accumulate in large amounts and leak into the surroundings. Monitoring well systems must be installed and sampled regularly by methods designed to detect contamination of the groundwater within poorly defined statistical limits.

Any leakage or contamination of ground or surface water will trigger remedial action at the operator's expense, even outside the facility boundary. Remedial action usually means locating a zone of contaminated groundwater, defining its characteristics, pumping, treating, and disposing of the water until that which remains meets the standards imposed by the regulatory enforcement staff. Many remedial action programs have required the removal of the waste from the site and its redeposit in some other location. There are some specific prohibitions affecting land-disposal facility operations. Reflecting Congress's determination to eliminate all hazardous waste disposal in landfills, indeed to eliminate all landfills, a number of particular materials are banned from landfill disposal, at least until they are "treated" to reduce their hazardous nature. Organic solvents were a prime target of this banning legislation. Free liquids, even in containers, are substantially barred from landfills, and the so-called "kitty litter treatment" in which adsorbents are added to liquids to make them act as solids is generally not allowed.

There are some chemical and physical systems for solidifying or binding liquids so they cannot be squeezed out or released by degradation of the solidifying agents. These methods are often the basis for a process called "delisting," by which a particular waste from a particular location and process can be shown to be nonhazardous through a petition process and therefore disposable in an ordinary landfill.

A complete description of all the responsibilities of a generator who chooses to treat, store, and dispose of hazardous waste, if it could be written at all, would require a book substantially longer than this one. Repeating earlier advice, those who choose this route will need the best advice they can get. Beware of shortcuts or of snap decisions based on a little knowledge such as that contained in this book. As we shall see later, mistakes can have such serious consequences that only unusually careful attention to all the details of the requirements will avoid major business and personal risk.

TRANSPORTATION OF HAZARDOUS WASTE

Anyone who engages in the transportation of hazardous waste outside the site on which the waste is generated is a transporter under regulatory control. This broad definition includes the common carriers operating on roads, rails, and even with aircraft, and also covers the private company or even an individual who occasionally moves waste off-site using his own vehicles. If the waste being moved is required to be manifested, even a few hundred feet of travel on a public road is enough to trigger the full regulatory program on the transporter.

As we have said earlier, the generator of waste is under obligation to prepare a proper Manifest, to assure himself that the transporter has a current EPA Identification Number, and to designate a destination TSDF which will accept the waste and subject it to the best available technology for treatment and/or disposal.

Once the transporter accepts the waste by signing the Uniform Hazardous Waste Manifest, he becomes subject to intensive regulation by both the EPA and the Department of Transportation. These two Federal agencies have agreed on joint regulatory procedures and standards. Many States have transportation controls that may be far more restrictive than the Federal rules. Restrictions on movement of hazardous materials, of which wastes are merely a part, may limit use of particular streets or roads during specified hours, particularly if the routes include tunnels or bridges, or heavily populated districts. Specialty haulers have long had to cope with this web of controls, and a generator hauling his own waste had better learn all about them.

A simple, uncomplicated trip from generator to disposer requires only that the trucker, for example, assure himself that the load is properly packaged and labeled and that the vehicle is placarded as required under DOT regulations. Then, the Manifest must be kept with the waste at all times. When custody is transferred, the original transporter obtains a signature from the recipient on the Manifest, retains a copy, and sends the rest of the copies along with the waste or leaves them with the disposal site operator.

If the transporter finds that the destination facility or transporter will not accept the waste, he must contact the generator for instructions, revising the Manifest if need be. Along the way a transporter may store waste for up to ten days without an RCRA storage permit.

Operating heavy vehicles on rail or highway, transporters can get into serious trouble in several (maybe many) ways. If the transporter for some reason mixes hazardous wastes in a single container (like a tank car or truck) when they were shipped under two or more different DOT classifications, he becomes a generator and must comply with all the regulations we have just described. Similarly, if quantities of wastes in excess of the 1,000 kilo or 100 kilo per month limitations accumulate in transportation equipment, it is subject to the generator rules when removed. Cleaning facilities for containers and tanks generally are generators and often are storage facilities under the regulations.

Finally, if there is an incident, usually accidental, that results in a "release" of a hazardous material into the environment from a transport vehicle, the Department of Transportation and EPA regulations generally require the transporter to notify the National Response Center by phone at 800-424-8802 or 800-426-2675 and to file a written report.

Although the transportation company is expected to take reasonable steps to minimize the damage from a release, the Federal and State agencies will expect the generator to accept responsibility for providing to the public safety personnel on the scene information about the hazards created by the release and the measures available for fire fighting, for example. A release of any kind triggers the need for abatement of the hazards and for remedial action to restore the damaged environment and natural resources to an acceptable level of quality.

A major transportation accident which releases hazardous substances, particularly if the cargo is a classified waste, creates a major crisis for local public safety personnel, the generator, and the EPA and its State allies. The official procedure calls for the EPA to provide an On-Scene Coordinator, an experienced person in the management of releases and their consequences. The coordinator has authority to make substantial expenditures if need be to mitigate the effects of the release and to provide for public safety. These expenses will be recovered from the generator or owner of the waste in due time. Most large companies engaged in the manufacture and distribution of hazardous materials have established teams available for immediate dispatch to the scene of a release. All generators must have someone available on call to answer questions that arise during the emergency.

There are special additional requirements for anyone transporting hazardous waste for use as fuel. Used lubricating oil can be delivered only to a facility that is licensed and permitted for recycling used oil.

Finally, the small business must be very careful not to step over the line and become a transporter unknowingly. Putting any quantity of a waste required to be manifested into a personal car or truck and moving it over a public road is an unlawful act, regardless of its destination. No landfill operator can legally accept such waste, and unauthorized disposal can make one subject to severe criminal penalties.

An example of how easy it is to fall into this trap arose recently when neighbors complained about seeing old drums and smelling bad smells from a small suburban-rural woodlot west of Minneapolis. When Federal investigators went on the property and found that indeed some illegal disposal had occurred, they learned that the property owner was the maintenance man at a local manufacturing firm. He said that he was attempting to help his employer by loading drums of this waste into his pickup and taking them home where he buried them on the property.

While the owner of the business denied responsibility for directing his employee to take this improper action, the court found him liable for violating the State and Federal regulations. Criminal sanctions including prison and heavy fines were imposed, and the business was destroyed by the financial liabilities which resulted. Good intentions are not enough, although many in the community thought this particular case might have had an unfair outcome.

PENALTIES FOR VIOLATIONS OF THE SOLID WASTE LAWS

In any conversation or review of the solid waste laws and regulations, the question will inevitably be raised as to legal penalties which may be imposed for violations. In this discussion, we will focus on the sanctions applicable under RCRA and the 1984 Amendments (Hazardous and Solid Waste Amendments—HSWA).

Most citizens are accustomed to a legal, or at least a criminal structure that assesses a defined penalty for a specific act or failure to act—a crime. Thus, overparking draws a fine of, say, ten dollars or at worst a towing charge, storage, etc., but we know what our risk is and most of us simply pay up. Similarly, many people believe that a person convicted of a specific crime is likely to receive a specific punishment—a fine of "not more than"—or imprisonment of "not more than"—a stated term. As evidence of these expectations we just have to read the letters to the editor after a sentencing in a notorious criminal case.

On this basic quid-pro-quo scheme, the penalty for violation of the RCRA law or any regulation issued under its authority (specifically, of the management rules issued under Subtitle C) is a civil penalty of *$25,000 per day of violation*. This penalty can be assessed even if the agency has issued no notice of violation or compliance order. The rules are self-implementing. Any time the enforcing agency (Federal or State) determines that a violation has occurred or is occurring, an administrative order can be issued requiring the violation be stopped immediately or at some future time. Alternatively, the agency can seek a civil injunction against the violator in a U.S. District Court. Approved State programs also provide for injunctive relief in the appropriate State Court.

A "person" subject to regulation under RCRA and HSWA is expected to know the laws and regulations and to comply. This is the meaning of the self-implementing or self-executing regulation. The real person or organization cannot wait for notification that a violation has occurred and then act to eliminate the violation. He can be convicted whether or not a compliance order was ever issued.

The element of "knowing" violation received great stress in the original RCRA, but even more in the 1984 HSWA. Today "knowing" violations of the regulations under Title C can give rise to criminal, not merely civil charges. The criminal liability rises to $50,000 and/or two years' imprisonment for knowingly doing or not doing a long list of actions enumerated in Section 3008 of the amended act. Among these are:

1. Treating, storing, or disposing of identified hazardous waste without a permit
2. Making false statements in a permit application, report, or required record
3. Knowingly shipping hazardous waste without a proper manifest
4. Exporting hazardous waste without complying with the rules for that activity

A second offense under the rule is subject to doubled penalties, both the fines and the allowed term of imprisonment. Much more important, any knowing violations are felonies, conviction for which results in loss of personal rights and forfeiture of one's eligibility for any office of trust unless pardoned—a very high price to pay for an error in judgment or a moment's greedy lapse from grace.

The really new element introduced in the 1984 Amendments (HSWA) is a newly defined crime of *knowing endangerment*. This innovation basically says that anyone engaged in any aspect of the management of hazardous waste, while in violation of any of many regulations, who places another person in "imminent danger of death or serious bodily harm" is subject, on conviction, to a fine of up to $250,000 or up to 15 years in prison, each doubled for a second conviction.

The legal rules for defining "knowing," "endangerment," etc., are beyond the scope of this discussion. It is at best a whole new field for litigation which will be before the courts for many years. Corporations can be fined up to one million dollars on conviction, but the major effect of such an event would probably be the validation of civil lawsuits against the corporation by those affected. Clearly, in developing the "knowing endangerment" concept, the Congress had in mind situations in which irresponsible people or their organizations mishandled, improperly stored, or disposed of waste that caused serious illness, debility, or death in an affected population. Such circumstances have been alleged in a number of well-publicized cases—Love Canal, the Stringfellow Pits, and others.

It is important, though, in thinking about penalties and punishments for violations of the rules, to know that the first conviction obtained under the knowing endangerment clause did not affect the general population and had none of the characteristics of a highly publicized public risk. Obviously, people were endangered, some fatally, but in a context that did not appear to fit the circumstances Congress had in mind.

A company and two of its managers were convicted of knowing endangerment when several employees allegedly became ill and at least one died from exposure to hazardous waste while handling it in the course of their regular work. Conviction was based on evidence that the employees were not properly trained or supervised and that adequate personal protective clothing, respirators, and other gear were not provided or not required to be used. The managers were subjected to personal fines and were sent to jail. The company paid fines of over $4 million and, of course, is vulnerable to civil lawsuits that could total much more.

The point of this discussion of knowing endangerment is that Congress and the courts have woven a web of punitive laws around the crime of mishandling or improperly disposing of hazardous waste. The web is often invisible even to the knowledgeable attorney, and prosecutors have very wide latitude in choosing cases and authorities for prosecution and even whether to bring any action at all. Thus, to say that violation of the rules leads to specific exposures of persons and companies is just not possible.

In his invaluable book *Federal Regulation of Hazardous Waste: A Guide to RCRA*, John Quarles says: "Because the potential exposure to both personal and corporate punishment for noncompliance is so great, it is critical that all persons responsible for assuring compliance with the regulatory scheme fully appreciate the range of enforcement options open to the government." Quarles goes on to discuss the general acts and omissions that can constitute noncompliance with the self-executing standards of performance promulgated by the EPA for generators, transporters, and TSDF operators. All are required to comply with the standards whether the inspector has been around or not, as we have already mentioned.

Failure to have a permit or operating in violation of a permit requirement is an obvious instance of noncompliance. A less known but high-risk area of exposure to enforcement action is the idea of an "imminent hazard." The law grants the EPA authority to require persons (and companies) responsible for creating an imminent hazard to take corrective action under threat of severe legal sanctions. An imminent hazard is loosely defined to include circumstances which "may present" a hazard, and in fact, in cases in which no damage has occurred and even when none is expected, the rule can be applied.

An example of the reach of this authority is this: a company received a notice that it was potentially responsible for toxic metal waste *believed* to have been placed in a landfill. The company had no record of any operation in the region, much less of one that generated the type of waste described. After a frantic search, it was found that a subsidiary company, since liquidated, had (before becoming a part of the existing corporation) operated a plant in the region that might have produced the type of waste that was the source of the complaint. This one had a happy ending. Some pure good fortune and good local police work turned up credible witnesses who described the detoxification of the waste before disposal, and no further action was needed. Many other combinations of circumstances can give rise to the need for quick action to obtain facts and to plan remedial action if needed.

Prosecutors have access to many other provisions of Federal law: making false statements to any government agency is a crime; doing so on the telephone or by mail may constitute wire or mail fraud; the conspiracy statutes have been invoked to punish unlawful private agreements that frustrate the purpose of the laws.

In summary, we can say that the severity of the legal sanctions for violation of the hazardous waste management rules is so great that no responsible manager or individual can afford to do less than the maximum good-faith effort needed to comply. Not only is today's action important—hazardous wastes can and do return from the grave to create new and larger problems than they would have in their original appearance if they had been conscientiously managed.

UNDERGROUND STORAGE TANKS

In its deliberations on amendment of RCRA before the 1984 legislation, Congress was confronted with a substantial public interest in reducing damage to groundwater and other values from leakage of underground storage tanks, largely from automotive gasoline retailing facilities. Although the problem has almost no relationship to the problem of hazardous waste management addressed by RCRA, it was decided to create a new regulatory program and attach it to RCRA as Subtitle I. Responsibility for implementation was placed on the Administrator of the EPA.

Congress wanted the EPA to find out how many tanks existed, what the risk to the public was, and to find ways to reduce the hazard in the future. Further, the agency was to obtain remedial action at any sites that were found to need it. As always, the definition of an underground storage tank presents the first problem. Congress chose to define such tanks as any which, including their associated piping, were 10% or more below ground.

From a legal and political viewpoint, it was necessary to exclude from this broad class some tanks or at least containers that otherwise would meet this broad definition. These excluded things are not the target of the program. The list of things not included in the term "Underground Storage Tank" is interesting. The term does not include any—

(A) farm or residential tank of 1,100 gallons or less capacity used for storing motor fuel for noncommercial purposes,
(B) tank used for storing heating oil for consumptive use on the premises where stored,
(C) septic tank,
(D) pipeline facility (including gathering lines) regulated under—
  (i) the Natural Gas Pipeline Safety Act of 1968,
  (ii) the Hazardous Liquid Pipeline Safety Act of 1979, or
  (iii) an intrastate pipeline facility regulated under State laws comparable to the provisions of law referred to in clause (i) or (ii) of this subparagraph,

(E) surface impoundment, pit, pond, or lagoon,
(F) storm water or waste water collection system,
(G) flow-through process tank,
(H) liquid trap or associated gathering lines directly related to oil or gas production and gathering operations, or
(I) storage tank situated in an underground area (such as a basement, cellar, mine working, drift, shaft, or tunnel) if the storage tank is situated upon or above the surface of the floor.

The definition then goes on to exclude any pipes connected to the tanks listed in subparagraphs (A) through (I).

Having with unusual precision set up the target, the new program first sought information. It defines a group of "regulated substances" including liquid petroleum products and all those defined in the Superfund law (CERCLA) Section 101(14), but *excluding* any substance regulated as a hazardous waste under RCRA!

To obtain the information base for the program, every owner of an underground storage tank was required to report by November 8, 1985, the existence of any tank covered by the regulation and its size, age, type, location, and uses. Figure 14 shows the form used which is still required for a new tank before it can be used.

More demandingly, the owner of any tank taken out of service after January 1, 1974, was required to report a lot of detail about the idle (often vanished) tank. Tanks taken out of service before 1974 were not to be reported.

Any new tank had to be announced to the regulatory authority within 30 days of its inaugural use. As a means of further encouraging reporting, suppliers of petroleum liquids and regulated substances were required for the first 18 months to notify a tank owner of his reporting responsibility when anything was deposited in the tank.

Financial Responsibility

The law then directs the Administrator of the EPA to develop comprehensive regulations for detection, prevention, and correction of "releases" (leaks) from tanks, allowing classification of tanks by age, location, use, etc., and requiring owners to show their financial capability for doing those things when and if required.

In proposing regulations under the new law, the EPA recognized that major impacts on small businesses, especially gasoline retailers and owners of private underground storage tanks, are to be expected. In the period since the proposed scheme has been under review, a great many small business people have complained bitterly about the difficulty of dealing with the financial responsibility requirement. At least one Western town has experienced a threatened shutdown by gasoline retailers protesting the unreasonable requirements. Opinions vary about the effectiveness of the action, but it did get some press coverage and therefore accomplished its major objective.

The difficulty for the regulatory agencies (States are required by law to implement the Federal regulations) is that a leaking underground storage tank can contaminate a lot of soil and affect groundwater over a large area—from a few thousand square feet to several city blocks. It is common for the first symptom of a leak to be the emergence of gasoline vapors into someone's basement or a sewer—both extremely dangerous situations.

Cleanup of contaminated soil and pumping of polluted groundwater for treatment can be a difficult and thus expensive venture. (Refer to Chapter 7, "Waste Management: Land Disposal and Protection of Groundwater.") The EPA's analysis indicates that many motor fuel retailers have less than ten employees and a net worth of $100,000 or less. Cleanup costs from a major tank leak in vulnerable soil can add up to many times $100,000. Indeed, the EPA believes most will cost in the neighborhood of three times that. The EPA's concept of the process of corrective action is shown in Figure 15. To fully appreciate what this chart means, one must realize that the time period covered between the upper left entry to the procedure and the final exit at the lower right may be as long as thirty years.

To ensure that corrective action will be paid for if and when needed (we will discuss release detection and other triggering mechanisms for corrective action later), the owner of an underground petroleum storage tank must somehow obtain a guarantee that he or someone in his behalf can pay up to $1 million if he owns from 1 to 12 tanks, the amount increasing to $6 million for an owner of 341 (*sic*) or more "individual containment units" or tanks meeting the definition and not exempted under any of the many exemptions.

The perception of unfairness in this rule is natural, given the requirement for only 6 times as much guarantee for 50 to 300 times as many tanks, as against requirements for the small business. It is also true that the rules will permit a large corporation to provide its own guarantee based on its large net worth.

If an owner is using the tank to store any material regulated under RCRA Subtitle C (Hazardous Solid Waste), then the financial responsibility requirements for a TSDF will apply—they are more complex.

Many different financial mechanisms may be used to guarantee the level of responsibility defined for any particular owner. Insurance against "sudden" and "nonsudden" releases can sometimes be obtained. The cost of such insurance is difficult to pin down and depends on the particular tank and its circumstances.

Figure 14, page 1

Notification for Underground Storage Tanks

FORM APPROVED
OMB NO. 2050-0049
APPROVAL EXPIRES 6-30-88

STATE USE ONLY

I.D. Number

Date Received

GENERAL INFORMATION

Notification is required by Federal law for all underground tanks that have been used to store regulated substances since January 1, 1974, that are in the ground as of May 8, 1986, or that are brought into use after May 8, 1986. The information requested is required by Section 9002 of the Resource Conservation and Recovery Act, (RCRA), as amended.

The primary purpose of this notification program is to locate and evaluate underground tanks that store or have stored petroleum or hazardous substances. It is expected that the information you provide will be based on reasonably available records, or, in the absence of such records, your knowledge, belief, or recollection.

Who Must Notify? Section 9002 of RCRA, as amended, requires that, unless exempted, owners of underground tanks that store regulated substances must notify designated State or local agencies of the existence of their tanks. Owner means—

(a) in the case of an underground storage tank in use on November 8, 1984, or brought into use after that date, any person who owns an underground storage tank used for the storage, use, or dispensing of regulated substances, and

(b) in the case of any underground storage tank in use before November 8, 1984, but no longer in use on that date, any person who owned such tank immediately before the discontinuation of its use.

What Tanks Are Included? Underground storage tank is defined as any one or combination of tanks that (1) is used to contain an accumulation of "regulated substances," and (2) whose volume (including connected underground piping) is 10% or more beneath the ground. Some examples are underground tanks storing: **1.** gasoline, used oil, or diesel fuel, and **2.** industrial solvents, pesticides, herbicides or fumigants.

What Tanks Are Excluded? Tanks removed from the ground are not subject to notification. Other tanks excluded from notification are:

1. farm or residential tanks of 1,100 gallons or less capacity used for storing motor fuel for noncommercial purposes;
2. tanks used for storing heating oil for consumptive use on the premises where stored;
3. septic tanks;
4. pipeline facilities (including gathering lines) regulated under the Natural Gas Pipeline Safety Act of 1968, or the Hazardous Liquid Pipeline Safety Act of 1979, or which is an intrastate pipeline facility regulated under State laws;
5. surface impoundments, pits, ponds, or lagoons;
6. storm water or waste water collection systems;
7. flow-through process tanks;
8. liquid traps or associated gathering lines directly related to oil or gas production and gathering operations;
9. storage tanks situated in an underground area (such as a basement, cellar, mineworking, drift, shaft, or tunnel) if the storage tank is situated upon or above the surface of the floor.

What Substances Are Covered? The notification requirements apply to underground storage tanks that contain regulated substances. This includes any substance defined as hazardous in section 101 (14) of the Comprehensive Environmental Response, Compensation and Liability Act of 1980 (CERCLA), with the exception of those substances regulated as hazardous waste under Subtitle C of RCRA. It also includes petroleum, e.g., crude oil or any fraction thereof which is liquid at standard conditions of temperature and pressure (60 degrees Fahrenheit and 14.7 pounds per square inch absolute).

Where To Notify? Completed notification forms should be sent to the address given at the top of this page.

When To Notify? 1. Owners of underground storage tanks in use or that have been taken out of operation after January 1, 1974, but still in the ground, must notify by May 8, 1986. **2.** Owners who bring underground storage tanks into use after May 8, 1986, must notify within 30 days of bringing the tanks into use.

Penalties: Any owner who knowingly fails to notify or submits false information shall be subject to a civil penalty not to exceed $10,000 for each tank for which notification is not given or for which false information is submitted.

INSTRUCTIONS

Please type or print in ink all items except "signature" in Section V. **This form must by completed for each location containing underground storage tanks.** If more than 5 tanks are owned at this location, photocopy the reverse side, and staple continuation sheets to this form.

Indicate number of continuation sheets attached ☐

I. OWNERSHIP OF TANK(S)

Owner Name (Corporation, Individual, Public Agency, or Other Entity)

Street Address

County

City | State | ZIP Code

Area Code | Phone Number

Type of Owner *(Mark all that apply ☒)*

☐ Current
☐ Former
☐ State or Local Gov't
☐ Federal Gov't (GSA facility I.D. no. ____________)
☐ Private or Corporate
☐ Ownership uncertain

II. LOCATION OF TANK(S)

(If same as Section 1, mark box here ☐)

Facility Name or Company Site Identifier, as applicable

Street Address or State Road, as applicable

County

City (nearest) | State | ZIP Code

Indicate number of tanks at this location ☐

Mark box here if tank(s) are located on land within an Indian reservation or on other Indian trust lands ☐

III. CONTACT PERSON AT TANK LOCATION

Name (If same as Section I, mark box here ☐) | Job Title | Area Code | Phone Number

IV. TYPE OF NOTIFICATION

☐ Mark box here only if this is an amended or subsequent notification for this location.

V. CERTIFICATION (Read and sign after completing Section VI.)

I certify under penalty of law that I have personally examined and am familiar with the information submitted in this and all attached documents, and that based on my inquiry of those individuals immediately responsible for obtaining the information, I believe that the submitted information is true, accurate, and complete.

Name and official title of owner or owner's authorized representative | Signature | Date Signed

CONTINUE ON REVERSE SIDE

Figure 14, page 2

Owner Name (from Section I) ______________ Location (from Section II) ______________ Page No. ____ of ____ Pages

VI. DESCRIPTION OF UNDERGROUND STORAGE TANKS *(Complete for each tank at this location.)*					
Tank Identification No. (e.g., ABC-123), or Arbitrarily Assigned Sequential Number (e.g., 1,2,3...)	**Tank No.**	**Tank No.**	**Tank No.**	**Tank No.**	**Tank No.**
1. Status of Tank *(Mark all that apply ☒)*					
Currently in Use	☐	☐	☐	☐	☐
Temporarily Out of Use	☐	☐	☐	☐	☐
Permanently Out of Use	☐	☐	☐	☐	☐
Brought into Use after 5/8/86	☐	☐	☐	☐	☐
2. Estimated Age (Years)					
3. Estimated Total Capacity (Gallons)					
4. Material of Construction *(Mark one ☒)*					
Steel	☐	☐	☐	☐	☐
Concrete	☐	☐	☐	☐	☐
Fiberglass Reinforced Plastic	☐	☐	☐	☐	☐
Unknown	☐	☐	☐	☐	☐
Other, Please Specify	____	____	____	____	____
5. Internal Protection *(Mark all that apply ☒)*					
Cathodic Protection	☐	☐	☐	☐	☐
Interior Lining (e.g., epoxy resins)	☐	☐	☐	☐	☐
None	☐	☐	☐	☐	☐
Unknown	☐	☐	☐	☐	☐
Other, Please Specify	____	____	____	____	____
6. External Protection *(Mark all that apply ☒)*					
Cathodic Protection	☐	☐	☐	☐	☐
Painted (e.g., asphaltic)	☐	☐	☐	☐	☐
Fiberglass Reinforced Plastic Coated	☐	☐	☐	☐	☐
None	☐	☐	☐	☐	☐
Unknown	☐	☐	☐	☐	☐
Other, Please Specify	____	____	____	____	____
7. Piping *(Mark all that apply ☒)*					
Bare Steel	☐	☐	☐	☐	☐
Galvanized Steel	☐	☐	☐	☐	☐
Fiberglass Reinforced Plastic	☐	☐	☐	☐	☐
Cathodically Protected	☐	☐	☐	☐	☐
Unknown	☐	☐	☐	☐	☐
Other, Please Specify	____	____	____	____	____
8. Substance Currently or Last Stored in Greatest Quantity by Volume *(Mark all that apply ☒)*					
a. Empty	☐	☐	☐	☐	☐
b. Petroleum					
Diesel	☐	☐	☐	☐	☐
Kerosene	☐	☐	☐	☐	☐
Gasoline (including alcohol blends)	☐	☐	☐	☐	☐
Used Oil	☐	☐	☐	☐	☐
Other, Please Specify	____	____	____	____	____
c. Hazardous Substance	☐	☐	☐	☐	☐
Please Indicate Name of Principal CERCLA Substance	____	____	____	____	____
OR					
Chemical Abstract Service (CAS) No.	____	____	____	____	____
Mark box ☒ if tank stores a mixture of substances	☐	☐	☐	☐	☐
d. Unknown	☐	☐	☐	☐	☐
9. Additional Information (for tanks permanently taken out of service)					
a. Estimated date last used (mo/yr)	/	/	/	/	/
b. Estimated quantity of substance remaining (gal.)	____	____	____	____	____
c. Mark box ☒ if tank was filled with inert material (e.g., sand, concrete)	☐	☐	☐	☐	☐

Figure 15

Corrective Action Process for Petroleum Underground Storage Tanks

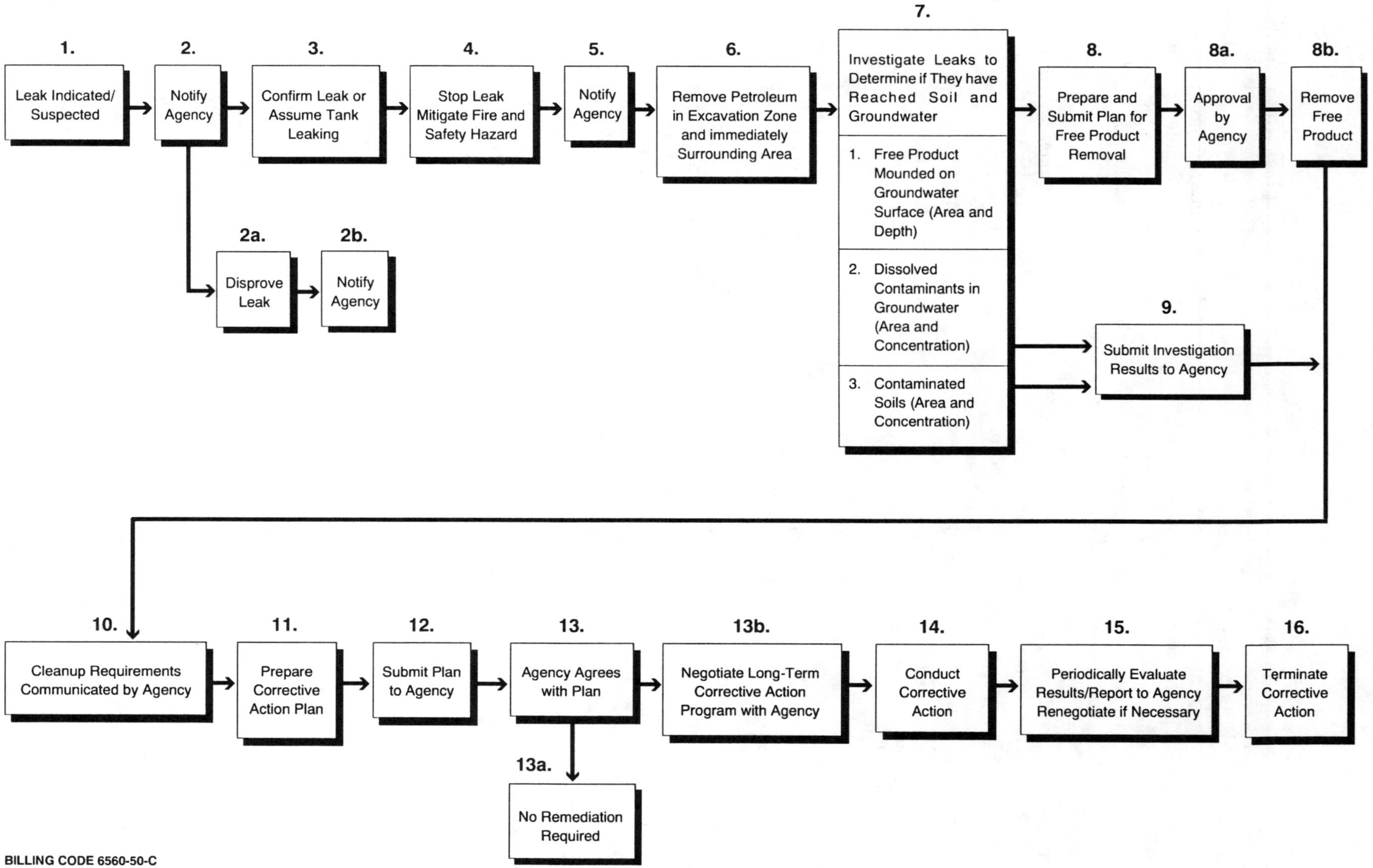

BILLING CODE 6560-50-C

Source: USEPA

The cost of insurance is universally described by small businesses as "prohibitive," and the EPA expects real difficulty in finding coverage in the commercial insurance market. Because of this insurance problem, a complicated and to some an unreliable mechanism has been provided by which whole classes of tank owners can seek a "suspension of enforcement" of the financial responsibility rule by showing that no access to outside guarantees is available. It is said, in the Regulatory Impact Analysis, to require an hour's "managerial time" per year to obtain this suspension—a likely story! The same analysis estimates that 555 firms may close because of this particular requirement.

Businesses 50% or more owned by larger companies have access to corporate guarantees in several forms to meet their needs under this Section 9003(d). If their net worth is ten or more times the required level of coverage, corporate guarantees (self-insurance) can be acceptable. The rules for this mechanism are very strict, with the language of the guarantee spelled out word for word in the regulation and a routine requirement of an additional guarantee by an independent trustee.

Finally, when the Superfund Amendment and Reauthorization Act (SARA) was adopted in 1986, a new section was added to RCRA's underground storage tank program establishing a Leaking Underground Storage Tank Trust Fund similar to the one already established by taxation of petroleum and chemicals and commonly called Superfund. Under the new fund, the financial responsibility amounts required of owners become a "kind of deductible," and the EPA is authorized to use funds from the trust for the costs of corrective actions needed above and beyond the amounts provided by the guarantees. The backing of this fund is not likely to be helpful to the owner/operator, since the agency will almost certainly come after the responsible party for any costs it may incur for corrective action, both short and long term.

The Nature of Underground Tanks

In the Preamble to the proposed Regulations on Underground Storage Tanks published in the *Federal Register* beginning at page 12,662 on April 17, 1987, the EPA estimated that the rule would affect about 1,400,000 tanks in use at about 500,000 facilities. Facilities range from a small service station with one or two tanks to a petroleum refinery or distribution point with many tanks in service. The EPA believed that 95% of all these tanks were in service for storing petroleum and that only 4% were used for storing hazardous substances.

As an important sidelight, few of these owners had or have a choice as to the placement of their tanks underground. Nearly all local fire regulations follow the National Fire Protection Code in demanding that gasoline and other volatile flammable liquid storage tanks be underground unless they are located on large sites allowing several hundred feet between the tank and adjacent buildings. This is another example of the complexity of decisions in the area of human and environmental safety. Fire is an immediate and familiar hazard. Leaks into soil, groundwater, and even basements and sewers are quiet and not always easy to find.

In the approach to the regulations, the EPA describes six classes or kinds of UST systems (*U*nderground *S*torage *T*anks):

1. Steel UST systems not protected from corrosion ("bare steel tanks")
2. Steel UST systems protected with a non–corrodible coating
3. Steel UST systems with cathodic protection
4. Steel UST systems with a noncorrodible coating and cathodic protection
5. UST systems made totally from noncorrodible material
6. UST systems made of noncorrodible materials or protected steel using secondary containment systems

Much of this is self-explanatory. Noncorrodible coatings or systems imply plastics or other organic coatings like coal tar and asphalt. Cathodic protection is accomplished by the use of one of several means of connecting a corrodible metal object like a tank or pipeline to a "sacrificial" metal called an anode, usually made of magnesium, which corrodes faster than the tank (or cathode) and thus protects the metal of the tank. Much more can be said about this, and it is important in the future installation and use of underground storage tanks. A good reference book in corrosion engineering is suggested.

There are clearly a lot more tanks to be found below the ground surface in the United States than the regulations cover. Remember that many (perhaps half) are exempted either because of their ownership (farm fuel tanks), their specific location (on the floor of a basement or mine working), or use (in-process tanks, storage of hazardous waste regulated under Subtitle C of RCRA), or one of several other special classes.

At any rate, the regulations are aimed very specifically at underground motor fuel storage systems, both past and present. Reading down the classes of tanks just given pretty much leads one through a chronology of the technology of motor fuel storage since the automotive age began. This leads to a very good assumption: the older a fuel tank is, the more likely it is to have leaked. Thus the regulatory approach is designed to deal first with the most likely sources of leakage—releases.

As a simplification, it is very significant that tanks "taken out of service" before 1974 are ignored. This is unfortunate in many ways, particularly since the rumored existence of an abandoned tank can play hob with a real estate transaction. Most State programs are seeking at least to find and classify these abandoned facilities.

Existing tanks are required to install leak-detection systems and to maintain records proving the integrity of the system either by careful inventory-keeping; by pressure-testing under independent supervision; or by actual sampling of soil and water near the tank installation. All existing systems are to meet the standards for installation and operation of new tanks within ten years after the new-tank regulations are promulgated.

Many petroleum-product suppliers are willing to help their retailer-customers understand and comply with the regulations. For reasons of exposure to liability for the consequences of a leak or release, most suppliers will not accept responsibility for the independent retailer's situation. Some States offer technical assistance, but in many if not most circumstances, the small businessperson is on his own with entrepreneurs and expert professional people willing to give sales pitches or expert advice, but he does not understand either. The problem is that his old business has to operate in a new world. Perhaps his best hope is his trade association, since at least all the other members are in the same position.

New Underground Storage Tanks

The new law contained an "interim prohibition" which, after May 8, 1985, allows new installation of UST's to be done only if three conditions are met:

1. The UST system can be shown to prevent releases because of structural failure or corrosion for the life of the system;
2. The system is cathodically protected against corrosion, is made of noncorrodible material, or otherwise designed to prevent releases; and
3. Is compatible with the substances to be stored.

These interim prohibitions were extended when the Proposed Rules for Underground Storage Tanks were released April 17, 1987, and essentially remain in force until the effective dates (which are many for different classes of tank owners) specified in the Final Rules promulgated September 23, 1988.

An attentive reader will quickly recognize that any fool would design a system that will not leak for the life of the system—after all, the system life is at an end when leakage becomes a problem, at least until repairs are made. All this really does is to place any owners of new systems in a position such that, when they fail, their judgment in the design will likely be challenged in the blazing light of authorized second-guessing.

The standard for new Underground Storage Tanks occupies about 200 pages in the *Federal Register* of September 23, 1988. The rules include a requirement for notification to the regulatory authority (a State Agency) that the tank exists, including details of its construction, operation, and provisions for containing releases and for detecting them when they occur. The notification form almost amounts to a license application and is subject to all the penalties for errors and omissions common to those which apply to statements to governments.

The regulations for new tanks will rely on leak detection based on the use of secondary protection barriers. A secondary protection system can be based on building a tank with two walls—basically a tank for the actual storage within another tank separated by a significant space—several inches at a minimum. Another secondary barrier can be constructed by setting the storage tank within a box or vault made of concrete or other impervious material. Often the box is backfilled with gravel or sand so it can be covered in a service station driveway, for example.

In any case, leak-detection equipment must be installed between the tank (the primary containment) and the outer tank or vault (the secondary containment) that is sufficiently sensitive to detect a release of a defined amount over a specific period of time.

Inventory records seeking to detect a release are notoriously unreliable because of human error, errors in meters and pumps, and spills that are not accounted for. The development and demonstration of devices for leak detection is in full swing, with significant new technology being force-fed to meet the need at reasonable costs.

Sampling for free petroleum inside the containment is an old method that has many difficulties attached. For example, free product can be spilled near the tank during filling or even during routine sampling. Gas analysis equipment can unfailingly detect the presence of gasoline leaks and spills but is hard to use in interpreting the quantity of a release. A very small amount of liquid can fill a large volume of secondary containment space with vapors. A new approach using specially sensitized chemical sensors attached to fiber optic cable systems and detectors offers promise of improving sensitivity, selectivity, and reliability of leak-detection systems.

In summary, then, the new UST regulatory program, like all the others, deals with real problems that occur a significant number of times in the country. The EPA estimated in 1987 that 25% or more of the bare steel tanks in service were leaking. Finding these sources of "releases" to air, soil, and groundwater will be difficult and expensive. The experience of studying and designing remedial projects will require new and expensive methods of construction and installation of new motor-fuel retailing and distribution facilities. How much will it cost?

The EPA suggested to Congress that the cost of cleaning up a plume from a release would "seldom" exceed $281,000 (in 1987 funds, presumably). At a failure rate of 5% per year, we can expect around 28,000 tanks to be the source of a release, at least until all are fitted with secondary containment. As an upper limit, around $9 or $10 billion per year might be needed, not counting new and replacement installations.

Each citizen must make his/her own decision as to the worth of the program. As we suggested at the beginning of this chapter, Congress rarely has much information on costs at the time it acts.

Chapter 13

Superfund and Its Extensions

BACKGROUND OF CERCLA

Any expression of concern, fear, or anger about "toxic waste" or "hazardous dumps" likely arises from a situation which the Superfund law was intended to correct.

During the public and congressional debates about waste management during the 1970s, it was apparent to many that the RCRA law (see Chapter 12 regarding that act) was intended to prevent future problems like the Love Canal and its notorious relatives. It was equally clear, given the Constitutional doctrine prohibiting *ex post facto* laws, that something special in the way of legislation was needed to give government the authority to see that the old contaminated sites were cleaned while avoiding the political accusation that tax money would be used to correct problems caused by private interests—that the program would violate the sacrosanct principle that the Polluter Pays.

The result of this search for a solution is something special indeed—the Superfund Law, more precisely known as the Comprehensive Environmental Response, Compensation, and Liability Act of 1980, CERCLA, since amended by the Superfund Amendments and Reauthorization Act of 1986, universally called SARA. A main lead in one of the principal journals of the field cried "SARA is no Lady"—which the reader will have to judge as we proceed.

STRUCTURE OF CERCLA

In its essentials, the act is straightforward in its approach:

- It seeks to define its target—contamination of the "environment" in the most general terms.
- Methods of discovering the contamination are presented, with penalties for those who actively or through failure to act conceal the existence of a problem.
- It provides authority needed by the assigned government agency to evaluate the problem and to define a remedy.
- It provides ways to force those responsible for the contamination to clean it up at their expense.
- Failing the identification of a non-Federal actor to pay for the cleanup, the law provides money for remedial action through Superfund, now officially known as the Hazardous Substance Superfund, which is provided by special taxes and general revenues at a level of about $8.5 billion for the period 1986–1991.

Thus, the simple structure of the law: define the danger, find out where and when it has or does exist, then rectify it. The trouble arises, as usual, when trying to implement the ideas in legal and technical ways that can withstand challenges in the courts on Constitutional and equitable grounds.

Under intense political pressure, Congress had to craft a statute that could avoid conflict with many of our most cherished ideas of civil liberties while accomplishing the "greater good" of abating a perceived and often all-too-real danger to public health arising from unwise, careless, and outright dangerous disposal of hazardous materials in the past in all sorts of places.

The problem for the bill's authors was that many of the conditions addressed by the bill were not unlawful at the time they were created, with the possible exception of situations in which there was proof that the disposal injured someone sufficiently for a civil court to award economic damages on the basis of the common law—a difficult legal route to pursue. The drafters of Superfund came up with a number of novel ideas to achieve the ends of the legislation. They must have worked, because in its first five years, none of the main concepts of CERCLA was overturned by the courts.

Novel Ideas in the Law

This book is obviously not a lawyer's treatise on the intricacies of Superfund or any other law. Even for laypeople, however, it is necessary to have some feel for some of the ways in which our American tradition of instinctive negative reaction to government intrusion has been pushed aside in pursuit of the "larger good" in this field.

The first novel idea is that this law is not about wastes as we commonly know them. Neither is it solely about abandoned sites that get all the publicity. CERCLA deals with hazardous *substances*, hazardous *chemicals*, and *extremely hazardous substances* as explicitly defined in the act. The administrator is given the authority to expand the lists of hazardous substances pretty much at will. The definitions, combined, are much broader than any definition of hazardous waste to be found in RCRA and its regulatory extensions.

There is one broad and, to some, curious exclusion from the provisions of Superfund: petroleum and its related products and fractions are completely excluded from regulation under this particular statute and the regulations implementing it. The extent of the exemption is shown by a recent court decision that otherwise-hazardous substances, when used as additives in motor gasoline, were not subject to the Superfund regulations. The explanation for this sweeping exclusion is that the Clean Water Act contains a regulatory structure for oil spills that compares in many ways to the spill provisions of Superfund. The difference is that the Clean Water Act does not contain many of the severe penalties available to the government through administrative action that are unique to Superfund.

Leaving out petroleum, then, what are the substances addressed by Superfund? A short answer is: everything imaginable. A longer answer starts with the general and proceeds to the specific:

1. Any element, compound, mixture, solution, or substance which, when released to the environment, may present substantial danger to public health or to the environment itself.
2. Any substance listed under applicable sections of the Clean Water Act [Sections 311(b) (2) (A) and Section 307(a)].
3. Any substance classified hazardous because of characteristics listed under RCRA Section 3001. This provision excludes those wastes excluded from RCRA / SWDA by Congress.
4. All "listed" wastes under RCRA.
5. Any hazardous air pollutant listed in Section 112 of the Clean Air Act.
6. Any "imminently hazardous" chemical substances or mixture for which the government has taken regulatory action under Section 7 of the Toxic Substances Control Act. Section 7 is one of several authorities granted to the Administrator of the EPA to take summary action under "imminent hazard" conditions.

Reading this definition in the light of the comprehensive nature of the other statutes incorporated by reference makes it clear that just about anything that is placed, spilled, or emitted into the water, land, or air of the United States comes under the regulatory reach of Superfund.

The unlimited range of the materials covered is still more emphasized when one considers that all that is needed is that something "may" create a hazard. Later, we shall see that the government does not have to prove that a "release to the environment" actually occurred or that anyone or anything is actually endangered. It is only necessary to show that a possibility of a release and of endangerment or harm exists.

The second novel idea that occurs in the Superfund law is the concept of a "release" and, by extension, of a "continuing release."

A release is anything that occurs that results in some material appearing in the environment, presumably as opposed to its being confined in some form of containment, although this is not said. The words that are used to describe "release" include spilling, leaking, pumping, pouring, emitting, emptying, discharging, injecting, escaping, leaching, dumping, and disposing. Included are the abandonment or discarding of barrels or other closed containers containing any hazardous substances, pollutants, or contaminants.

Releases excluded from regulation under Superfund are those regulated under other programs and laws: in-workplace releases under OSHA (the Occupational Safety and Health Act); vehicle exhaust emissions and pipeline pumping station engine exhausts; and most nuclear releases. Under other provisions, most Federally permitted discharges under the other environmental statutes are generally exempt from regulation as releases.

The idea of a "continuing release" is fundamental to the whole function of Superfund as it applies to old sites in existence before the law was passed. Anyone who paid attention in his civics classes "knows" that the government cannot pass an *ex post facto* law declaring what you did yesterday a crime. If it was not already a crime on the day you committed the act, you cannot be prosecuted.

Recognizing that many of the actions involved in existing abandoned dump sites and other contaminated soil and water areas were not unlawful at the time they were done, Congress had a problem. The legal construct of a continuing release was the solution. First, the statutory decision was made that each day on which something escaped from containment—was released—into soil, groundwater, or air constituted a separate release. Contamination in soil, presumed to spread slowly into adjacent soil (or groundwater, the logic applies) is seen to contaminate new soil each day, thus a new "release" may occur each day.

All that remains to circumvent the Constitutional provision is a passage that allows the government to avoid the need for establishing any factual base for the continuing release or any harm that might result by inserting the words "may" or "might" or "potential hazard." This done, the Federal courts have upheld the regulatory program's retroactive responsibility provisions.

Lest anyone who has not followed this field of law carefully feel tempted to challenge many of the elements of the regulations, a word of caution is in order. The authors of the *Superfund Handbook*, a copyrighted publication of ERT, Inc. and the law firm of Sidley and Austin, report that in the first five years of Superfund, the government filed about 200 lawsuits pursuant to the provisions of Superfund. This litigation mostly concerned the Constitutionality and other legal tests of the 1980 statute, and was largely successful from the government point of view. The 1986 amendments (SARA) codified the outcome of many of these suits as well as the operating policies adopted during the period by the EPA, giving them the force of law.

Finally, another novel idea was the approach to the responsibility for contamination and the resulting required cleanups. This structure is essential to the implementation of Superfund as we know it.

The law has a vindictive tone that often requires so much effort to find and punish the perpetrators of releases that the actual cleanup may be delayed for years. The 1986 amendments, applying an even heavier hand, attempted to avoid this undesirable side effect by authorizing the government to decide on the required cleanup, to get it done, and then to collect its costs from those it deems responsible, with only very limited authority for the courts to intervene in the process.

Having set up and financed the Superfund out of taxes on the oil and chemical industries, the law and the intent of Congress set out to avoid the use of Superfund money at all costs, or if it is used, to pursue private parties relentlessly for reimbursement, wielding severe financial and criminal sanctions as weapons. This approach is believed to be justified by estimates that expenditures of as much as $160 billion will be needed to remedy the damage at upwards of 25,000 sites scattered across all the States. Other estimates range as high as $600 billion.

Even the process of negotiated settlements between prosecutors and responsible parties is restricted by the statute, and the government has authority to pursue any one party while saying to him that it is his problem to recover from the other parties because the government doesn't need to parcel out responsibility—although it can and sometimes does.

The hardest thing for a nonspecialist or nonattorney to handle intellectually about the Superfund law and its implementation is the continuing impression that he and his company have no rights in these cases. Furthermore, the government can, seemingly arbitrarily, create a criminal case or at least one involving large institutional and personal financial risk arising out of actions taken, sometimes long ago, by others no longer around.

When the business community applauds a court finding that bankruptcy is an effective economic shield against claims under Superfund, we can be sure that a new world has dawned for the business executive or owner.

Releases Old and New—Reporting Requirements

Under the logical assumption that the person in the best position to report releases is the one who created the release, Superfund sets up two kinds of required reporting: Spill Reports and Facility Notification.

Spill Reporting. This is a shorthand way of describing an obligation to report a release of any hazardous substance (other than those occurring under a Federal permit) from a transportation vehicle or vessel or a facility in a quantity larger than the reportable quantity for that particular substance. Reportable quantities may be set at 1, 10, 100, 1,000, or 5,000 pounds per day. Reportable quantity for most things is set at 1 pound unless the EPA publishes a larger amount with justification based on a risk analysis. This system obviously places a premium on knowledge of materials being handled and their regulatory status.

Because of the exclusion of oil from the Superfund system, reporting requirements for oil spills are different, and those handling oil need to study that set of requirements.

For everything but oil, the reportable quantities are listed in Volume 40 of the Code of Federal Regulations, Section 302.5. There are some exemptions other than oil spills. For example, things like engine exhaust emissions, routine discharges under the Clean Air Act and the Clean Water Act, the routine application of fertilizer (and other farm chemicals, in practice), and other releases that must be reported under other regulatory systems. Radioactive spills are not reportable to the National Response Center under the pollution laws but to the Nuclear Regulatory Commission.

The "person in charge" of a vessel, transportation vehicle, or a facility is responsible for the reporting of spills. Reports are made by telephone to the National Response Center at Coast Guard Headquarters in Washington, D.C. The phone number is (800) 424-8802. In most cases you will be required to notify an appropriate State agency also and, independently, the National Response Center. Remember that reporting to the National Response Center cannot be avoided by reporting to any other agency—do not accept anyone's assurance that they will report for you. Some States have no minimum reportable quantities—any spill of a hazardous substance (their lists of such substances may be different, also) must be reported to the State office in that case. Most local telephone directories list the State agency numbers in their emergency number sections.

Spill notices must be given "immediately." Penalties have been imposed for delays of one or two hours. It is safe to say that a spill that brings response from local emergency forces like fire, police, and medical or paramedical personnel had better be reported as soon as the immediate call for emergency help has been entered. When in doubt over the need for a report or whether others may have already called, report anyway. You may later be found to be a "person in charge" and subject to the severe penalties which apply.

The surest way of avoiding confusion and ensuring that timely reports are filed is the clear designation of a person to report for any facility or transporter in the event of an accident or a spill. Truck drivers are required to be trained in this process before transporting hazardous loads under Department of Transportation rules. Never, no matter what the situation, allow a load of hazardous material on the road unless the driver is properly trained and experienced. Department of Transportation regulations require proper placards on the vehicle and specific identification of the materials on container labels. A major reason for these requirements is the protection of local firefighters and police responding to a transportation emergency.

Failing to report a spill you know about when you are responsible for reporting, or filing of false or misleading information is a crime under Federal law and under most State law, also. The Federal rules for these acts of omission or commission can result in penalties to individuals of three to five years in prison and a fine of up to $250,000. Companies can be fined up to $500,000 for failure to report and are subject to very large liabilities to the Federal government for permitting the spill to occur and for the costs of cleanup.

To make it easy to find out about unreported spills, the law permits payment of a $10,000 reward for information leading to an arrest and criminal conviction of a person for violation of CERCLA requirements.

Over and above the risk of criminal penalties, the EPA can impose administrative civil penalties up to $25,000 per day of violation, even more for repeated offenses.

Facility Notification. This is the second kind of notice required under Superfund. Owners or operators are required to report the existence of any facility where *hazardous wastes* have been treated, stored, or disposed of unless the facility has already been reported under the RCRA notification procedure. Note carefully that this is one of the few instances in which *hazardous wastes* as opposed to hazardous substances are the subject of Superfund activity.

This notice is filed by the use of an EPA form: "Notification of Hazardous Waste Site," which was required to be completed and sent to the EPA Regional Office having jurisdiction by October 9, 1981. Any newly discovered sites should be reported as soon as they become known. It would probably be wise to contact the State hazardous waste regulatory office to find out how to make these reports. A letter to the appropriate Regional Headquarters of the Federal Environmental Protection Agency would also be indicated.

It is important that this notification, unlike spill notices, applies only to sites involving hazardous *wastes*. It requires identification of the types and amounts of hazardous waste thought to be there and any known or suspected releases from the facility. You must file if you presently own such a facility, owned or operated such a facility (presumably in the past) at the time of handling, treatment, or disposal of hazardous wastes, or if you accepted hazardous waste for transport and selected such a facility for treatment, storage, or disposal.

Records of facility notifications as to location and title of the facility and of the kind, amounts, origin, etc., of any hazardous wastes contained or deposited at the facility must be retained by the reporting owner/operator for 50 years. Few companies have adequate records storage for a 50-year time span. The requirement cannot, however, be dismissed as frivolous. Failure to maintain the records in usable form or any alteration, modification, etc., is a crime leading to a $10,000 fine or a year's imprisonment.

Anyone can report the existence of a suspected disposal site to the State agency having jurisdiction or to the EPA Regional Office for the area. Some environmental public-interest groups have active "hunt-the-dump" programs and can assist the citizen who feels a report is needed but is uncertain how to get it done.

All that is really needed is a phone call or letter to the Solid Waste Division of your State environmental control agency. Such calls should not ordinarily be made to the National Response Center or other "hotline" numbers. These numbers can be reserved for acute emergencies. Calling them for response to a long-abandoned site may interfere with their availability for reporting spills and other emergency releases. In 1986, from many sources, the EPA and the Coast Guard received 2,700 notices of releases. The *EPA Journal* for January/February, 1987, indicated that up to 2,500 of these reported releases would require cleanup.

Oil spills involving crude oil and petroleum fractions such as gasoline, heating oil, and the like, can be a source of confusion and misunderstanding. Petroleum and fractions are exempt from regulation under Superfund; so oil spills are regulated under the Clean Water Act, Section 311, sometimes called the "Oil Spill Prevention and Control Act."

Because of the basic difference in regulatory authority, penalties are not the same as under Superfund. Methods of cleanup and disposal of cleanup waste may be permissible that might not be allowed under the other solid waste laws, and there may be other differences. Reports of oil spill incidents must be made to the same National Response Center as are similar reports under RCRA and Superfund. Observers and responsible parties should always call the Center unless they are certain that the incident has already been called in.

Once the government has information indicating that a release or a "threatened" release is in existence, it has broad powers to obtain information from a landowner, operator, or even an adjacent property owner to determine the need for a response or even to make a decision on how to respond to the reported actual or threatened release. It is vital to understand that no actual release has to have occurred—only that there may be a "threat" of some release—to trigger this right to enter and inspect property and to demand information relating to the need for some kind of response action.

The President, through his delegates, is empowered to require the following kinds of information:

1. The identity, nature, and quantity of materials generated, treated, or disposed of at, or transported to, a facility or vessel;

2. The nature or extent of a release or threatened release of a hazardous substance, pollutant, or contaminant at or from a facility or vessel; and

3. The ability of the facility or vessel owner or operator to pay for or perform a cleanup.

To confirm this information, government officials may, after giving reasonable notice, inspect and copy documents they believe are related to the information demanded, and enter and inspect the property. While there is some provision for protection of trade secrets, in general, any factual data on samples and analyses cannot be protected as confidential and become public information.

As we said earlier, it is the very essence of the Superfund law that the President has broad, even seemingly unlimited powers to respond to actual or threatened releases of hazardous materials. The breadth of this authority is such that no actual release has to occur or to have occurred in the past. Furthermore, the definition of hazardous materials, pollutants, or contaminants is very broad. Just about anything that *may* cause death, disease, behavioral abnormalities, cancer, genetic effects, morbidity, or deformities can be the target of a Superfund removal or remedial action program or requirement. In dealing with a spill, it is expected that the government will assume the worst about the spilled material until that material is fully identified by those on the scene.

There are some limits to these powers, largely introduced in SARA, but they are so specialized as to have little impact. This specific one may be important: asbestos in common use is ruled off limits for Superfund response action, presumably because it is covered under other laws.

As we have repeatedly said, this is not a book of legal advice. However, anyone who intends to challenge the right of government to seek and demand information or who insists on his Constitutional protections in a Superfund investigation had better retain a qualified attorney at the outset and follow that attorney's advice very carefully. Many of those kinds of issues may be regarded as settled by higher courts to the point that one's objections can be considered frivolous and to constitute expressions of bad faith that can lead to even more stringent punishment at the will of the administrative officers or prosecutors.

Cleanup—The National Contingency Plan

Concerned about the possibility of massive oil spills into navigable waters, Congress mandated in the Clean Water Act the establishment of a National Oil and Hazardous Substances Pollution Contingency Plan. This Plan was, and is, intended to spell out the structure of the response to environmental emergencies resulting from spills.

For historic reasons, the Coast Guard has a prominent role in reacting to releases from vessels and off- and on-shore facilities. It is natural, then, that the National Response Center remains a function located at Coast Guard Headquarters in Washington.

Congress had good reason to worry. The enormous spills caused by the wrecks of several large and superlarge oil tankers like the *Torrey Canyon* and the *Amoco Cadiz*, although occurring in distant waters, showed the potential for such disasters on the American coasts and perhaps in the extensive inland waterways of North America. The massive oil spill from the *Exxon Valdez* in Alaska's Prince William Sound indicated that Congress's worst nightmares could actually come about in the real world.

As concern for contamination of land and air grew, the National Contingency Plan became more comprehensive. Today it provides the basic framework of organization, personnel assignment, and standards which are necessary to respond to releases and threatened releases of hazardous substances, oil, pollutants, or contaminants (as broadly defined) anywhere within the territory of the United States, including the territorial seas.

The Plan (usually, in EPA jargon, called the NCP) can be found in the Code of Federal Regulations, Volume 40, Chapter 1, Subchapter J, Part 300. Since it is over 100 pages long, it is impractical to reproduce it here. More important, it is altered regularly.

The Code of Federal Regulations prints all important regulatory information in current effect at the Federal level. Anyone wanting to get involved in understanding and complying with the regulatory system needs to know what the CFR is all about and how to use it. Copies are on hand at all Federal Depository libraries across the country, either in hard copy printed form or a microfiche.

Major city and some university libraries are depository libraries and can be consulted. A call to your local library (usually the government Documents Collection) will almost always produce information on the location of the nearest collection of government documents for reference. Frequently, even a library having no copies itself can obtain material from a depository collection for a modest charge. Single copies or the entire Code can be obtained from the Superintendent of Documents in Washington.

The scope of the National Contingency Plan is, as we might expect, formidable. Table 13 shows the Table of Contents, Table 14 the official Scope and Application excerpted from the Plan.

As ordered by Congress, the Plan does provide a policy and organizational framework for the government response to releases of all kinds into the environment, either of an emergency nature or of a long-term nature, a response required to reduce or eliminate damage to natural resources through remedial action. Both kinds of action are needed, and the National Plan has to address the whole problem. Restoration of natural resources after damage by a release or releases is encouraged by a system of appointment of "trustees" of specific natural resources who can behave as owner-advocates in ensuring their recovery to the maximum extent possible and compensation to the public for any losses by those ultimately responsible.

No one can really know whether application of the full weight of legal sanctions available to the Federal government under the Superfund law (instead of the application of the Clean Water Act, as actually had to be done in Alaska) would have made the *Exxon Valdez* cleanup faster or better. From outside the immediate group concerned with the remedial action, it appears that problems of organization, logistics, and the sheer size of the spill made it nearly impossible to clean up by human intervention. Prevention may be the only course that will work—if it is possible in an international environment of distrust and unfair competition.

The National Priority List

One of the vital parts of the National Contingency Plan is a list, required by statute, of the major hazardous substance contamination sites in the nation. While usually called "hazardous waste sites" in the common language of the work, the contaminants at many sites were not "waste" as we know it but were commercial chemicals until released to become pollutants or contaminants. A common case arises from the discovery of abandoned materials left in an abandoned mining or manufacturing building—whose owners may have intended to return and use them up but never did. "Dumps" are by no means the only important items on the National Priority List (NPL).

The National Priority List, updated regularly, is supposed to comprise, beginning with the worst, a tabulation of at least 400 sites considered most important for cleanup. In 1977 the list had about 500 sites grouped into

Table 13

Partial Table of Contents for the National Contingency Plans

National Oil and Hazardous Substances Contingency Plan
(40CFR Ch. 1, Subch. J)

Subpart A—Introduction

Section	
300.1	Purpose and objectives
300.2	Authority
300.3	Scope
300.4	Application
300.5	Abbreviations
300.6	Definitions

Subpart B—Responsibility

300.21	Duties of President delegated to Federal agencies
300.22	Coordination among and by Federal agencies
300.23	Other assistance by Federal agencies
300.24	State and local participation
300.25	Nongovernmental participation

Subpart C—Organization

300.31	Organizational concepts
300.32	Planning and coordination
300.33	Response operations
300.34	Special forces and teams
300.35	Multi-regional responses
300.36	Communications
300.37	Special considerations
300.38	Worker health and safety
300.39	Public information
300.40	OSC reports

Subpart D—Plans

300.41	Regional and local plans
300.42	Regional contingency plans
300.43	Local contingency plans

Subpart E—Operational Phases for Oil Removal

300.51	Phase I—Discovery and notification
300.52	Phase II—Preliminary assessment and initiation of action
300.53	Phase III—Containment, countermeasures, cleanup, and disposal
300.54	Phase IV—Documentation and cost recovery
300.55	General pattern of response
300.56	(Reserved)
300.57	Waterfowl conservation
300.58	Funding

Subpart F—Hazardous Substances Response

300.61	General
300.62	State role
300.63	Discovery or notification
300.64	Preliminary assessment for removal actions
300.65	Removals
300.66	Site evaluation phase and National Priorities List determination
300.67	Community relations
300.68	Remedial action
300.69	Documentation and cost recovery
300.70	Methods of remedying releases
300.71	Other party responses

Subpart G—Trustees for Natural Resources

300.72	Designation of Federal trustees
300.73	State trustees
300.74	Responsibilities of trustees

Subpart H—Use of Dispersants and Other Chemicals

300.81	General
300.82	Definitions
300.83	NCP Product Schedule
300.84	Authorization of use
300.85	Data requirements
300.86	Addition of products to schedule

Appendix A to Part 300—Uncontrolled Hazardous Waste Site Ranking System: A User's Manual

Appendix B to Part 300—National Priorities List

Table 14

Scope of the National Contingency Plan
(40CFR Ch. 1, Subch. J, § 300.3)

300.3 Scope

(a) The Plan applies to all Federal agencies and this Plan is in effect for:

1. Discharges or substantial threats of discharges of oil to or upon the navigable waters of the United States and adjoining shorelines, for the contiguous zone in connection with activities under the Outer Continental Shelf Lands Act or the Deepwater Port Act of 1974, or which may affect natural resources belonging to, appertaining to, or under the exclusive management authority of the United States (including resources under the Magnuson Fishery Conservation and Management Act) (see Sections 311(b) (1) and 502 (7) of the CWA).

2. Releases or substantial threats of releases of hazardous substances into the environment, and releases or substantial threats of releases of pollutants or contaminants which may present an imminent and substantial danger to public health or welfare.

(b) The Plan provides for efficient, coordinated, and effective response to discharges of oil and releases of hazardous substances, pollutants, and contaminants, in accordance with the authorities of CERCLA and the CWA. It provides for:

1. Division and specification of responsibilities among the Federal, State, and local governments in response actions and appropriate roles for private entities.

2. The national response organization that may be brought to bear in response actions, including description of the organization, response personnel, and resources that are available to respond.

3. The establishment of requirements for Federal regional and Federal local contingency plans, and encouragement of preplanning for response by other levels of government.

4. Procedures for undertaking removal operations pursuant to Section 11 of the CWA.

5. Procedures for undertaking response operations pursuant to CERCLA.

6. Designation of trustees for natural resources for purposes of CERCLA.

7. National policies and procedures for the use of dispersants and other chemicals in removal and response actions.

(c) In implementing this Plan, consideration shall be given to the Joint Canada/U.S. Contingency Plan, the U.S./Mexico Joint Contingency Plan, and international assistance plans and agreements, security regulations and responsibilities based on international agreements, Federal statutes, and executive orders. Actions taken pursuant to this Plan shall conform to the provisions of international joint contingency plans, where they are applicable. The Department of State should be consulted prior to taking any action which may affect its activities.

classes in decreasing order of severity or importance. Ten years later, the list had grown to over 1,500 sites.

The NPL is developed from required facility reports, from information provided to the National Response Center by responsible parties or the public, and by nomination by the States. The law requires that at least one site—the highest-priority one by State criteria—in each State must be included on the list. It appears that at least one site from every State is included in the top 100.

The decision to add or delete a site from the NPL is made after a process comparable to a rule making. A site *may* be included if it meets one of three requirements:

1. A health advisory has been issued by the Agency for Toxic Substances and Disease Registry which recommends that persons stay away from the site;

2. The EPA makes a determination that the site poses a significant danger to public health; or

3. The EPA decides to approach cleanup as a "remedial action" as opposed to a "removal action." (This distinction will be discussed later.)

If a proposed site has at least one of the three characteristics just listed, then a preliminary assessment and/or a site investigation is conducted by or for the EPA. Contractors, State or local agencies, or EPA staff conduct a quick evaluation based on available information from affected persons and, often, the press.

If a site is prominently discussed, is controversial, or attracts the press to a high level of public concern, much of the formal process may be bypassed and a site investigation, often complete with photo opportunity and press conference, will be conducted. Such a level of public activity will almost always result in at least a proposal to score the site carefully with an eye toward a proposal for addition to the NPL. This decision requires public notice and a hearing before the site can be added. Among the essentials of the decision-making process is the score of the site on the Hazard Ranking System. It is important to know that site visits and on-scene assessments are not essential to the classification and listing system. Such decisions can be and are made for less-known and noncontroversial sites on the basis of information provided to someone in an EPA office who never sees the site.

Citizens can petition the EPA to do a preliminary assessment. If this work shows that the site may be threatening to human health or the environment, the EPA will do scoring under the Hazard Ranking System. State agencies also do the scoring, particularly if they disagree with the EPA's priority decisions.

The Hazard Ranking System, originally developed by MITRE, a contract research firm, is itself controversial. The SARA Amendments to CERCLA require it to be modified. The wanted changes broaden the elements of risk that should be included in the ranking process. The original Hazard Ranking System rates the threat of a site through three pathways of exposure: surface water, groundwater, and air pollution. Congress wanted some consideration of the indirect threat to the human food chain through damaged natural resources.

All of these subscores are developed with a great deal of subjective judgment for the simple and largely unavoidable reason that useful data for most of the input do not exist. The worksheets contain explicit directions for scores to be used when there is no data. While such methods are often useful to the practitioner faced with a decision as to the allocation of limited resources, one problem is that they give the appearance of objectivity—a number score—that is often not supported by any facts and may well be in conflict with an evaluation by experienced professionals.

The really important thing about all this investigation, scoring, and bureaucracy is that no site is eligible for full evaluation or final cleanup using Superfund dollars unless it is first listed on the National Priority List. We still do not know how many sites will eventually be eligible for Superfund cleanup. More than 6,000 sites are known and a total of perhaps 25,000 is often discussed. Of all these, the NPL lists just over 1,500. This smaller number certainly represents the most notorious locations. Only time will tell whether it represents the most important ones.

Mitigating this bureaucratic tangle is the fact that the EPA has three authorities under Superfund, supplemented by provisions of the Clean Water Act and some others, to clean up oil and hazardous waste contamination:

1. Under the oil spill provisions of the Clean Water Act;

2. The hazardous materials spill provisions of the Superfund law; and

3. The disposal-site remediation provisions of the Superfund law and RCRA, which together give the EPA almost unlimited authority to clean up both active (RCRA) and abandoned (CERCLA) sites.

Behind and beyond all these formal powers, the laws include a broad authorization for the EPA to identify "imminent hazards" to human health or the environment and to take whatever steps are required in the judgment of the Administrator of the EPA to eliminate, mitigate, or control the situation. These provisions are so sweeping in their authority to define the imminent hazard and to determine appropriate action to eliminate or mitigate the hazard that it is hard to imagine a situation in which a reasonably bold administrator would lack power to act—with little restraint on the costs and side effects of his actions.

Listing of a site is important in establishing the urgency and therefore the probability of some action on the site. Absence from the list does not foreclose action—it may simply mean that the EPA or the State plans to deal with the matter under some other authority—perhaps a law with less red tape like the Clean Water Act or State statutes, many of which the EPA has the power to enforce as if they were Federal laws when it chooses to do so.

Finally, the owner of a site not yet on the NPL can take action to remove the source of contamination and to clean up the site on his own initiative. In many cases, such voluntary action by the owner will show a commitment or concern that is important in future dealings with the regulators and may well save the owner some money as well as show the community a willingness to play a constructive part in the solution of what is essentially a community concern.

The Phased Response System

The National Contingency Plan, as already discussed, organizes the total national response to release of hazardous substances. At the head of the organization is the National Response Team, made up of representatives of at least 12 Federal agencies. This large and unwieldy group "assures" preparedness across the entire nation to respond to "significant" discharges of oil or hazardous substances. Counterpart people form the Regional Response Teams in each standard Federal Region and act to help implement the National Response through regional and local contingency plans.

The EPA representative on each of these teams is the Chairperson, and the Coast Guard representative is Vice Chairperson.

The Coast Guard (in coastal areas) and the EPA (in inland areas) appoint On-Scene Coordinators (OSC) for main areas of each region. This person directs Federally funded response efforts at any Superfund site and coordinates all other Federal efforts on the scene. The powers of the OSC are potentially very broad. His/her decisions often are not reviewable by courts or higher authority, at least when made under emergency conditions.

Lest anyone be intimidated by this formidable array of agencies and individuals, only rarely does anyone not employed full time in handling hazardous materials encounter this large structure of response capability. To the teams already named, we can add Regional Strike Teams intended to assist the On-Scene Coordinator and the EPA's own Environmental Response Team.

Generally, these backup teams of experts are biologists, chemists, on-site safety specialists, hydrologists, engineers, or people specially trained in the use of advanced cleanup technology and equipment. These are often the first people to appear on the scene of an emergency or even an old site—dressed in "moon suits" that impress the danger on anyone watching.

The full response system is designed to be operated in seven phases, for hazardous substance releases. The oil response has only four phases.

Phase I	Discovery and notification
Phase II	Preliminary assessment of the hazard
Phase III	Immediate removal action. Actions taken to mitigate acute threats of harm to human health or the environment. This phase may be bypassed to Phase IV if conditions warrant.
Phase IV	More detailed assessment of the threatened harm and more deliberate planning of either removal or remedial action. May be a follow-on to Phase III after the emergency is past, and may lead to listing on the National Priority List.
Phase V	Planned removal actions even where a continuing emergency exists. In this type of removal, the State must participate in the costs.
Phase VI	Long-term remedial action. This phase includes the Remedial Investigation/Feasibility Study and all long-term actions designed to clean up the site to abate any migration of hazardous substances from the site in harmful quantity.
Phase VII	Actions by the EPA to gather data for cost recovery

Cleanup—The Approach to Major (NPL) Sites

Any time the EPA (or its delegates, the States) finds that a site presents some imminent hazard stemming from a potential for fire or explosion or severe hazard from direct contact of persons, livestock, etc., with hazardous materials, short-term *removal actions* can be conducted. In this context, *removal* implies collecting, excavating, gathering, or any other action that eliminates the dangerous or suspect material from the site for safe disposal, often to another location, but occasionally to a different spot on the same property. Removal may encompass a very wide range of actions. Fencing of the site is a common one, while physical removal of drums, piles of material, contaminated soil, and even liquids found in pits is often accomplished as rapidly as possible after discovery of a site presenting an imminent hazard.

Not so common, but illustrative of the EPA's broad authority, is the decision to provide safe drinking water to hundreds or even thousands of people who are deemed to be threatened by damage to their normal supplies. Emergency or even permanent evacuation of affected people can be part of a removal action, including the provision of temporary housing and assistance in obtaining new permanent housing.

Removal actions do not require the site to be listed on the National Priority List. Expenditure of Superfund effort is limited to $2,000,000 or one year's time unless the site becomes listed on the NPL. More important, the government can order private interests responsible for the site to undertake any or all of the actions it believes are necessary. For such orders, neither the $2,000,000 limit nor the time limit applies!

Often while the removal action, if any, is taking place, the EPA will order a Remedial Investigation and Feasibility Study prepared for the site. Before 1986, policy encouraged private owners and other potentially responsible parties to design and carry out these studies under the direction and with the agreement of an EPA Regional Office or a delegated State agency. This policy resulted in a good deal of technical interaction in informal discussion of the results of ongoing studies and, often, modification of the plan as new data indicated the desirability of changing. Under the 1986 Amendments, the EPA has the sole responsibility for the preliminary studies and for the design of the remedy for a site. Potentially responsible parties may be allowed to participate but only if the EPA wishes and on the EPA's terms.

Generally, if a Potentially Responsible Party (PRP) wishes to conduct the Remedial Investigation and Feasibility Study (RI/FS), the PRP must agree to do the work within a scope-of-work prepared by the government. There may be some negotiation about the scope but not much. Actually, the SARA Amendments provide for the government to give notice to the parties it believes may be responsible for the site 90 days before it starts on the preparation of the RI/FS. The private interests can respond by offering a work plan for approval and indicating a willingness to proceed when the work plan is approved. If the PRP's do not respond, the government will see to it the work is done and then sue the PRP's to recover its costs.

As will be more fully discussed later in this chapter, the U.S. Army Corps of Engineers, acting through its Division of Civil Works, is often hired by the Environmental Protection Agency either to accomplish or to contract with private engineering firms the performance of technical work needed to prepare the studies it finds necessary. Often, the Corps is given the task, on behalf of the EPA, of managing the contracting and construction-type activities involved in a remedial action.

This relationship is based on Congress's long familiarity and confidence in the competence of the Army Engineers in the management of complex engineering and technical programs throughout the Federal establishment. It has, thus far, been decided that the EPA does not need a construction contracting and supervisory capability of its own.

Remedial Investigation and Feasibility Study

Formally, the Remedial Investigation is a separate job from the Feasibility Study. The Remedial Investigation is entirely a data-gathering and fact-finding field investigation. Normally, extensive sampling of soil, water, air, and even of sensitive plants and other biota on and near the site are conducted. Extreme care in sampling techniques and in maintaining the identity and integrity of the samples is always required in compliance with a Quality Control/Quality Assurance Plan incorporated into the scope-of-work at the outset. Analyses are numerous and frequently expensive to conduct. Only a few laboratories have the equipment, expertise, and government certifications required to participate in these studies. More important, the work must be directed by highly qualified geotechnical engineers, geologists, and other specially trained persons in order to obtain meaningful results.

It is in the Remedial Investigation that many (sometimes literally hundreds) of testing wells are drilled to track contaminants underground. Each investigation is unique, based on unique conditions of contamination and unique factors at the site, like soil types, vertical distance to impervious zones or bedrock, the kinds of rock, rates of migration of groundwater, and many, many other factors. Close cooperation between the regulatory agency, its contractors (usually geotechnical engineering firms), and potentially responsible parties yields the best results.

The Feasibility Study, often conducted nearly simultaneously with the Remedial Investigation, but formally separate, is an engineering study. On the basis of the data produced in the Remedial Investigation, experience with the behavior of environmental contaminants, and professional judgment, the available options for remedying the contamination are explored. As defined in CERCLA Section 101, a remedy or remedial action includes those activities consistent with a permanent remedy which are done in place of or in addition to removal action (as just discussed) to prevent or minimize the release or threat of release so that hazardous substances *will not migrate* to cause substantial danger to present or future public health, welfare, or the environment. This is quite an order, since no one really knows the future.

Under this sweeping concept of remedy, many on-site activities like excavation, treatment of soil or water, and groundwater recovery, purification, and monitoring are easily included. Further inclusions are permanent ways of limiting public access to the site, evacuation and housing of threatened people, provision of alternate water supplies, and emergency assistance under the Disaster Relief Act.

The design of remedial action is done under policy constraints contained in the National Contingency Plan and specific intent-of-Congress statements. The former bias in many cases toward excavation followed by disposal in a secure landfill has been specifically rejected on the reasonable ground that merely moving contamination to another site is logically unsound as a remedy. Congress has called for "permanent" remedies using treatment or resource recovery technologies to the "maximum extent practicable." In reality, this means that if such technologies are not applied, then the failure to use them must be justified.

These new remedial actions, coming under the heading of "source control," are intended to eliminate hazardous substances from a site. If, as was common under earlier practice, hazardous substances are merely contained or sealed, but remain, then the EPA must reinvestigate every five years—not an appealing thought.

Other remedial actions are taken to reduce or eliminate contamination that has migrated from the site. By far the most common form of this "residual contamination management" is the removal or pumping of groundwater for treatment or disposal. The recovered groundwater is expected to carry with it any contaminants that may have migrated into it from a release site.

How Clean Is Clean?

In Superfund cleanup activities, the source of greatest controversy and of greatest difficulty in the design of remedial action is the problem of knowing when cleanup is sufficient and can be considered complete. One of the unavoidable laws of nature is that removal of dissolved or dispersed materials is less and less effective as the removal proceeds; thus, the amount left, while getting smaller and smaller as the cleaning proceeds, will always be more than zero. In the first years of Superfund, the answer tended to be based on risk analysis (see Chapter 6, "Risk Assessment and Management") and a statutory concern about cost-effective use of the Fund money. Recently, the courts seem to be establishing a doctrine that the extent of cleanup is solely a decision for the On-Scene Coordinator, a relatively low-level official who runs the field operation. One thing on which there seems to be agreement among many is that a cleanup in which the government does not expect to recover significant funds from nongovernment responsible parties is never as complete as one in which private organizations with "deep pockets" can be found to finance it.

Some interests seek a cleanup to "background" levels of soil and water contamination. The difficulty here is that there are few or no data on background levels of environmental contamination with which the results of a cleanup can be compared. The current approach (SARA) requires the identification of legally *applicable or relevant and appropriate requirements* (called ARAR's, naturally) from all other Federal environmental regulations or from more stringent State ones.

This direction from Congress causes a sometimes nightmarish search for the ARAR for a particular site, environmental medium, and substance migrating from the site. A brief list of the relevant Federal programs includes those developed under the Clean Air Act; Resource Recovery and Conservation Act; Safe Drinking Water Act; Clean Water Act; Toxic Substances Control Act; Marine Protection, Research and Sanctuaries Act; and others. In general, State regulations are permitted to impose more stringent limits than Federal rules, and the Feds are required to enforce them when they apply.

More specifically, the SARA Amendments require cleanup of groundwater to the Maximum Contaminant Levels developed under the Safe Drinking Water Act and/or water quality criteria published under the Clean Water Act. Again, State standards may be more demanding than the Federal ones, and if so, they must be applied. Realistically, many of the required standards for ARAR's for chemicals do not exist and may not be established for years, if ever.

Although it has not been heard much in recent times, there remains a certain pressure for cleanup standards based on "nondegradation." This is the same as requiring achievement of background levels and presents the same problem—lack of data on the prerelease situation.

Few sites can ever be cleaned up to meet every ARAR. Compliance with this requirement can be waived under certain circumstances. One major rule affecting waivers of ARAR's is based on a judgment about cost-effective application of Superfund money at a particular site and the need at other sites. This cost-effectiveness analysis is not available in private party cleanups, creating a clear possibility of a double standard—one for government cleanups, another and more stringent one for privately funded ones.

Remedial Action, Public Participation, and the Administrative Record

In general, when the EPA does, or contracts for, a Remedial Investigation and Feasibility Study, it specifically instructs the engineering contractor to avoid selecting or even recommending a particular remedial action plan in its report. Privately prepared plans often include such recommendations.

The RI/FS is a public document and is part of the Administrative Record for the site. During the progress of the RI/FS studies, interested parties may place information bearing on the agency's decision as to a remedy into the Record by following a relatively simple procedure. At some point in the process, the EPA is required to have a community relations program for the site. Public meetings to explain the findings of the RI/FS program are held after announcement in the local press. All interests, from responsible parties to individual members of the public, are entitled to appear and to express their views and concerns at these meetings.

In response to the common concerns that the major actors, both government and private, have access to the best professional assistance in reviewing the proposed action and preparing detailed comments, the EPA can provide grants of up to $50,000 to citizens' groups, etc., to help pay for the costs of professional help which they may need in the process.

At any rate, at the end of the process and based on the Administrative Record which it develops, the EPA develops a Record of Decision that defines the remedy to be applied along with all the other requirements to be met by whoever cleans up the site.

The Administrative Record and the Record of Decision are not just more bureaucratic paper—although they look and sound as if they are. The Federal courts have held that they will not review any action by the EPA on a Superfund site except on the basis of the Administrative Record alone. Any court review, if granted at all, will not allow introduction of data or information not included in the Administrative Record. Furthermore, any court review of the remedy described in the Record of Decision cannot be had until the cleanup is complete and the government sues to recover its costs from responsible parties.

A thorough understanding of the unique status of administrative decision making under Superfund is essential for any potentially responsible party. These rules and practices demand that anyone seriously involved in these matters retain experienced and competent professional engineers and legal advisers from the earliest indication that a company or organization may be involved with a hazardous waste site.

In a similar way, any individual or group of citizens needs to ensure that their input finds its way into the Administrative Record. Nothing that is considered relevant to the eventual decision on a remedial action plan should be too small to be the subject of a letter to the office directing the RI/FS work, with a specific request that the letter be made a part of the Record in the case.

Citizens have access to other powers under the Superfund law through the provisions for citizens' suits against the government. We will discuss this process and its limits later in this chapter.

CLEANUP

The Actual Work

When J. Winston Porter was Assistant Administrator of the EPA for Solid Waste and Emergency Response, he characterized Superfund cleanups as "major unstructured engineering projects." An unstructured engineering project is one in which only the most general idea of the desired objective is available, without useful guideposts which the engineer can rely upon to judge the nature, acceptable cost range, or even the need for a solution. A structured project might direct the engineer to design a bridge over a specific body of water at a particular location. An unstructured project which might have a similar outcome would direct the engineer to evaluate and design means to allow traffic to move between the parts of the area lying on both sides of the water. It is easy to see that the structured assignment calls for less effort and even less understanding of the total context of the problem than does the unstructured one.

In remedial action, the RI/FS reports are the result of an unstructured assignment for which the engineer, with selected assistants having needed expertise, is expected to develop the data on the characteristics of the site, to investigate and understand the behavior of the chemical substances present, both natural and from any release, and to understand the various technical means available for achieving permanent removal of the undesirable substances from the site, or at least prevent their migration. In addition, all this is to be done in a cost-effective manner—that is, achieving the best possible results for a reasonable expense. The work has to be done in anticipation of challenge and controversy, and subject to intensive public review.

Even when the remedial action plan is agreed upon, a major site cleanup can be compared only to a large-scale construction project in its impact on the site and the surrounding community. Recognizing this, the EPA often hires the U.S. Army Corps of Engineers Division of Civil Works to manage the cleanup contract on behalf of the On-Scene Coordinator. At least in the beginning, the OSC's were rarely experienced in management of large-scale contracts, and the Corps can share its long experience and expertise in bidding, selection of contractors, and supervision of the fieldwork.

Most important, no two cleanups are ever alike, even if they are superficially similar. It would be a waste of space here to attempt to describe "the process" because each site must have its own process developed for its conditions alone. The *EPA Journal* for January/February, 1987, contains some good case studies, and the general literature in the field of hazardous materials and waste management frequently has descriptive articles on the more highly publicized cleanups.

Probably the most difficult part of a cleanup is its lack of believability for the public. Places like Love Canal remain suspect even when restoration is certified as satisfactory by prestigious bodies like the Centers for Disease Control and the Agency for Toxic Substances and Disease Registry. Having choices, people simply avoid these sites and any possible problem for a long time. Many States now require the actual or former presence of waste materials on a site to be permanently recorded in the deed of ownership of the real property. There is no real cure for the credibility gap—except time, perhaps.

Costs, Liabilities, and Settlements

Volumes have been written for attorneys and other professionals involved in the enforcement and cost recovery aspects of Superfund. Any person or business who is under serious scrutiny as a Potentially Responsible Party for a site that may be contaminated *must* seek qualified legal advice immediately. Often, as we shall see, action taken by an owner/operator *before* the EPA takes formal notice can have important effects in reducing the costs and time demands of a response action.

There are two basic ways of becoming involved with Superfund enforcement and cost-recovery actions: as the owner/operator of a site or facility which comes under review either because of required reports or by means of some publicity or public concern; and as a contributor to a disposal site failure where several or even hundreds of people and businesses contributed wastes to the site, often over many years. These two ways of becoming involved both grow out of a long-term disposal site operation; so the enforcement process does not necessarily cover the site of a transportation accident after the emergency is over. However, the eventual treatment of accident sites is in many ways so similar to the ones we will discuss that we do not single them out here.

In the event that the EPA expresses some interest in your site or facility, usually by a letter requesting information, the first step after retaining counsel should be the beginning of your involvement in creating the Administrative Record of the case. Particularly, look for any information that the EPA has that might be favorable to your involvement—expressions of good faith, for example, or of satisfaction from the requesting office with your reporting efforts can be of value in the future.

If the EPA determines that a period of negotiation is likely to produce action by the Potentially Responsible Party (you!) that will result in faster agreements to take response action and will expedite remedial action, it can declare a moratorium on response action. At the same time, the EPA must inform the recipients of the notice of what it knows about the quantity and nature of the contaminating substances believed to be present and some estimates of the relative amount when it believes more than one contaminant is present. Names and addresses of all other known PRP's must also be supplied.

The moratorium period is 60 days, and during that time the PRP or the group of PRP's has an opportunity to make a proposal to conduct or to pay for a Remedial Investigation/Feasibility Study. While there is no guarantee that the government will accept such offers, and indeed the SARA Amendments seem subtly to shift the preference to strictly government-conducted studies, most practitioners seem to find the offer to be a worthwhile step. A good serious proposal will allow PRP's to participate in discussions about the RI/FS process for the site and may result in substantial savings of time and money. Reimbursing the government for the cost of a study conducted under its procurement regulations and under the direction of its technical bureaucracy can become very expensive. There are reports that privately conducted RI/FS programs have been done for one-third of the government costs.

Since the RI/FS for even a limited groundwater-contamination site can cost hundreds of thousands of dollars and the work on a large-scale site many millions, it is essential that the persons and companies that are expected to pay stay close to the work and the regulations at all times. Hostility and failure to cooperate can only lead to even greater application of the government's formidable power in these matters.

Whoever does the studies, eventually the government, usually through the Regional Administrator of the EPA, will issue a Record of Decision (ROD) defining the work to be undertaken to accomplish the cleanup. At this point, a proposal to conduct the cleanup can be made by the PRP and should be considered carefully as an alternative to simply watching the work proceed in the expectation that you will be sued for all the government's costs when it is complete. Remember, your only chances to appeal the requirements of the cleanup occur before the Record of Decision is signed and after the cleanup is complete. The latter course seems uncertain at best.

A cleanup proposal must be a serious one that illustrates an understanding of the responsibility one is asking to assume and showing that the work will accomplish all requirements spelled out in the Record of Decision. A general offer is not likely to lead to reasonable negotiations with the regulators. A specific offer needs to consider the ability of the PRP to gain access to all of the property needed for well sites, for example, or for off-site disposal of any hazardous waste generated in the cleanup itself. Superfund gives the President the authority to enter on adjacent property for the purpose of obtaining information needed in a cleanup or investigation, but private parties do not gain this power directly.

State solid waste officials have a major interest in promoting private financing of response activities. Under the basic provisions of Superfund, States must pay 10% of the costs of response actions funded by Superfund dollars, and 50% if the response is required on State-owned land or land owned by a political subdivision like a city or county. While the States, too, can sue for reimbursement, few want to wait, and some have no sources of funds for their share of cleanup costs.

All of these factors encourage responsible parties to settle with the government, either when it sues after the fact to recover costs, or, as newly provided in SARA, under an Administrative Order issued on the basis of the Record of Decision in the case. Under the settlement concepts presently in use, the responsible party has an opportunity to have input to the whole process, often can do the work at less cost than the government could, avoid litigation, and can control the selection of professional advisers to ensure high-quality work.

Remember that once issued, the Administrative Order is not reviewable in court except in extraordinary circumstances. Any failure to comply with the Order can lead to an administrative penalty of up to three times the cost of the remedy! High-grade negotiation beforehand and strict compliance with an Order once issued is obviously the way to go.

Large Sites

To this point, our discussion has focussed on the type of site for which one legal entity is basically responsible and is acting to protect its own interests, often with historic knowledge and control over the site. Another class of large sites, where many entities sent substances for disposal, often without even knowing where the site was or who operated it, is a feature of the most-publicized activities of implementing Superfund.

The sites in this category that are the subject of cleanup activity are usually ones that were occupied for varying lengths of time by persons and companies offering disposal and recycling services to manufacturers and processors of commercial chemicals and solvents. Many of them had associated landfill operations which accepted any kind of waste for disposal. Some had been the site of spectacular fires and explosions and were well known so that they rapidly gained a position on the National Priority List.

From the beginning, in responding to these kinds of sites, Superfund incorporated the legal concept of *joint and several liability* as applying to the users. This concept of law generally means that any person or company which is found to have "contributed" hazardous substances to a site where a release has occurred or is threatened to occur is responsible for the consequences *as if he were the only contributor*. Each contributor is potentially liable for all of the response actions and costs for the site even if hundreds of contributors were involved.

In dealing with a large site like a leaking landfill once operated by a disposal contractor, the government will, based on its assessment of the site and on records it can find, identify some contributors and send them notice of their liability. The notice, as we have already discussed, has to say why the recipient is believed to be a PRP and who some others are, as well as give some indication of the extent of the notified company's involvement.

In such cases, it is expected that the major contributors will organize a committee or consultative group to negotiate jointly with the EPA toward a settlement. Remember the EPA holds each one fully responsible for the entire cleanup and leaves it to the parties to work out a sharing of costs by negotiation or by filing lawsuits for "contribution" on PRP's not directly pursued by the EPA. In some cases, the groups of PRP's have obtained agreement that they can pursue the remedial action, but under present thinking, it is more likely that the government will proceed with the cleanup while settlement negotiations continue. It is not the EPA's practice nor has it been held to have any obligation to pursue all the PRP's. Courts have supported the placing of that responsibility on the notified PRP's if they want to gain other contributions to reduce their own liability.

The SARA amendments allow the EPA to use its subpoena power to gather information on relative allocations of responsibility for a multiple-contributor site. This can result in a document called a Nonbinding Allocation of Responsibility (NBAR), but this document has no official standing in any lawsuit and remains, at this writing, of uncertain value in the minds of attorneys and other representatives of PRP's. If prepared and issued, the document represents at least a preliminary attempt by the government to balance quantity of waste against "quality" or hazard and allocate the responsibility on that basis. It's easy to see why the NBAR is not an easy document to prepare and is bound to lead to endless controversy. Few government servants want to get involved in voluntarily preparing a document which will become instantly controversial.

There is a special provision which allows the EPA to reach final settlement with contributors whose involvement at a site is very small compared to the total amount of waste and hazardous materials found there. Under this *de minimis* settlement, the government offers a consent decree in Federal court or an Administrative Order under consent, usually requiring payment of some amount, often of seemingly arbitrary size. Small contributors are often assessed $25,000, and settlements of this kind over $500,000 require so much red tape for the government that they will be rare. The group of major PRP's organized on a major site may offer a similar deal to minor contributors, but in considering such offers, one must ensure that one can obtain a meaningful release from future liability. One way to do this is to insist on being sued so the agreement is approved by a court having competent jurisdiction.

Any contributor who has an opportunity to reach a *de minimis* settlement in one of these cases should seriously consider it, naturally with competent legal advice. One of these matters is a bad place to insist on one's legal rights or the protection of one's good business name—the risks are just too high in almost all cases.

The End of the Process

Strictly speaking, remedial actions have no final physical end point. In groundwater recovery and treatment, for example, if the desired target quality is reached, often the system can be shut down. Still, monitoring must continue for long periods (thirty years is common), and resumption of the remedial action may be required if contamination returns or if standards change.

For most, the end of the legal process is a most important achievement. Usually, when a settlement is agreed upon either as closure to a lawsuit or as a response to an Administrative Order, the desired end is a Covenant Not to Sue, granting release from further liability to the government for the site or the part of the site under consideration. The covenant can be granted if it is deemed in the public interest under a strict review process, if it expedites the response action consistent with the National Contingency Plan, and if the recipient of the covenant is in full compliance with the approved response action.

One must understand that all such agreements can and will be "reopened" if the government finds new information or a new release occurs which results from some condition not known when the covenant is granted. One advantage of a *de minimis* settlement is the absence of a reopener in its Covenant Not to Sue. There are other circumstances in which the covenant cannot be reopened, as an attorney can advise when these unusual opportunities arise.

After all this, the responsible party or parties must complete the response action as agreed. This may take years, particularly if it involves removing contaminants from soil by aeration or from groundwater by pumping and disposal. Monitoring reports are required regularly, and the mechanical and chemical systems must be operated and maintained as any complex process system must. Performance of the systems must be consistent over the life of the project, often considerably longer than the economic life of a manufacturing system.

Few companies have any provision within their management structure or financial controls for the long-term maintenance of remedial systems. Before top management enters into a consent order, these needs should be fully addressed. Failure to comply with a remedial action plan is usually grounds for very substantial penalties, often without court action or review. Most Administrative Orders, including Orders under consent, will contain provisions in which the responsible party has agreed to the penalties in advance. It would not be unusual for the cost of a breakdown of a remedial system, if a result of poor maintenance, to exceed substantially the cost of effective management of the remedial system over the long pull.

CITIZENS' INTERVENTION—A NEGLECTED SOURCE OF POWER

Throughout the processes we have been discussing, the citizen as an individual or as represented by a civic association or even a neighborhood group has the opportunity to be heard as a matter of right at many points. We have already mentioned that grants up to $50,000 are available to such groups to allow them to cover costs of effective intervention. Even without this money, individual citizens, particularly if directly affected by the decisions being made, can and should appear at the public meetings and hearings that must be announced in advance in local papers and can often be found on community bulletin boards, etc.

In appearing at such meetings, citizens just have to remember that, despite the fancy language and formidable terminology used in these matters, the regulatory agencies are ultimately dependent on public support for their effectiveness and indeed their existence. Few public officials can last long in office if they attract too much unfavorable attention from the public.

As a last resort, the SARA amendments created a right in Superfund matters for citizens to sue the Administrator of the EPA for certain failures to perform tasks which the citizen believes he is supposed to do under the law. This right, while rather tightly restricted, is a potent motivator for the regulators to perform their proper functions as ordered by Congress.

Citizens' suits can be brought only to have the Administrator and his delegates perform duties required under law. Where the law gives the Administrator discretion as to whether to act or what action to take, citizens' suits are not maintainable. Nevertheless, this provision is a major source of power for the public-interest group striving to be heard.

TITLE III OF SARA—THE COMMUNITY RIGHT TO KNOW

In the aftermath of several major tragedies, most notably those at Soveso, Italy, and at Bhopal, India, in which toxic materials escaped from industrial facilities and caused injury and loss of life in the vicinity, Congress added Title III to the SARA Amendments to Superfund.

This addition creates a new program of still-unknown dimensions. The basic objectives are laudable: persons living near industrial facilities which use or store and handle toxic materials ought to know what hazards are associated with sharing the community with such operations.

To accomplish this objective, Title III sets up an elaborate system of local, regional, and State planning agencies which are supposed to coordinate a massive information program based on mandatory reports to various levels of government as to the nature and quantity of hundreds of chemicals on hand at the facilities or used there. In particular, local emergency forces, the fire and police functions, are given specific rights to information which they may have had difficulty getting before.

This program does not particularly apply to hazardous waste, although it may have some overlap, as do nearly all these programs. Its requirements are all in addition to the regulations on hazardous waste management. For these reasons and because it would take a book the size of this one to discuss fully the Community Right to Know regulations, we will explore this no further here. Those who wish to understand the program and its impact on them and their community should contact their municipal officials, who are required by the law to provide information to the public.

SOME REFLECTIONS ON SUPERFUND

After wading through this long description, readers will have some feel for the vindictive tone of the Superfund law. The assumption throughout is that these releases, whether today or remaining from actions of many years ago, were careless, at best thoughtless, acts which were largely motivated by an unwillingness to spend money at the time to find alternatives. In short, by greed.

While there can be no doubt that many sites result from the careless and selfish acts of people who just wanted to get rid of a problem, the huge array of enforcement tools and the failure to allow for good faith by the circumvention of historic rights has to be examined for what it is. Due process, when represented by equal subjection of everyone to unfair laws, is not enough. Posturing members of Congress demanding impossible rates of progress in correcting long-past errors will not help.

The Environmental Protection Agency has many dedicated, hardworking, and experienced professionals working to abate the real hazards to life, health, and the environment which result from a very few really dangerous sites. Ways should be found to let them work with deliberate speed toward resolution of these problems by reasonable means and with reasonably effective use of their valuable time and effort.

Epilogue

This book, with its underlying drumbeat of warnings about the severe consequences of errors and omissions in the management of hazardous wastes, might have been written from a completely different perspective.

In a more perfect society, all actors in the modern world of productive use of resources for the material (and spiritual) benefit of the population would cooperate in gaining and using a better understanding of the effects of our common decisions on the natural world and its life forms. Using that better understanding, we could have avoided at least the worst cases of short-sighted and irreversible environmental degradation and abuse.

In the real world of the United States, our worst natures have instead been revealed.

When, in the 1960s, environmental philosophy began the necessary conversion to environmental politics, the first casualties were reason and reflection. In our system, it is expected that partisan attitudes will develop and that "we" will oppose "them" with a minimum of listening, resulting from a maximum of talking and shouting.

Any time the American people are told that something has gone wrong and that someone may have been injured, there is a reflex: "Who's responsible for this? How can he be punished? Who's going to pay to fix it?"

In response, there comes a chorus: "Not me! I never did anything illegal!" The idea that actions or inaction might be bad judgment or a result of a failure to understand the situation can be dealt with only in a vindictive and punitive way.

So it happened that the environmental laws were made criminal statutes. Some people maintain that basic safeguards in the Constitution were evaded with participation of the courts, whose judges also read the newspapers and are believers in the ancient laws of retributive justice. Whether this is true is moot now, since all these issues are regarded as settled. At any rate, given the deep distrust of government and its agents that is endemic in the American psyche, it probably had to be this way.

This approach has important costs to the society and to the environment that the laws are intended to protect. The most obvious of these costs is loss of candor. No responsible executive of any company can ever say, "Yes, we understand our operations have resulted in a problem to the community." Instead his attorney will have pointed out that such a remark may be an admission of criminal liability, which he is entitled to avoid under the Fifth Amendment to the Constitution.

Thus, instead of looking his neighbors and friends in the eye and starting a discussion about remedies, he will say, "Talk to my lawyer—I can't discuss the matter." The same attorney will often advise that any action taken to abate an insult to the environment may be viewed as an admission of guilt, so the problem may go unremedied for long periods, even when it's clear what action is needed.

It is less likely today, but it was common in the 1970s for corporate financial officers to maintain that spending capital funds for environmental protection was improper use of stockholders' funds unless a court had ordered the "unproductive" investment.

It seems clear that the criminalization of undesirable environmental behavior has had a major inhibiting effect on the agencies designed to develop the regulatory schemes and the enforcement practices demanded by these complex laws. It is accepted throughout the regulatory agencies that nothing they do is final until some court has ruled on it. Frequently a rule or action of some kind actually takes effect without litigation, but there is still a feeling among the regulators that some judge will always be looking over their shoulder and second-guessing their product.

Since the courts have become accepted as the final arbiter of all actions, public and private, we perhaps should not be surprised at this effect. Still, there are tremendous time-lapses between legislation and regulation, which arise to a large extent from regulation-writers' concern that some court will make them look bad by ordering the work redone. Worse, members of Congress and their anonymous staff persons may, by commenting on the unfinished work or even on rumors, derail the most dedicated civil servant from his task.

Under pressure from public interest groups, Congress created the idea of citizens' suits, which authorize "citizens" to bring suit against the Administrator of EPA to force him or her to take action that the "citizen" believes should have been taken but was not. Generally, these suits cannot demand that the Administrator do anything in a matter for which the law grants him the authority to use his judgment. They therefore can only be used to get a court to order the Administrator to do something that he is already required to do by his oath of office! In practice, such suits, brought by public-interest organizations on behalf of some citizen, usually result in some private settlement negotiations under the eye of the court. In these negotiations, often whole new sections of law and/or regulations are developed, frequently to supply a perceived omission by the Congress or to replace sections of the law that are unworkable. These "settlements," which are agreed between the litigating parties and the regulatory agency under the imprimatur of the court, may be of enormous effect. Two major ones were the creation of the whole scheme of regulation of toxic pollutants under the Clean Water Act and the sweeping system of nondegradation and of increments of pollution permitted under the Clean Air Act. Both were later incorporated into the laws.

It is tempting simply to consider much of this legalism and harassment of the bureaucrats part of the American Way of Life in a litigious society. This must be considered in the context of the behavior of our European and Japanese cousins and competitors in the industrialized world, who have never subscribed to the punitive concept of environmental protection. Indeed, recent disclosures have shown that the Communist approach to environmental protection was to ignore it completely, even at the price of severe impacts on the health of the population. Western Europe and Japan never fell into this trap. However, those countries quietly did their work by negotiation and by the granting of public subsidies when the costs of environmental controls were agreed to be too heavy for their industries to bear. While giving lip service to the "Polluter Pays" principle, the need for maintaining competitive conditions has never been far from the minds of civil servants and politicians of those regions.

Today, with "green politics" emerging as a force in Europe, the situation is changing, but few native activists can be found even now.

More than twenty years have passed in the battle to find better ways of managing the affairs of people to gain the unquestioned goal of a sustainable world. In such a world, the quality of human life will be balanced by reasonable demands on the resources of the earth. As a participant in many aspects of this struggle, one is saddened by the huge drain on the limited resources of ingenuity, intellect, and capital that continues because of the lack of a decision-making process more effective than that provided by courts of law. The issues needing decision, incorporating sound science, economical technology, and an effective politician's grasp of the aspirations of people in a society, demand our best. Too often, they receive only a narrow legalistic evaluation by judges and others who may be nearly illiterate about science and engineering and who, when they seek counsel from the scientific community are met by a cop-out: "Everything is possible, but you'll have to decide what's important because we pure scientists can't comment. After all, we don't know all the facts!"

The courts never know all the facts either, but they are accustomed to making decisions anyway. In the adversarial legal system, truth is what the judges say it is. In the real and complex web of environmental interdependence, absolute truth is hard to find and even harder to prove.

APPENDIX

GLOSSARY OF SELECTED TERMS USED IN THIS BOOK

The following is intended to provide clarification of some terms which have been given unusual or specialized meanings when used in connection with environmental activities. Most of those here are, of course, used in the regulation and discussion of hazardous materials and of hazardous wastes.

In general, if the ordinary or "dictionary" use of a term is that used in the hazardous waste "business," you will not find it here. Beware of those terms which look familiar, but have meanings that can be connected with their ordinary use only by a tortured interpretation of language.

Abate (also as abatement, etc.). To take action to reduce or eliminate a hazard or a violation of a regulation or standard.

Above-Ground Storage Facility. A tank or container, the bottom of which is not more than six inches below the surface of ground on which it stands. If the tank and its associated piping have more than 10% of the total volume of the system below ground level, it is an underground facility.

Acidity. The property of a solution which is determined by a pH measuring less than 7 or by determining the amount of a base that is required to neutralize the solution.

Activated Carbon. A special form of carbon, often derived from charcoal and treated to make it able to adsorb and retain certain chemical substances.

Acute (as in exposure). Exposure to a toxic agent over a brief period (24–96 hours).

Acute Dermal LC_{50}. Amount of toxic agent which, when applied to the skin of a test animal, results in the death of half of the test population. Expressed as milligrams of agent per kilogram body weight.

Acute Oral LC_{50}. Same as above, but the dose is given by mouth.

Acute Toxicity. Having toxic effects (see *Toxicity*) that are expressed within a short time after exposure, usually within one to four days.

Adsorption. The holding of chemical substances on the surface of a solid, usually by the formation of chemical bonds.

Ames Test. A screening test for certain forms of carcinogenic effects in which selected strains of bacteria are exposed to toxic agents and then evaluated for changes in chromo-patterns that may indicate damage to genetic material of the cells. It was developed by Dr. Bruce Ames and his collaborators, for whom it is named.

ANPR. When a regulatory agency (the EPA, etc.) wishes to indicate its intent to issue a rule, but wants to solicit public input on the subject matter without the need for formal hearings and many of the restrictions on informal discussions of the subject, an Advance Notice of Proposed Rulemaking is issued in the *Federal Register* giving the rationale for the proposed rule and soliciting comments in detail.

Aquaclude. A layer of water-resistant soil or rock that prevents water from moving between aquifers (which see).

Aquifer. A geological formation (layer.of sand, gravel, or rock) or part of a formation which is capable of yielding useful amounts of water to a well or spring.

Asbestos. Any fibrous form of any of the minerals chrysotile, crocidolite, amosite, anthophyllite, tremolite, or actinolite.

Asbestos-Containing Material. Any solid material, whether in use or a waste, that contains 1% or more of asbestos.

Asbestos-Containing Waste. As applied to demolition and renovation activities, this term includes only *friable* asbestos waste and asbestos waste from air pollution control devices.

Askarel. A trade name applied to compositions containing chemical compounds derived from biphenyl and chlorine, often used interchangeably with PCB.

Attenuation. In groundwater, any decrease in the amount or concentration of a pollutant which results from chemical, physical, or biological action on the water as it moves through the ground.

Bioaccumulation. A buildup of a chemical compound or element in a living organism which occurs when the organism takes in more of the chemical than it can get rid of in the same length of time and stores the chemical in its tissues, etc.

Bioassay. Measuring the effect(s) of environmental exposures by intentional exposure of living organisms to a chemical or to a condition.

Biochemical Oxygen Demand. An index number which measures some kinds of water pollution. It is based on the amount of oxygen used by bacteria that grow in a water sample under prescribed conditions for a fixed time, usually five days. The symbol for the result is BOD_5.

Biodegradation. Decomposition of a substance into simpler substances by the action of microorganisms, usually in soil. It may or may not detoxify the material which is decomposed.

Biomass. Total amount of living matter found within a prescribed volume or space.

Biota. All living organisms which are to be found within a prescribed volume or space.

Bottom Ash. Combustion residues which do not leave an incinerator with the gaseous products, but are recovered from the bottom of the combustor or furnace.

BTU. Abbreviation for British Thermal Unit, which is the amount of heat energy needed to increase the temperature of a pound of water by one degree Fahrenheit under specified conditions.

Carcinogenic. Capable of causing cancer.

Casing (of a well). Pipe or tubing inserted into a well bore to prevent caving of the earth or rock and to allow, through seals or other devices, only the desired fluids to enter the well.

COD. Abbreviation for the term "chemical oxygen demand," a measure of the amount of oxygen which can react with a sample of waste water under specified conditions.

Chemical Waste Landfill. Generally, any landfill in which wastes from chemical and related industries are or have been deposited. The term is specifically used in connection with PCB disposal in landfills.

Chlorinated Hydrocarbons. Chemical group in which chlorine is substituted for hydrogen in common derivatives of petroleum or natural gas. Misused in some regulations to apply to certain pesticides that are more complex than hydrocarbons.

Chlorinated Solvents. An inexact term applied to those chlorinated hydrocarbons which are often used as cleaning and degreasing agents.

Closure. All the activities involved in taking a hazardous waste facility out of service and securing it for the time required by the regulations.

Combustible Waste. Any waste which will burn. If it begins to burn under special test conditions, it may be "ignitable hazardous waste."

Confined Aquifer. A porous underground formation containing water which cannot escape because of water-resistant layers of soil or rock above and below.

Confining Bed. See *Aquaclude.*

Contaminate. In groundwater usage, the act of releasing a substance so as to create a concentration in excess of the maximum contamination level set by a regulatory authority.

Contingency Plan. A document required to be kept at a hazardous waste treatment, storage, or disposal facility. The document describes in detail a plan of action to be followed in the event of a fire, explosion, or release of hazardous waste constituents.

Conversion (of solid waste). Resource recovery by application of biological, chemical, or physical methods to change solid waste into usable forms.

Cost-effectiveness Analysis. An economic study designed to find which of several waste management methods or systems will achieve regulatory requirements at least cost.

Dermal Toxicity. Toxic effects developed from application of a substance to the skin of an animal or human.

Designated Facility. A hazardous waste treatment, storage, or disposal facility operating under appropriate permits which is shown as a destination on a hazardous waste manifest.

Disposal (of solid waste). Any action or failure to act that results in solid waste, hazardous waste, or any constituents of such wastes entering the environment.

Dose. In toxicology, the amount of a chemical or substance to which an organism is exposed.

Dump. A site used for disposal of solid wastes without environmental controls.

Engineered Storage. A "disposal" method which is designed to keep solidified wastes safely for a long period of time in the hope that a permanent disposal method will be developed. One feature of the design is that the wastes must be retrievable at any time.

Environmental Compliance Audit. This describes a complete review of all operations of a facility, intending to determine whether that facility is in compliance with all the environmental regulations that apply to that facility.

Environmental Fate. The movement of a chemical through the environment by transport in air, water, and soil and the resulting exposures of living organisms.

Environmentally Sensitive Areas. In EPA usage, these are natural features such as wetlands, flood plains, and critical habitats of endangered species, etc.

Exposure. Any contact with a chemical or physical agent. The amount of the agent is the dose.

Exposure Assessment. Measurement or estimation of the extent of exposure of people to a chemical through environmental routes.

EP Toxicity. An estimate of the potential of a solid waste to contaminate groundwater when disposed of in the soil. It is based on an Extraction Procedure.

Facility (in solid waste usage). A very broad term including buildings and equipment, piping, wells, pits, ponds, lagoons, ditches, landfills, containers, even motor vehicles, rail equipment, or aircraft. The term also is used for a site or area where hazardous waste is or has been. Does *not* include any consumer product or any watercraft or vessel.

Flash Point. The lowest temperature at which a liquid produces vapors which can be ignited by a flame under specified test conditions.

Fly Ash. Solid material which is carried out of a furnace or incinerator by the gases.

Free Liquids. Liquids which easily separate from solid waste without the application of pressure or heat.

Fugitive Dust. Strictly, airborne soil particles from unpolluted areas such as roads, soil piles, etc. Often used, however, to describe any airborne solid matter released or emitted from any source other than a chimney or stack.

Groundwater. Generally, any water found in a liquid state below the surface of the ground.

Halogen. Generic term for the elements chlorine, bromine, and iodine.

Hazard. The chance that a particular substance will have an adverse effect on human health or the environment in a particular set of circumstances which creates an exposure to the substance.

Hazard Ranking System. A mathematical formula or "model" used by the USEPA to establish priority of a reported site under Superfund in comparison to other sites. It is used for the development of the National Priority List and is described in the Appendix to the National Contingency Plan.

Hazardous Substance. In general, any substance which can cause harm to human or environmental health whenever excessive exposure occurs. For regulatory purposes, consult the chapter in this book entitled "How to Recognize a Hazardous Waste" and relevant statutes.

Hazardous Waste. Waste materials which are listed by the USEPA or a State regulatory authority or possessing characteristics defined by the USEPA under authority of the Resource Conservation and Recovery Act (RCRA) as those of hazardous wastes.

Heavy Metals. Refers to members of a group of metallic elements which are recognized as toxic and generally bioaccumulative. The term arises from the relatively high atomic weights of these elements.

Hydrolysis. Chemical reaction between water and another substance that results in the breakdown of the reactant into a different chemical substance.

Ignitability. A characteristic which defines certain hazardous wastes which will catch fire under specific conditions. See Chapter 4, "How to Recognize a Hazardous Waste."

Incompatible Waste. Waste which cannot be disposed of or handled at a particular facility because it may damage the equipment or cause harm to the system. Also describes waste which, when mixed with other materials without controls, may create fire, explosion, or other severe hazards.

Infectious Waste. Broadly, all wastes generated in the treatment of communicable disease or other pathology. Includes blood and blood fractions, tissues, human wastes, and most disposable supplies when discarded by hospitals, clinics, medical offices, and similar facilities. For a more extensive definition, see the Code of Federal Regulations, Volume 40.

Lagoon. A pond or lake used for treatment of wastes which is shallow to allow penetration of sunlight and oxygen by natural contact for treatment of wastes.

Landfarming. Application of wastes to land, with or without incorporation into the soil. The term includes the use of waste as a soil improver or fertilizer.

Landfill Cell. A section of a hazardous waste landfill which is isolated from the rest of the landfill volume by liners, etc.

Land Treatment Facility. Place where hazardous wastes are applied or incorporated into the land surface. If any wastes remain after closure, it is a disposal facility.

LC_{50}. A concentration of a test substance in surrounding air or water which results in the death of half of the test organisms subjected to the material within a specified time and under other controlled conditions.

LD_{50}. A dose (quantity) of a toxicant which causes death of half of the test organisms within a specified time and under other controlled test conditions. The dose is usually measured as a weight of the test material per unit body weight of the organism, for example, as milligrams per kilogram (mg/kg).

Leach. To transfer materials from solids into liquids which are in contact with the solids. An example is the removal of soluble materials from wastes in a landfill by rainwater which penetrates into the solid materials.

Leachate. The liquid, including suspended solids, which is the product of leaching.

Lethal Dose. Often used as an equivalent for LC_{50}.

Love Canal. A name popularly applied to a waste disposal site near Niagara, New York. The problems arising from conditions at this site were influential in the development of Federal laws regulating hazardous wastes.

Manifest. In hazardous waste parlance, the document which is used to describe the type and amount of hazardous wastes in a shipment and to direct the wastes from origin to destination in accordance with regulations issued by the USEPA, the USDOT, and numerous state agencies (see Chapter 2).

Monitoring. Sampling activity to determine the conditions in the environment around a particular location. May include air, water, soil, and groundwater sampling and evaluation.

Monitoring Well. A well used to obtain groundwater samples for evaluation and/or to determine the elevation of groundwater in specific geologic strata.

Mutagen. A substance which causes genetic changes in an organism exposed to it which result in inheritable change in subsequent generations of the exposed organism.

National Contingency Plan. A comprehensive document which defines the Federal response to oil and hazardous material spills, required to be prepared under the Clean Water Act.

Natural Resources. Defined in CERCLA to include land, fish, wildlife, biota, air, water, groundwater, drinking water, and other such resources under Federal stewardship.

Polyaromatic Hydrocarbons. Chemicals of which each molecule is constructed with two or more benzenoid rings. Often considered to be carcinogenic.

Protocol. In environmental usage, the outline of the approach and methods to be used in conducting a test or experiment.

Reactivity. A "characteristic" defining a class of hazardous wastes. See Chapter 4, "How to Recognize a Hazardous Waste."

Recoverable Resources. Used materials which still have value for additional use or recycling.

Release. In hazardous waste usage, any action or failure to act which results in any thing or substance entering the environment. Releases can be either instantaneous or continuing, and many common occurrences are excluded, for example, any discharge of pollutants which is permitted under the Clean Air Act or the Clean Water Act. Releases are discussed in Chapter 13. For an exhaustive definition, refer to Title I, Section 101 (22) of the Comprehensive Environmental Response, Compensation, and Liabilities Act (Superfund).

Remedial Action. Generally, those actions taken at a hazardous waste site which are consistent with but not to be considered the eventual "permanent" remedy described in CERCLA.

Resource Conservation. Reduction of the amounts of solid waste that are generated, reduction of overall resource consumption, and utilization of recovered resources.

Risk. The probability of occurrence of unwanted consequences of an action or exposure.

Risk Assessment. Methodology combining exposure assessment with health and environmental effects data to estimate risks to human or environmental target organisms which result from exposure to pollutants.

Sanitary Landfill. A waste-disposal facility at which solid wastes are spread and compacted under a soil cover applied at least daily.

Saturated Zone. Geological layer or stratum in which all spaces between solids are filled with water.

Scoping. The public-input process conducted to define the information to be sought, issues to be addressed, and alternative actions to be considered in the preparation of an Environmental Impact Statement.

Sole Source Aquifer. When an aquifer is designated thus by the Administrator of the EPA, development may be restricted or controlled to avoid risk to the quality of drinking water drawn from the aquifer.

Special Waste. Term applied to certain wastes which are generated in very large volumes but are considered to pose low risks. Examples are mining wastes, utility boiler ash and scrubber sludge, and oil and gas drilling waste solids.

Stringfellow Acid Pits. A notorious Superfund site in Riverside county in southern California.

Superfund. 1. The word commonly used to refer to the complex of laws and regulations designed to clean up old and abandoned waste disposal and other contaminated sites. 2. Shorthand reference to the Hazardous Substance Response Trust Fund, the actual point of deposit in the United States Treasury for the taxes collected for funding the remedial action, etc., that is required by the CERCLA statute (which see).

Teratogen. Term applied to a toxic substance which causes malformation of offspring, but the malformations or defects are not inheritable.

Thermal Treatment. Generic term for processes which treat hazardous waste primarily by the use of high temperatures to cause changes in the waste. Examples include incineration, pyrolysis, calcination, etc.

Times Beach, Missouri. A small town that gave its name to a controversial Superfund site when its streets were treated with waste oil containing dioxin.

Toxic Substance. Any material or mixture that is capable of causing an unreasonable threat to human health or the environment.

Toxicity. The property of a substance to cause any adverse physiological effects (on living organisms).

Toxicity Characteristic Leaching Procedure. A defined test protocol for deciding whether a material is a hazardous waste as defined by the characteristic of toxicity.

Transporter. Under RCRA, any person who transports hazardous waste off the site of generation by any means.

Treatment. In hazardous waste, application of technological means to a waste to render it nonhazardous or less hazardous or more suitable for resource recovery.

Uniform Hazardous Waste Manifest. A document required to accompany any shipment of hazardous waste.

Valley of the Drums. A notorious Superfund site in Kentucky.

Zone of Aeration (vadose zone). Geological stratum or part of one that lies above the saturated zone. The space between solids in this zone is filled with a mixture of water and atmospheric gases.

SIMPLIFIED GLOSSARY OF OFTEN-USED ENVIRONMENTAL ACRONYMS

Note: Most of these acronymns and the meanings here associated with them come from the regularly published *Glossary of EPA Acronyms.* There are many hundreds in common use and not listed here. This list focuses on the popular acronyms among the staff persons within the USEPA who work on solid and hazardous waste issues.

Many of these terms find their way into all kinds of public notices and regulatory rules. They are constantly changing as the work of the agency changes.

If you don't know what an acronym means, *ask*! It seems likely that a request to your EPA Regional Office would get you a copy of the current *Glossary.*

AA	Assistant Administrator (of USEPA)
AEA	Atomic Energy Act
ALJ	Administrative Law Judge
ANPR	Advance Notice of Proposed Rulemaking
AO	Administrators Office (of USEPA)
APHA	American Public Health Association
API	American Paper Institute ("little API")
API	American Petroleum Institute ("big API")
ARAR	Applicable or Relevant and Appropriate Requirements
ASHAA	Asbestos in Schools Hazard Abatement Act of 1984
ASTHO	Association of State and Territorial Health Officials
ASTSWMO	Association of State and Territorial Solid Waste Management Officials
ATA	American Trucking Association
ATSDR	Agency for Toxic Substances and Disease Registry (HHS, which see)
BAC	Biotechnology Advisory Committee
BOD	Biological Oxygen Demand (measure of water contamination)
BTU	British Thermal Unit (measure of quantity of heat)
CAA	Clean Air Act
CAFO	Consent Agreement/Final Order
CAG	Carcinogen Assessment Group
CCU	Correspondence Control Unit (in OECM, which see)
CDC	Centers for Disease Control (HHS)
CDD	Chlorinated dibenzo-p-dioxin
CDF	Chlorinated dibenzofuran
CEAS	Office of Criminal Enforcement and Special Litigation
CED	CERCLA Enforcement Division (in OSWER, which see)
CED	Criminal Enforcement Division (in OECM)
CEQ	Council on Environmental Quality
CERCLA	Comprehensive Environmental Response, Compensation, and Liability Act of 1980 (Superfund)
CERCLIS	Comprehensive Environmental Response, Compensation, and Liability Information System
CERI	Center for Environmental Research Information
CFC	Chlorofluorocarbons
CFC	Combined Federal Campaign (the Federal United Fund)
CFR	Code of Federal Regulations

CMA Chemical Manufacturers Association
COD Chemical Oxygen Demand (measure of water contamination)
COE Corps of Engineers, US Army
CPO Certified Project Officer
CPSC Consumer Product Safety Commission
CRS Congressional Research Service
CSI Clean Sites, Inc. (a nonprofit corporation)
CWA Clean Water Act (also FWPCA, which see)
CWTC Chemical Waste Transportation Council
DA Deputy Administrator (of USEPA)
DBCP Dibromochloropropane
DO Dissolved Oxygen
DOC Department of Commerce
DOD Department of Defense
DOE Department of Energy
DOI Department of the Interior
DOJ Department of Justice
DOL Department of Labor
DOT Department of Transportation
DRA Deputy Regional Administrator (of USEPA)
DRE Destruction and Removal Efficiency
EA Enforcement Agreement
EA Environmental Assessment (under NEPA, which see)
EAG Exposure Assessment Group
ECU Environmental Crimes Unit (in DOJ)
EDB Ethylene Dibromide
EDF Environmental Defense Fund
EEC European Economic Community
EIS Environmental Impact Statement
ELI Environmental Law Institute
EPA Environmental Protection Agency (always US, Federal)
EPCA Energy Policy and Conservation Act of 1975
ERD Emergency Response Division (in OSWER)
ERNS Emergency Response Notification System
ESA Endangered Species Act
FATES FIFRA and TSCA Enforcement System
FDA Food and Drug Adminstration
FIFRA Federal Insecticide, Fungicide, and Rodenticide Act
FOIA Freedom of Information Act
FR *Federal Register*
FWPCA Federal Water Pollution Control Act
FWS Fish and Wildlife Service
FY Fiscal Year
GAC Groundwater Activated Carbon
GAO General Accounting Office (US Congress)
GEA *Glossary of EPA Acronyms*
GLP Good Laboratory Practices
GW Groundwater
HHS Department of Health and Human Services
HMTA Hazardous Materials Transportation Act
HRSD Hazardous Response Support Division (in OSWER)
HSCD Hazardous Site Control Division (in OSWER)
HSWA Hazardous and Solid Waste Amendments of 1984
HUD Department of Housing and Urban Development
HWDMS Hazardous Waste Data Management System (in OSWER)
HWED Hazardous Waste Enforcement Division (in OECM)
HWERL Hazardous Waste Engineering Research Laboratory
HWGTF Hazardous Waste Groundwater Task Force
IAG Interagency Agreement
ICC Interstate Commerce Commission
IG Inspector General
IGD Inspector General Division
IRM Interim Remedial Measures (under CERCLA)
ISD Interim Status Document (under RCRA)
LDC London Dumping Convention
LDR Land Disposal Restrictions
LUST Leaking Underground Storage Tanks
MCL Maximum Contaminant Level
MEK Ketones
MMT Million Metric Tons
MPRSA Marine Protection, Research, and Sanctuaries Act
MSDS Material Safety Data Sheet
NACA National Agricultural Chemicals Association
NAM National Association of Manufacturers of the USA
NAS National Academy of Sciences
NBAR Nonbinding Allocation of Responsibility
NCA Noise Control Act
NCHS National Center for Health Statistics
NCI National Cancer Institute
NCP National Contingency Plan (under CERCLA)
NEPA National Environmental Policy Act
NGA National Governors' Association
NHTSA National Highway Traffic Safety Administration

NIH	National Institute of Health
NIOSH	National Institute for Occupational Safety and Health
NOAA	National Oceanic and Atmospheric Administration
NOEL	No Observable Effect Level
NPDES	National Pollutant Discharge Elimination System (Under CWA)
NPL	National Priority List (under CERCLA)
NPRM	Notice of Proposed Rulemaking
NRC	Nuclear Regulatory Commission
NRDC	Natural Resources Defense Council
NRT	National Response Team
NSF	National Science Foundation
NSWMA	National Solid Waste Management Association
NTP	National Toxicology Program (in HHS)
NWF	National Wildlife Federation
NWPA	Nuclear Waste Policy Act
OALJ	Office of Administrative Law Judges
OCE	Office of Criminal Enforcement
OCL	Office of Congressional Liaison
OCSLA	Outer Continental Shelf Lands Act
ODW	Office of Drinking Water (in OW)
OECM	Office of Enforcement and Compliance Monitoring
OEET	Office of Environmental Engineering and Technology
OERR	Office of Emergency and Remedial Response
OGC	Office of the General Counsel
OGWP	Office of Groundwater Protection
OIA	Office of International Activities
OIG	Office of the Inspector General
O&M	Operations and Maintenance
OMB	Office of Management and Budget (of White House)
OPMS	Office of Program Management and Support (in OSWER)
OPPM	Office of Policy and Program Management (in OSWER)
ORD	Office of Research and Development
OSHA	Occupational Safety and Health Act
OSHA	Occupational Safety and Health Administration
OSM	Office of Surface Mining (in DOI)
OSTP	Office of Science and Technology Policy (in White House)
OSW	Office of Solid Waste (in OSWER)
OSWER	Office of Solid Waste and Emergency Response
OTA	Office of Technology Assessment (of US Congress)
OTS	Office of Toxic Substances
OUST	Office of Underground Storage Tanks
OW	Office of Water
OWPE	Office of Waste Programs Enforcement
PAH	Polycyclic Aromatic Hydrocarbons
PAT	Permit Assistance Team
PBB	Polybrominated biphenyl
PCB	Polychlorinated biphenyl
PCDD	Polychlorinated dibenzodioxin
PCDF	Polychlorinated dibenzofuran
PIC	Product of Incomplete Combustion
ppb	Parts per billion
ppm	Parts per million
ppt	Parts per trillion
PRP	Potentially Responsible Party (under CERCLA)
PSPD	Permits and State Programs Division (in OSWER)
QA	Quality Assurance
QC	Quality Control
RA	Regional Administrator
RA	Remedial Action
RCRA	Resource Conservation and Recovery Act
R&D	Research and Development
RD	Remedial Design
RDF	Refuse-Derived Fuel
RE	Reportable Event
RED	RCRA Enforcement Division
RIA	Regulatory Impact Analysis
RICO	Racketeer Influenced and Corrupt Organizations Act
RI/FS	Remedial Investigation/Feasibility Study
RIP	RCRA Implementation Plan
RMCL	Recommended Maximum Contaminant Levels
ROD	Record of Decision
RPM	Remedial Project Manager
RRT	Requisite Remedial Technology
RTP	Research Triangle Park (EPA laboratory site in North Carolina)
SAB	Science Advisory Board
SAP	Scientific Advisory Panel
SARA	Superfund Amendments and Reauthorization Act
SBA	Small Business Administration
SCAP	Superfund Comprehensive Accomplishments Plan
SCRP	Superfund Community Relations Program
SCSA	Soil Conservation Society of America
SDWA	Safe Drinking Water Act
SFFAS	Superfund Financial Assessment System

SMCRA	Surface Mining Control and Reclamation Act of 1977
SNARL	Suggested No Adverse Response Level
SPCC	Spill Prevention, Containment, and Countermeasures
SQG	Small Quantity Generator
SWDA	Solid Waste Disposal Act
SWERD	Solid Waste and Emergency Response Division (of OGC)
TANSTAFL	There Ain't No Such Thing as a Free Lunch
TAP	Technical Assistance Program (usually found in State offices)
TCDD	Dioxin (Tetrachlorodibenzo-p-dioxin)
TCE	Trichloroethylene
TCLP	Toxicity Characteristic Leaching Procedure
TLV	Threshold Limit Value
TOC	Total Organic Carbon
TPY	Tons per Year
TOSCA	Toxic Substances Control Act
TSDF	Treatment, Storage, and Disposal Facility
TWMD	Toxics and Waste Management Division (of EPA Regional Office)
UIC	Underground Injection Control (under SDWA)
UMTRCA	Uranium Mill Tailings Radiation Control Act
UNEP	United Nations Environment Program
UN/NA	International identifying number established by the United Nations for a hazardous material
USC	United States Code (codified laws of the United States)
USCA	United States Code Annotated
USDA	US Department of Agriculture
USDHHS	US Department of Health and Human Services
USDW	Underground Source of Drinking Water
USEPA	US Environmental Protection Agency
USGS	US Geological Survey
UST	Underground Storage Tanks
VOC	Volatile Organic Compounds
WAP	Waste Analysis Plan
WIC	Washington Information Center (of EPA)
WHO	World Health Organization

U.S. ENVIRONMENTAL PROTECTION AGENCY REGIONAL OFFICE CONTACTS, HAZARDOUS WASTE

Region I
(Connecticut, Maine, Massachusetts, New Hampshire, Rhode Island, Vermont)
Director
Waste Management Division
JFK Federal Building
Boston, MA 02203
(617) 223-5186

Region II
(New Jersey, New York, Puerto Rico, Virgin Islands)
Director
Air and Waste Management Division
26 Federal Plaza, Room 1009
New York, NY 10278
(212) 264-2301

Region III
(Delaware, Maryland, Pennsylvania, Virginia, West Virginia, District of Columbia)
Director
Hazardous Waste Management Division
841 Chestnut Street
Philadelphia, PA 19107
(215) 598-8181

Region IV
(Alabama, Florida, Georgia, Kentucky, Mississippi, North Carolina, South Carolina, Tennessee)
Director
Waste Management Division
345 Courtland Street, NE
Atlanta, GA 30365
(404) 257-3454

Region V
(Illinois, Indiana, Michigan, Minnesota, Ohio, Wisconsin)
Director
Waste Management Division
230 South Dearborn Street
Chicago, IL 60604
(312) 886-7579

Region VI
(Arkansas, Louisiana, New Mexico, Oklahoma, Texas)
Director
Waste Management Division
First International Building
1201 Elm Street
Dallas, TX 75270
(214) 729-2730

Region VII
(Iowa, Kansas, Missouri, Nebraska)
Director
Waste Management Division
726 Minnesota Avenue
Kansas City, KS 66101
(913) 757-2850

Region VIII
(Colorado, Montana, North Dakota, South Dakota, Utah, Wyoming)
Director
Waste Management Division
One Denver Place
999 18th Street
Denver, CO 80202
(303) 564-1720

Region IX
(Arizona, California, Hawaii, Nevada, Pacific Islands)
Director
Toxics and Waste Management Division
215 Fremont Street
San Francisco, CA 94105
(415) 454-7640

Region X
(Alaska, Idaho, Oregon, Washington)
Director
Hazardous Waste Division
1200 Sixth Avenue
Seattle, WA 98101
(206) 399-1352

STATE AGENCY OFFICES RESPONSIBLE FOR HAZARDOUS WASTE MANAGEMENT

Alabama
Chief, Land Division
Alabama Department of Environmental Management
1751 Federal Drive
Montgomery, AL 36130
(205) 271-7730

Alaska
Program Manager, Hazardous Waste and Superfund Programs
Alaska Department of Environmental Conservation
Environmental Quality Operations
Air and Solid Waste Management
Pouch O
Juneau, AK 99811
(907) 465-2666

Arizona
Manager, Office of Waste and Water Quality Management
Arizona Department of Health Services
Division of Environmental Health Services
1740 West Adams Street
Phoenix, AZ 85007
(602) 257-2304

Arkansas
Chief, Solid and Hazardous Waste Division
Arkansas Department of Pollution Control and Ecology
8001 National Drive
P.O. Box 9583
Little Rock, AR 72219
(501) 562-7444, ext. 504

California
Chairman, Waste Management Board
California Waste Management Board
1020 Ninth Street, Suite 300
Sacramento, CA 95814
(916) 322-3330

Colorado
Director, Waste Management Division
Colorado Department of Health
4210 East 11th Avenue
Denver, CO 80220
(303) 320-8333, Ext. 4364

Connecticut
Director, Hazardous Materials Management Unit
Connecticut Department of Environmental Protection
165 Capitol Avenue
Hartford, CT 06106
(203) 566-5712

Delaware
Supervisor, Solid and Hazardous Waste Management Branch/Superfund
Delaware Department of Natural Resources and Environmental Control
Division of Air and Waste Management
Solid Waste Management Branch
89 Kings Highway
P.O. Box 1401
Dover, DE 19901
(302) 736-4781

District of Columbia
Chief, Pesticides and Hazardous Waste Management Branch/Superfund
Department of Consumer and Regulatory Affairs
5010 Overlook Avenue, S.W., Room 114
Washington, DC 20032
(202) 767-8422

Florida
Administrator, Solid and Hazardous Waste Section
Florida Department of Environmental Regulation
2600 Blair Stone Road
Tallahassee, FL 32301
(904) 488-0300

Georgia
Chief, Land Protection Branch
Georgia Department of Natural Resources
Environmental Protection Division
270 Washington Street, SW, Room 723
Atlanta, GA 30334
(404) 656-2833

Hawaii
Hawaii Department of Health
Environmental Protection and Health Services Division
P.O. Box 3378
Honolulu, HI 96801
(808) 548-4139

Idaho
Manager, Solid and Hazardous Waste Section
Department of Health and Welfare
Division of Environment
450 West State Street
Boise, ID 83720
(208) 334-4107

Illinois
Manager, Division of Land Pollution Control
Illinois Environmental Protection Agency
2200 Churchill Road
Springfield, IL 62706
(217) 782-6760

Indiana
Director, Division of Land Pollution Control
Indiana State Board of Health
1330 West Michigan Street
Indianapolis, IN 46206
(317) 243-5026

Iowa
Coordinator, Hazardous Waste Program
Iowa Department of Water, Air, and Waste Management
900 East Grand Street
Henry A. Wallace Building
Des Moines, IA 50319
(515) 281-8925

Kansas
Manager, Bureau of Waste Management
Kansas Department of Health and Environment
Forbes Field
Topeka, KS 66620
(913) 862-9360, Ext. 290

Kentucky
Director, Division of Waste Management
Department for Environmental Protection
18 Reilly Road
Frankfort, KY 40601
(502) 564-6716, Ext. 214

Louisiana
Louisiana Department of Environmental Quality
Office of Solid and Hazardous Waste
P.O. Box 44066
Baton Rouge, LA 70804
(504) 342-1216

Maine
Director, Bureau of Oil and Hazardous Materials Control
Maine Department of Environmental Protection
State House–Station 17
Augusta, ME 04333
(207) 289-2651

Maryland
Director, Waste Management Administration
Office of Environmental Programs
201 West Preston Street
Baltimore, MD 21210
(301) 225-5647

Massachusetts
Director, Division of Hazardous Waste
Department of Environmental Quality Engineering
Division of Solid and Hazardous Waste
One Winter Street, Fifth Floor
Boston, MA 02108
(617) 292-5589

Michigan
Chief, Hazardous Waste Division
Department of Natural Resources
P.O. Box 30038
Lansing, MI 48909
(517) 373-2730

Minnesota
Director, Division of Solid and Hazardous Waste
Minnesota Pollution Control Agency
520 Lafayette Road North
St. Paul, MN 55155
(612) 296-7282

Mississippi
Division Director
Solid Waste Management/Superfund
Department of Natural Resources
Bureau of Pollution Control
P.O. Box 10385
Jackson, MS 39209
(601) 961-5062

Missouri
Program Director, Waste Management
Department of Natural Resources
Division of Environmental Quality
117 East Dunkin Street
P.O. Box 1368
Jefferson City, MO 65102
(314) 751-3241

Montana
Chief, Solid and Hazardous Waste Bureau
Department of Health and Environmental Sciences
Solid Waste Management Bureau
Room B-201, Cogswell Building
Helena, MT 59620
(406) 444-2821

Nebraska
Division Chief
Department of Environmental Control
Land Quality Division
P.O. Box 94877, Statehouse Station
301 Centennial Mall South
Lincoln, NE 68509
(402) 471-2186

Nevada
Administrator, Division of Environmental Protection
Department of Conservation and Natural Resources
Capitol Complex
Carson City, NV 89710
(702) 885-4570

New Hampshire
Division of Public Health
Department of Health and Welfare
Health and Welfare Building
Hazen Drive
Concord, NH 03301
(603) 271-4611

New Jersey
Director, Division of Waste Management
Department of Environmental Protection
P.O. Box CN026
32 East Hanover Street
Trenton, NJ 08625
(609) 292-1250

New Mexico
Division of Solid Waste
Health and Environmental Department
Solid and Hazardous Waste Management Program
P.O. Box 968
Santa Fe, NM 87504
(505) 827-2779

New York
Director, Division of Solid and Hazardous Waste
Department of Environmental Conservation
50 Wolf Road
Albany, NY 12233
(518) 457-6603

North Carolina
Head, Solid Hazardous Waste Management Branch
Department of Human Resources
P.O. Box 2091
Raleigh, NC 27602
(919) 733-2178

North Dakota
Director, Division of Environmental Waste Management and Special Studies/Superfund
State Department of Health
1200 Missouri Avenue, Room 302
P.O. Box 5520
Bismarck, ND 58502
(701) 224-2366

Ohio
Chief, Division of Solid and Hazardous Waste Management
Ohio Environmental Protection Agency
361 East Broad Street
Columbus, OH 43215
(614) 466-7220

Oklahoma
Waste Management Service
Oklahoma State Department of Health
P.O. Box 53551
1000 NE 10th Street
Oklahoma City, OK 73152
(405) 271-5338

Oregon
Administrator, Solid Waste Division
Department of Environmental Quality
Hazardous and Solid Waste Division
522 SW Fifth Avenue
P.O. Box 1760
Portland, OR 97207
(503) 229-5913

Pennsylvania
Director, Bureau of Solid Waste Management
Department of Environmental Resources
P.O. Box 2063
Harrisburg, PA 17120
(717) 787-9870

Rhode Island
Chief, Division of Air and Hazardous Waste Management
Department of Environmental Management
Solid Waste Management Program
204 Cannon Building
75 Davis Street
Providence, RI 02908
(401) 277-2797

South Carolina
Chief, Bureau of Solid and Hazardous Waste Management
Department of Health and Environmental Control
J. Marion Sims Building
2600 Bull Street
Columbia, SC 29201
(803) 758-5681

South Dakota
Administator, Office of Air Quality and Solid Waste
Department of Water and Natural Resources
Room 217, Foss Building
523 E. Capitol
Pierre, SD 57501
(605) 773-3153

Tennessee
Director, Division of Solid Waste Management
Customs House, Fourth Floor
701 Broadway
Nashville, TN 37219
(615) 741-3424

Texas
Division Director,
Hazardous and Solid Waste Division
Texas Water Commission
P.O. Box 13087, Capitol Station
1700 North Congress
Austin, TX 78711
(512) 463-7760

Utah
Director, Bureau of Solid Waste Management/Superfund
Division of Environmental Health
P.O. Box 45500
State Office Building
Salt Lake City, UT 84140
(801) 533-4145

Vermont
Chief of Operations
Agency of Environmental Conservation
Department of Water Resources and Environmental Engineering
Waste Management Division
State Office Building
Montpelier, VT 05602
(802) 828-3361

Virginia
Director, Division of Solid and Hazardous Waste Management
Department of Health
Monroe Building, 11th Floor
101 North 14th Street
Richmond, VA 23219
(804) 225-2667

Washington
Supervisor, Solid and Hazardous Waste Management Division
Department of Ecology
Office of Land Programs
Mail Stop PV-11
Olympia, WA 98504
(206) 459-6316

West Virginia
Branch Head,
Solid and Hazardous Waste/Groundwater
Department of Natural Resources
Division of Water Resources
1201 Greenbrier Street, Second Floor
Charleston, WV 25312
(304) 348-3506

Wisconsin
Director, Bureau of Solid Waste Management
Department of Natural Resources
P.O. Box 7921
Madison, WI 53707
(608) 266-1327

Wyoming
Supervisor, Solid Waste Management Regulation
Department of Environmental Quality
Solid Waste Management Program
122 West 25th Street
Herschler Building
Cheyenne, WY 82002
(307) 777-7752

Suggested Reading

Allman, William F. "Staying Alive in the 20th Century." *Science 85* (October, 1985): 31.

American Institute of Chemical Engineers. *Environmental Progress* 7, no. 3 (1988). Special issue devoted to Waste Minimization.

Ames, Bruce N. "Environmental Pollution, Natural Carcinogens, and the Causes of Human Cancer: Six Errors." In *Important Advances in Oncology, 1989*, edited by V.T. DeVita, Jr., and S. Hellman. Philadelphia: J. B. Lippincott Co., 1989.

Ames, Bruce N., and Lou Sevirsky Gold. "Pesticides, Risk, and Applesauce." *Science* 244:755 (May 19, 1989).

Cherimisinoff, Paul N., and Kenneth A. Gigliello. *Leachate from Hazardous Waste Sites*. Lancaster: Technomic Publishing Co., Inc., 851 New Holland Ave., Lancaster, PA 17604, 1983.

Citizen's Toxic Protection Manual. Boston: National Toxics Campaign, 1988. 560 pp. loose-leaf format.

"The Compleat Worrier." *Science 85* (October 1985): 29–47. (Three articles in a special section of the issue)

The Conservation Foundation. *Risk Assessment and Risk Control*. Washington, D.C., 1985.

Crandall, Robert W., and Lester B. Lave. *The Scientific Basis of Health and Safety Regulation*. In a series: *Studies in the Regulation of Economic Activity*. Washington, D.C.: The Brookings Institution, 1981.

Davison, Gaynor W., and Basil W. Mercer. *Hazardous Waste Management*. New York: John Wiley and Sons, 1986.

Deisler, Paul F., ed. *Reducing the Carcinogenic Risks in Industry*. New York and Basel: Marcel Dekker, Inc., 1984.

Drabkin, Murray, James W. Morrison, and Lawrence S. Kirsch. "Bankruptcy and the Cleanup of Hazardous Waste: *Caveat Creditor*." *Environmental Law Reporter* 15 (1985). Note: Summary in *ELI Associates Newsletter*, a publication of the Environmental Law Institute, Washington, D.C.

The Dynamic Duo: RCRA and SARA Title III. Publication Office, 1525 New Hampshire Ave. NW, Washington, D.C. 20036.

Environmental Source Book (for current year). Center for Resource Economics/Island Press, 1718 Connecticut Ave., NW, Suite 300, Washington, D.C. 20009.

EPA Journal 13, no. 9 (November 1987). This issue is devoted to the subject of environmental risk in many aspects.

"EPA: A Regulatory Agency." *EPA Journal* 14, no. 7 (September 1986). Entire issue.

Freeman, Harry M. "Hazardous Waste Minimization." *Journal of the Air Pollution Control Association* 38, no.1 (January 1986): 59.

Freeman, Harry M., ed. *Standard Handbook of Hazardous Waste Treatment and Disposal.* New York: McGraw-Hill, Inc., 1989.

Guide to Hazardous Products Around the Home. Household Hazardous Waste Project, 901 South National Ave., Box 87, Springfield, MO 65804.

Gusman, S., et al. *Public Policy for Chemicals.* Washington, D.C.: The Conservation Foundation, 1980.

Household Hazards: A Guide to Detoxifying Your Home. New York State Department of Environmental Conservation, 50 Wolf Road, Albany, NY 12233.

Kroschwitz, Jaqueline I., and Melissa Wanamaker. *Chemistry—A First Course,* 2nd Edition. New York: McGraw-Hill Book Co., 1987.

National Association of Manufacturers of the United States. *Staying Out of Trouble: What You Should Know About the New Hazardous Waste Law.* Also see 1986 supplement. 1331 Pennsylvania Ave. NW, Washington, D.C. 20004-1703.

National Association of Manufacturers of the United States. *Preparing for Emergency Planning.* Washington, D.C., May 1987.

National Association of Manufacturers of the United States. *Waste Minimization: Manufacturers' Strategies for Success.* Washington, D.C., 1989.

"Protecting Our Drinking Water." *EPA Journal* 12, no. 7 (September 1986). Issue devoted to the special subject.

Quarles, John. *Federal Regulation of Hazardous Waste.* Washington, D.C.: The Environmental Law Institute, 1982.

RCRA Handbook, 2nd Edition, 1986. ERT, Inc., and Sidley and Austin, 696 Virginia Road, Concord, MA 01742.

Ruckelshaus, William D. (as Administrator of the U.S. Environmental Protection Agency). "Science, Risk, and Public Policy." Address to the National Academy of Sciences, June 22, 1983.

Sabatino, John. "Non-Obvious Hazardous Waste Determination." *Hazardous Waste Management Magazine* (January 1989): 30.

Specter, Michael. "Seeing Risk Everywhere." *The Washington Post* (May 7, 1989): A20.

Stanfield, Rochelle L. "Risky Business." In *Environmental Focus. National Journal* (January 25, 1986).

Superfund Handbook, 2nd Edition, 1987. ERT, Inc. and Sidley and Austin, 696 Virginia Road, Concord, MA 01742.

U.S. Congress, Congressional Budget Office. *Hazardous Waste Management: Recent Changes and Policy Alternatives.* Washington, D.C.: Government Printing Office.

U.S. Congress, General Accounting Office. GAO/RCED-87-27. *Hazardous Waste: Report to the Chairman, Subcommittee on Commerce, Transportation, and Tourism, House of Representatives.* Washington, D.C.: Government Printing Office, December 1986.

U.S. Congress, Office of Technology Assessment. *Serious Reduction of Hazardous Waste.* Washington, D.C.: Government Printing Office, 1987.

U.S. Congress, Office of Technology Assessment. *Protecting the Nation's Groundwater from Contamination.* Washington, D.C.: Government Printing Office, Stock No. 052-003-00966-8.

U.S. Environmental Protection Agency, Office of Solid Waste and Emergency Response. *The Hazardous Waste System.* Report released June 27, 1987, without document number.

U.S. Environmental Protection Agency, Office of Drinking Water. *Report to Congress: Injection of Hazardous Waste.* EPA 570/0-85-003, May 1985.

U.S. Environmental Protection Agency, Office of Ground-Water Protection. *Groundwater Protection Strategy.* Unnumbered report, August 1984.

U.S. Environmental Protection Agency, Office of Waste Programs Enforcement, Office of Solid Waste and Emergency Response. *RCRA Groundwater Monitoring Enforcement Guidance.* Unnumbered document ("Red Book"), August 1985.

U.S. Environmental Protection Agency, Robert S. Kerr Environmental Research Laboratory, Ada, OK 74820. *Leaking Underground Storage Tanks: Remediation with Emphasis on In-Situ Bioremediation*. EPA/600/2-67/008.

U.S. Environmental Protection Agency, Hazardous Waste EngineeringResearch Laboratory. *Waste Minimization: Opportunity Assessment Manual*. EPA/1625/7-88/003. Available as PB88-213004 from NTIS, 5285 Port Royal Road, Springfield, VA 22161.

U.S. Environmental Protection Agency, Report to Congress. *Wastes from the Extraction and the Beneficiation of Metallic Ores, Phosphate Rock, Asbestos, Overburden from Uranium Mining, and Oil Shale.* Document EPA/530-SW-85-033 from Supt. of Documents, U.S. Government Printing Office, Washington, D.C. 20402.

U.S. Environmental Protection Agency, Science Advisory Board. *Future Risk: Research Strategies for the 1990s*. SAB-EC-88-040. Also: Appendices A through E, separately bound.

U.S. Department of Health, Education, and Welfare, Public Health Service. *Fifth Annual Report on Carcinogens, Summary, 1989*. Document No. NTP 89-239.

World Environment Center, International Environment and Development Service. *Pollution Prevention Pays—An Overview by the 3M Company (USA) of Low and Non-Waste Technologies*. New York, 1985.

INDEX